SHOOTING INCIDENT
RECONSTRUCTION

SHOOTING INCIDENT RECONSTRUCTION

Lucien C. Haag

ELSEVIER

AMSTERDAM • BOSTON • HEIDELBERG • LONDON
NEW YORK • OXFORD • PARIS • SAN DIEGO
SAN FRANCISCO • SINGAPORE • SYDNEY • TOKYO

Academic Press is an imprint of Elsevier

Senior Acquisitions Editor: Mark Listewnik
Project Manager: Heather Furrow
Acquisitions Editor: Jennifer Soucy
Editorial Assistant: Kelly Weaver
Marketing Manager: Chris Nolin
Cover Design: Eric DeCicco
Composition: Integra Software Services
Cover Printer: Phoenix Color
Interior Printer: Maple Press

Academic Press is an imprint of Elsevier
30 Corporate Drive, Suite 400, Burlington, MA 01803, USA
525 B Street, Suite 1900, San Diego, California 92101-4495, USA
84 Theobald's Road, London WC1X 8RR, UK

This book is printed on acid-free paper. ∞

Library of Congress Cataloging-in-Publication Data
Haag, Lucien C.
    Shooting incident reconstruction / Lucien C. Haag.—1st ed.
      p. cm.
    Includes bibliographical references and index.
    ISBN-13: 978-0-12-088473-5   ISBN-10: 0-12-088473-9 (casebound : alk. paper)
    1. Forensic ballistics.   I. Title.
  HV8077.H22 2006
  363.25′62—dc22

                                            2005022278

British Library Cataloguing in Publication Data
A catalogue record for this book is available from the British Library

ISBN-13: 978-0-12-088473-5
ISBN-10: 0-12-088473-9

For all information on all Elsevier Academic Press publications
visit our Web site at www.books.elsevier.com

Printed in the United States of America
06 07 08 09 10 9 8 7 6 5 4 3 2

For Sandi, Matt and Mike for whom nearly every picnic or outing in our beautiful Arizona desert ended in gunfire.

And to the memory of Gene Wolberg.

CONTENTS

LIST OF COLOR PLATES

INTRODUCTION

At the time this introduction was written, I had been employed as a criminalist and forensic firearm examiner for more than 40 years; 17 of these with the Phoenix Arizona Police Department as a Criminalist, and later as Technical Director of that laboratory followed by another 23 years as a private consultant working for prosecutors, private attorneys, educational institutions, insurance companies, law firms, firearms manufacturers, and, on occasion, private individuals. I had always found the work interesting and challenging and still do. The concept of how science might aid the court and jury in determining what did and did not happen in the matter at trial is still an exciting one for me. Although many of us in the field of forensic science frequently disparage lawyers and the legal process, it is the anomalous trial outcome that gains our attention and generates our scorn. Most of the time, juries are able to grasp the evidence we present and that should be all that matters. What they do with that information may be, at times, disappointing to us personally but their decision is not ours to make, and it may often be made on some other basis than our observations and opinions derived from the physical evidence.

Working within the legal system is also fascinating. I suspect nearly all of us enjoy a good film about a trial. A trial can be high drama involving verbal and mental chess on the part of the lawyers and witnesses. Lives, careers, futures, personal freedom, and, in civil cases, large amounts of money are often at stake. The side that calls us as expert witnesses will usually praise our work but may also pressure us to extend ourselves beyond that which we should go in the furtherance of their cause. Our employer's cause must *not* become our cause. Our only advocacy must be for our analysis of the evidence carried out by scientifically sound means. Concurrently, the reader should remember that it is often our cross-examiner's mission to make us look like biased witnesses, fools, lackeys, mountebanks, or incompetents. The witness stand is a decidedly uncomfortable environment for most scientists and one best observed on the movie screen or television rather than from the actual site. It is, and should be, a stressful place but it is one that I have become used to and have even come to enjoy for the reasons stated earlier.

At the risk of seeming a bit immodest, it occurred to me that some readers might be interested in how I became gainfully employed (indeed, well-paid) shooting guns and shooting things for a living.

I grew up in the Midwest in the late 1940s and early 1950s. Guns—some of which were always loaded—were in almost every home and farmhouse I visited. My childhood friends all had access to guns and after school we could often be found in the field with a rifle or shotgun. This was with our parents' permission but without them necessarily being present. It was an age of trust on their part and personal responsibility on my part. At age 6 or 7, I had received my first Red Ryder BB gun from my father and my marksmanship training began. Neither I nor my friends ever considered using a gun to commit a crime or to endanger someone or damage their property. We certainly *never* discussed shooting at one of our classmates, our school, or our teachers.

My fondest memories of my father are of getting up before daybreak, having breakfast at some roadside truck stop then getting into the frosty woods at dawn with the sound of crunching autumn leaves under foot and with *my* rifle or *my* shotgun in hand. It did not much matter whether we got any squirrels or rabbits or whatever was the quarry of the day. We walked and talked and I learned of nature. He also taught me firearms safety and personal responsibility. I saw firsthand that firearms, even my diminutive 22 rifle, were capably of inflicting serious and fatal wounds. Guns were *not* toys or something to be handled carelessly. And my father trusted me with these guns. That meant a lot. I wish he were here to read this now. His lessons were ones that I have carried with me all of my life and have since passed on to my sons.

The use of guns in films of that time was typically portrayed as on the side of good. The Lone Ranger, Red Ryder, Roy Rogers, Gene Autry, and all the other lesser-known heroes of the Saturday afternoon matinee seldom had to shoot anyone because they were so competent and proficient in the use of their Colt single-action revolver or their Winchester rifle. They usually either shot the gun out of the bad guy's hand or simply got "the drop" on them through their superior proficiency in firearms handling. These were classic morality plays of good over evil in which firearms use was an integral part.

But today the blood-soaked films from Hollywood show guns creating as much death, destruction, and mayhem as possible in the shortest time possible. They are typically possessed by the psychologically flawed and unfit. It is difficult to think of a film in the past 20 years that depicts a gun on the side of right and in the hands of an honest person of character. It seems that we have forgotten that our special knowledge and proficiency with firearms is why we are citizens and not subjects. It is why we rightfully honor such men as Alvin York and Audie Murphy—men who grew up with firearms and used them for hunting, sport, and recreation and later used them so effectively in the defense of freedom.

In their day and in my youth firearms were *more* accessible and readily available with little or no restrictions (other than that imposed by our parents) than they are today. And there were no school shootings, gang shootings, drive-by shootings, or any of the other senseless acts of violence committed with firearms that we see today. As Hugh Downs (a well-known radio and television commentator) once pointed out in reference to the present day misuse of firearms— "It's a software problem, not a hardware problem."

But what of my life-long interest in firearms and how it relates to this book and this subject matter? I did bring home my share of rabbits and squirrels from the fields and woods of central Illinois but hunting was never a burning passion with me. I was more interested in how far and how accurately a bullet could be fired; what it looked like after it hit or went through something. Why did bullets make that fascinating whining sound when I would sit astraddle a railroad track and ricochet bullets off of the iron rail after an impact at low incident angle? I would shoot up a box of cartridges just to hear the sound that the departing bullets would make. I even heard some of these bullets impact the ground some distance downrange and subsequently searched many times but in vain in an effort to find one of these bullets just to see if its "new" shape corresponded to the gray elliptical smear of lead at the impact site on the rail. (These characteristic impact marks can be seen in Chapter 6.) While shooting at sticks floating down a slow-moving stream from an old covered bridge, I noticed that the sound of the bullet's impact with the water changed at a recurring point downrange and it became apparent that at the closer distances the bullets were entering the water whereas they were ricocheting at the greater distances. The phenomenon I was dealing with is known as the *critical angle*, I just did not know the name for it in 1952. In subsequent years I also fired many bullets vertically upward on calm days in deserts of California and Arizona with the misplaced hope of hearing one return to the ground. (I had previously measured the round trip time for BBs from my Red Ryder BB gun and a Crosman pellet gun in my back yard in Illinois.)

During my high-school years in Southern California I shot competitively on a church-sponsored rifle team. Yes, dear reader, churches and schools and colleges sponsored rifle teams and even supplied many of the guns! Even the University of California at Berkeley had a rifle club when I started there in 1961. Firearms and the people (including young people) who enjoyed shooting them had not yet been portrayed as they are today. I also became an avid handloader in my teenage years (and still am today) and many of my weekends during those years involved informal target practice in various remote locations in the Mojave Desert of California. All the time I was observing and learning things about firearms and ammunition that would become useful in later years and are now incorporated between the covers of this book.

After receiving my degree in Chemistry from Cal-Berkeley, I discovered the field of Criminalistics through several courses at California State University at Long Beach and realized for the first time that I could apply and utilize my interest in firearms in a professional way. I began interviewing and taking tests to join the staff of several crime laboratories in Southern California where I was living at the time. In 1965 a position for a second person in the then small Phoenix Police Crime Lab opened up. It was the classic case of being at the right place and the right time. During the years that I worked in the Phoenix Lab, I was able to apply my life-long interest in firearms to casework. I quickly became a member of the Association of Firearm and Tool Mark Examiners (AFTE) and began giving presentations at annual meetings and writing articles for the AFTE Journal. I started assembling handout materials for classes and workshops dealing with firearms evidence and the reconstruction of shooting incidents for various organizations.

Colleagues, students from these classes, and my wife, Sandi, all urged me to put these things together in the form of a book. This I have now done. But there is an additional reason and it arises as a consequence of my many years of reviewing the work of others who were most often employed by government laboratories. A very troubling change has been taking place in these laboratories over the last 30 years. They are taking on the properties of a clinical laboratory where the detective or investigator selects from a menu of tests, e.g., identify the fired bullet or cartridge case with the submitted gun, measure the trigger pull of the submitted gun, check the gun's safety system for proper operation, etc. In this strictly reactive role the forensic scientist is no longer functioning as a scientist at all. Rather he or she has been reduced to the role of a technician. Little or no discussion between the submitter and the laboratory examiner takes place regarding the details and issues associated with the case. The technician in this "clinical lab" organization is simply responding to the requests of the submitter. He or she may be doing the requested tests correctly and in accordance with some approved, standardized, certified, or accredited methodology but they are not fulfilling the true role of a forensic scientist.

It is the author's hope that this book will not only acquaint the reader with the many reconstructive aspects of firearms evidence but also inspire and reorient forensic scientists who examine such evidence. Firearms, expended cartridge cases, fired bullets, the wounds they inflict, the damage they produce, and the damage they sustain all tell a story. This book is intended to serve as a guide to understanding their language.

A couple of abbreviated quotes from G.G. Kelly, the first arms and ballistics officer for the New Zealand Police, say it all.

"The gun speaks . . . and the message of the gun is there to read by one who knows the language."

"The gun is a witness that speaks but once and tells its story with forceful truth to the interpreter who can understand the language."

"Everything that has a basis in physics is capable of being explained. All we have to do is to find the explanation."

REFERENCE AND FURTHER READING

Kelly, G.G., *The Gun in the Case*, Whitcombe & Tombs, Ltd, New Zealand, 1963.

CHAPTER 1

CASE APPROACH, PHILOSOPHY, AND OBJECTIVES

WHY THIS BOOK?

Many years ago the author was rigorously cross-examined by an excellent attorney who had put considerable thought and preparation into his questions. My work on the case was totally reconstructive in nature. My cross-examiner attempted to exclude my testimony on the basis that there was no such thing as "shooting reconstruction." He went on to claim that the term was something that I had made up. At the time I could not name a single textbook that was entitled *Shooting Reconstruction* that dealt specifically with shooting scene reconstruction or that had shooting reconstruction in its title. Neither could I name a forensic science textbook that even had a chapter devoted to this subject. To those who have familiarity with case law and tests of admissibility in the American legal system, his argument at the time was basically a *Frye* challenge.[1]

With the advent of the *Daubert* and *Kumho* decisions, future challenges are likely to be raised in American trials where reconstructive efforts have been undertaken in a shooting case and the results are offered in a trial. The idea for this book was the direct result of that cross-examination and is the product of nearly 40 years of applied research, casework, and trial experience in this specialized area of criminalistics.

---

[1] There was, in fact, a book that dealt almost exclusively with shooting incident reconstruction when the author was rigorously cross-examined on this subject some 20 years ago. The book, written by G.G. Kelly and first published in 1963, is entitled *The Gun in the Case*. It is long out of print but a good read if one can find a copy. G.G. Kelly was the Arms and Ballistics Officer for the New Zealand Police from 1929 to 1955. While I survived my cross-examiner's attack and my testimony was allowed in the trial alluded to, I nonetheless wished that I had known at the time of this wonderful and fascinating book.

RECONSTRUCTION—THE ULTIMATE GOAL OF CRIMINALISTICS

It may be useful to pause for a moment and consider the very concept of reconstruction and whether it is a legitimate function of forensic science. Probably the best quotes on this subject come from a contemporary textbook on criminalistics by DeForest *et al.* (1983) and are as follows:

> Physical evidence analysis is concerned with identification of traces of evidence, *reconstruction of events from the physical evidence record*, and establishing a common origin of samples of evidence. (p. 29)
>
> Reconstruction can assist in deciding what actually took place in a case and in limiting the different possibilities. (p. 45)

Eyewitnesses to events are notoriously unreliable even when they exist. People have trouble accurately remembering what they saw. It is even more of a problem if a complex series of events takes place suddenly and unexpectedly.

> Reconstruction may provide the only "independent witness" to the events and thus allow different eyewitness accounts to be evaluated for accuracy.
>
> Crime-scene reconstruction techniques are employed to learn what actually took place in a crime. Knowledge of what took place and how or when it happened can be more important than proving that an individual was at a scene.—A skilled reconstruction can be successful in sorting out the different versions of the events and helping to support or refute them. (p. 294)

Events that arise out of the use or misuse of firearms offer some very special and unique opportunities from a reconstruction standpoint. The wide variety of firearms and ammunition types, the relatively predictable behavior of projectiles and firearms discharge products, the chemistry of many of these ammunition-related products, and certain laws of physics may be utilized to evaluate the various accounts and theories of how an event took place. To some degree this is little different than the well-known principles of traffic accident reconstruction where the "ballistic" properties of motor vehicles give rise to momentum transfer, crush damage, and trace evidence exchanges. These phenomena are routinely used to reconstruct such things as the sequence of events, the location of one or more impacts, approximate speeds of vehicles, etc.

In summary, and in fact, there are many criminalists and forensic firearm examiners who perform various types of shooting scene reconstruction. A distance determination based on a powder pattern around a bullet hole is probably the simplest example of a reconstruction. A shotgun range-of-fire determination based on pellet pattern diameter represents another common example.

This book is an effort to describe the various principles of scene reconstruction as they relate to shooting incidents.

BASIC SKILLS AND APPROACH TO CASEWORK

From the very onset the true forensic scientist must be proactive by finding out what the case is about. From this, he or she must then make certain scientific assessments, define the important issues and questions in the case, ascertain what is in dispute, and then ultimately design a testing protocol based on the information derived from these previous efforts. He or she must focus on the *issues* in the case itself and not just the items of physical evidence within the case. One's first action should *not* be placing an evidence bullet on a scale to get its weight or test-firing a submitted gun to verify its operability. Rather it should and must be a reasoning process after making inquiry into the facts and issues in the particular case. This has always been and remains within the forensic scientist's control even in a laboratory that has been reduced to a clinical model. It simply requires that the analyst pick up the telephone and call the submitting investigator or attorney handling the case and ask a few key questions such as

- Tell me about this case. What are the issues?
- What do any witnesses to the incident say happened?
- Did the shooter provide an explanation?
- What is and what is not in dispute in this case?
- What are the competing hypotheses (theories)?
- What do you believe happened?
- What does the autopsy report* reveal? [*medical records if a gunshot wound without a fatality]
- What other evidence has been collected beyond that which has been submitted to the laboratory?

This last question is an important one that is often overlooked. It is not uncommon that investigators may select and submit only those items that they have concluded are relevant. This typically comes about from some restricted or narrow view that they have taken regarding the incident. The effect is to often blindside the laboratory analyst. It is scientific thinking, not the advanced technology now available in most laboratories, that is the means for solving problems. This book is about *thinking* and asking questions long before any effort is undertaken to answer them.

Those individuals addressing reconstructive issues must have good visualization skills and a fundamental understanding of firearms evidence, firearms design and operation, ammunition construction and basic ballistics (interior, exterior, and terminal), and the behavior of various materials when struck by projectiles.

A thorough study of the specific firearm(s) and ammunition involved in the case may be necessary. Once the issues have been defined in a particular case, the forensic scientist should begin by asking: "Is there anything about the firearm(s), its (their) operation, the ammunition, the purported events involved in this case that will allow the competing explanations or theories to be tested and evaluated?"

QUALIFICATIONS

Who should be doing this work and what should be their qualifications? In the author's view a degree in one of the physical sciences is desirable but not absolutely necessary. The advantage such a degree offers is a firm basis in scientific methodology and data evaluation but it does not insure that such an individual will use them. An individual who is both firearms-knowledgeable and interested in firearms *is* a requirement. For the proper and successful performance of this work, the analyst must have special knowledge and experience in the following areas in order to comprehensively reconstruct the wide variety of shooting incidents:

- the method of operation of the firearm(s) involved and the Class Characteristics of the firearm(s)
- small arms ammunition and projectile design characteristics critical to shooting reconstruction in general and to the case under investigation
- small arms propellants: their physical forms, basic chemical properties, and performance characteristics
- gunshot/powder residue pattern production, analysis, and interpretation
- fundamental exterior and terminal ballistics properties of projectiles to include:
    - "bullet wipe"
    - "lead splash"
    - bullet deformation due to impact
    - bullet destabilization due to intervening objects
    - bullet deflection due to ricochet and/or impact with intervening objects
    - cone fractures in glass and like materials
    - crater and/or spall production in frangible materials
- the nature of bullet perforation of thin materials such as sheet metal, glass, drywall, thin wooden boards, and vehicle tires

- bullet ricochet from:
  - yielding surfaces (soil, sand, water, sheet metal, wood, drywall)
  - frangible surfaces (cinder blocks, bricks, garden stepping stones)
  - non-yielding surfaces (concrete, stone, marble, heavy steel)
- the concept of *critical angle* as it relates to ricochet
- the examination and interpretation of ricocheted/deflected bullets
- the post-impact behavior of ricocheted/deflected bullets
- the recognition, examination, testing, and interpretation of bullet impact sites to include directionality determinations in non-orthogonal impacts through lead-in marks, lead splash, pinch-points, and fracture lines in painted metal surfaces
- trace evidence considerations and interpretation on recovered bullets and bullet impact sites
- the ability and skill to use various chemical reagents and tools associated with shooting incident reconstruction to include:
  - chemical tests for propellant residues and bullet metals (copper, lead, and nickel)
  - string lines
  - small, portable lasers
  - specialized dowel rods ("trajectory rods")
  - plumb bob and line
  - angle measuring devices (inclinometers, angle-finders, special protractors)
- methods for measuring and documenting the vertical and azimuth components of a projectile's path
- knowledge of basic trigonometric functions and calculations
- the proper use of the sodium rhodizonate test for lead and DTO and 2-NN tests for copper at or in suspected bullet impact sites
- cartridge case ejection behavior, factors affecting cartridge case ejection, interpretation and limitations associated with cartridge case location(s)
- contemporary shotshell construction
- the exterior ballistic performance of shot, wads, shotcups, and buffering material
- shotgun pellet pattern examination, extraction of pellet patterns on uneven surfaces and/or non-orthogonal impacts
- range-of-fire determinations in shotgun shootings
- contemporary exterior ballistics programs and the forensic application to include:
  - an understanding of the basic forces acting on a projectile in flight
  - the concept and use of ballistic coefficient with exterior ballistics programs
  - projectile flight path (trajectory profile)
  - line of sight vs. bullet path
  - the calculation of down range velocity
  - the calculation of flight time

- the concept of "lagtime"
- departure angle
- angle of fall
- the potential effect of environmental parameters on a projectile's flight
- the proper documentation of results and report writing.

GENERAL PHILOSOPHY

Question: What is it that we are setting out to prove in any case whether it has reconstructive aspects or is a simple comparison of a bullet to a submitted firearm? Before the reader spends much time pondering this question, I will answer it—*nothing*! I would urge every forensic scientist to heed the advice of two people: Dr P.C.H. Brouardel—a French medico-legalist (ca. 1880) who wrote:

> If the law has made you a witness, remain a man of science. You have no victim to avenge, no guilty person to convict nor innocent person to save—You must bear testimony within the limits of science.

The second is a statement by Dr Ed Blake, the well-known forensic serologist, who once said,

> If, in your analysis, you do not consider reasonable alternative explanations of an event, then what you are doing is not science.

Another useful approach to self-preservation in the courtroom is to contemplate your own cross-examination. As you work through the case, think of what questions you would ask if you were allowed to play lawyer-for-a-day and you wanted to expose any weaknesses or shortcomings in the analysis you conducted and the opinions you formed. After all, this *is* the basic mission of any attorney confronted with an opposing expert witness. Who knows best where you might have done a more thorough job than you? If the hypothetical cross-examination questions that you contemplate have merit and can be answered by some test or examination, you would be well advised to carry them out before issuing your report or appearing at trial. And if you have been thorough in this self cross-examination process, virtually any other questions that might be put to you at trial or deposition should pose no real challenge.

THE SCIENTIFIC METHOD

This whole discussion regarding a philosophy for casework quite naturally leads into a discussion of *The Scientific Method*. Since this *is* the approach we should be using in our evaluation and analysis it might do well to restate it.

(Besides, it can be surprisingly difficult to find a description of *The Scientific Method* when requested to explain it.) As a possessor of this book you will now have a ready source should the need arise. *The Scientific Method* is simply a way of thinking about problems and, ideally, solving them. In many instances the solution is so rapid and straightforward that the analyst who solved a particular problem may concede that he or she did not set down a written protocol before attacking the problem. In more complex situations, the analyst may be required to revise the hypothesis at the end of the process and modify the previous experiments or tests. This loop back to the initial steps of the Method may take place several times after the latter steps have been completed. Nonetheless, the steps of *The Scientific Method* will allow the problem, its analysis, and its solution to be explained in an orderly manner. *The Scientific Method* has at least five steps:

1. *Stating the Problem*: e.g., Can the distance from which a fatal shot was fired be determined?
2. *Forming a Hypothesis*: In doing so, the scientist considers what he or she knows about the problem, e.g., at close range gunshot residues (GSRs) will be deposited around the bullet hole or entry wound and, with appropriate materials and methodology, the characteristics of such residues can be used to establish the approximate muzzle-to-object distance.
3. *Experimentation and Observation* (Data Collection): Identifying and evaluating the effect of any variables that reasonably stand to affect the result is often an important initial consideration in the experimentation phase. In forensic science it is especially important that *all* observations be recorded or memorialized in some fashion so that the data can be reviewed by other scientists. This is, in part, because it may not always be possible to repeat the test or experiment with certain types of evidence after the passage of time or after certain types of tests are performed (e.g., powder patterns at selected distances with remaining evidence ammunition of a rare or unusual type).
4. *Interpreting the Data*: A careful study of the data (powder patterns from test-firings) provides the scientist with a means to evaluate the effect of the variables (distance) associated with the problem. The data should also provide a means of evaluating the reproducibility of the testing procedure or experiment (multiple shots at a fixed distance).
5. *Drawing Conclusions*: A conclusion regarding the problem stated in Step 1 may be drawn from the results of Steps 3 and 4. In some instances, a redesigning or modification of the test procedure or experiment may be deemed appropriate and additional data gathered before the scientist is able to draw meaningful conclusions.

The previous example of a distance determination is fairly straightforward: Question (Problem)—What was the distance from which a fatal shot was fired? Alternatively, the criminalist/firearm examiner may be presented with two conflicting accounts of the incident, e.g., the *shooter* says that he fired from distance "A" but an *eyewitness* says it was from range "B." Question—Can one of these accounts be refuted and the other affirmed? Or is either of these accounts supported by an analysis of the physical evidence?

From past experience and training the forensic scientist knows how GSRs are produced during the discharge of a firearm and how they behave with increasing distance between the muzzle and a struck surface (see the illustrations in the next chapter). We know how to set up and carry out test-firings with the responsible gun and like ammunition. The presence or absence of soot (smoke) deposits, the size of the powder pattern (diameter or radii) as well as the density of the powder pattern are all related to range of fire for a particular gun/ammunition combination.

These test patterns are compared with the GSR pattern of the decedent's clothing or other surface and the approximate muzzle-to-garment distance estimated. All of these matters are easy to set up, control, reproduce, document, and retain.

In summary, a forensic scientist should be able to describe the essential steps of *The Scientific Method*. A useful memory aid might be "PhD IC":

1. the *P*roblem
2. the *H*ypothesis
3. the gathering of *D*ata (experimentation/testing)
4. *I*nterpretation
5. *C*onclusions.

In addition to explaining *The Scientific Method*, he or she should be able to explain how his or her analysis conforms to this basic protocol.

This is, after all, the answer to the ultimate cross-examination question, "What method or procedure did you use in conducting your analysis and purported reconstruction of this incident?" *The Scientific Method* is not only an accepted method for carrying out any scientific inquiry; it is *the* method for all such inquiries. Carried out and documented properly, it allows reviewers, critics, opposing experts, and ultimately a court to evaluate your approach to the case at hand, your testing procedures, your data, your findings, and your subsequent conclusions. *The Scientific Method* supersedes all procedural "cookbooks" and rigid checklists that might exist for the routine examination of physical evidence. It is from *The Scientific Method* that all such procedures originated.

SPECIFIC CONSIDERATIONS

The reconstruction of shooting incidents may call upon one or more of the following:

- the presence of GSR deposits on skin, clothing, or other surfaces. Such deposits may be limited to sooty materials, vaporous lead deposits, or may include actual powder residue, unconsumed powder particles, and/or impact sites (stippling) produced by powder particles
- the pattern and density of such GSR deposits
- the physical form and/or chemical composition of the gunpowder in the ammunition associated with an incident and any powder present in a GSR deposit
- the chemical composition of the primer mixture used in the ammunition
- trace evidence around a bullet hole or at a bullet impact site (e.g., primer constituents, bullet lubricants, bullet metal)
- trace evidence *on* a recovered bullet (e.g., embedded glass particles, bone particles, paint particles, embedded fibers)
- the manufacturing features of the ammunition
- the design of a particular bullet
- the composition of a particular bullet (e.g., dead soft lead, lead hardened with antimony, lead alloys, copper jackets, brass jackets, aluminum jackets, steel jackets)
- trace evidence in or on a recovered firearm (e.g., blood and tissue in the bore)
- the cartridge case ejection pattern of a particular firearm (coupled with the location(s) of each expended cartridge case)
- the special exterior ballistic properties of shotgun ammunition (pellet patterns, wad behavior over distance)
- the terminal ballistic behavior of specific projectiles (orientation at impact, depth of penetration, and the degree and nature of deformation or expansion experienced by the projectile during penetration)
- the nature and distribution of secondary missiles generated during projectile perforation of intervening objects (may result in pseudo-stippling, satellite injuries, and damage to other nearby objects)
- ricochet behavior and characteristics of projectiles after impact with specific surfaces
- special attributes of some intervening objects that may permit the sequence of shots to be established (e.g., plate glass with intersecting radial fractures)
- special characteristics of projectile-created holes that allow the direction of the projectile's flight to be established
- the long range exterior ballistic performance of specific projectiles in long range shooting incidents

- visual considerations (e.g., the presence or absence of muzzle flash for a particular gun–ammunition combination)
- the nature and setting of the sights on a firearm (normally only of importance in long range shooting incidents)
- acoustical considerations (recorded gunshots, the sound of a bullet's arrival or passage at some downrange location, and "lagtime")
- the operational characteristics of the firearm to include any deficiencies or peculiarities
- the configuration of the firearm when found and recovered.

The fundamental concepts for the reconstruction of any shooting incident are:

- The relevant questions or issues must be identified early on and the potential reconstructive properties of the physical evidence recognized. Failing to do this may compromise or even obviate later efforts to reconstruct the incident.
- If you are to be a true forensic scientist, you *must* for the moment step out of your personal biases. (We *all* have them) Neither believe nor disbelieve the account provided by the shooter and/or eye and ear witnesses.
- Do not immediately accept or reject proposed explanations (hypotheses) offered by investigators, the prosecutor/plaintiff, the defendant's attorney, or the defendant. Listen attentively to any theory, account, or explanation. Taking some notes at this point might not be a bad idea. At some later time (probably while you are on the witness stand or in a deposition) you will be asked— "Did you consider the possibility that——?" or "Did you evaluate the account given by Mr. ——?" "No, I didn't" or "I wasn't asked to do that" may be truthful answers but they are not very good answers. "That's not my job" ranks no better. These answers will likely be followed with the question—"So you only did what you were asked to do by——" (fill in the blank with one of the following choices: *the police department, the prosecutor, the plaintiff's attorney, the defense attorney*).
Ask yourself these key questions:

  "What is in dispute and what is *un*disputed?"
  "What do we know about this incident?"
  "How might the physical evidence resolve (support/refute) the various accounts, explanations, (hypotheses) offered for the particular event?"
  "Is there anything about this gun, this ammunition, this recovered bullet, etc. that would allow the various accounts (or hypotheses) regarding this incident to be tested?"

- The physical evidence should be a sounding board against which to test or evaluate the various explanations offered. Plausible explanations will resonate; implausible and impossible explanations will not.

A strong skepticism and distrust of eyewitness accounts is both justified and encouraged. It is quite common for individuals with *no* reason or motive to favor one side or the other to be in error in one or more respects regarding their recollections of a shooting incident. Guns that never were there are "seen" and were often "fired." The description of the actual gun given by a witness or victim is frequently fraught with errors. The number of shots recalled is often incorrect. The timing of events, the sequence of events, positions and movements of participants, the distances involved are often not supported by the physical evidence. Shooters, victims, and witnesses frequently suffer temporal and auditory distortions when shootings occur. It is more often the exception than the rule that the physical evidence squares with the accounts of eye or ear witnesses in every respect. The degree of agreement between recollection and physical facts shows little if any improvement when one examines the accounts provided by the actual participants in a shooting incident. This includes law enforcement officers of long experience. The sincerity and seeming credibility of one or more witness/participants cannot be regarded as "the truth" of the matter. If this is the case, then what need do we have of the laboratory? It is not that you should regard the witness as incompetent, dishonest, or, worse, a liar. Rather, it goes to the very heart of a forensic scientist's role and that is to simply, objectively, and dispassionately test each account or hypothesis offered. It will also serve you well to think again of Dr Blake's warning and use your own intellectual skills in postulating any other *reasonable* alternative explanations for an event when you design your testing protocol for the matter under investigation.

It should also be recognized that each and every event in a shooting incident can seldom be completely reconstructed. The discharge of a firearm, the subsequent flight of a bullet over relatively short distances followed by its impact and penetration into a medium typically occur in very short time intervals amounting to a few hundredths or even thousandths of a second. These are time intervals much too quick to be observed by the human eye and recorded by the brain. However, the behavior of projectiles in flight and during the penetration of objects follows certain laws of physics and generates unique physical features and characteristics. These physical features and characteristics, preserved in the static aftermath of the incident, can often be utilized to reconstruct the flight path of the particular bullet. Such shot-by-shot reconstructive efforts in a multi-shot incident should be thought of like photographic snapshots where the object(s) struck appear to be stationary even though they might have been in motion at the time. Although the events taking place between shots can seldom be ascertained from these ballistic "snapshots" alone, many questions can be answered by integrating these "snapshots" with other information or evidence. It may be possible, for example to exclude certain theories or accounts of a shooting incident and support others. In the ideal case, it may be possible to eliminate all but one theory or explanation of an incident and arrive at a point where all the available physical evidence supports only one explanation or account of the incident. It should also be kept in mind that a thorough evaluation of an incident and examination of the physical evidence may also permit future questions or hypotheses to be answered.

Finally, I would remind the reader that the foregoing paragraph is nothing more than a restatement of *The Scientific Method*. For those looking for a simpler means of stating it, I might suggest the classic author Sir Arthur Conan Doyle and the Sherlock Holmes story, *The Sign of Four*, Chapter One. "Eliminate all other factors and the one which remains must be the truth," Holmes tells Dr Watson. And when Watson forgets this advice at a later point in the story (Chapter 6), Doyle, speaking through the fabled detective, says, "—apply my precept, when you have eliminated the impossible, whatever remains however improbable, must be the truth." Still good advice 100+ years later.

CONCLUDING COMMENTS

A considerable variety of interior, exterior, and terminal ballistic phenomena, reconstruction techniques, microchemical test procedures, trace evidence considerations, and laboratory examinations are presented in the subsequent chapters of this book. In one way or another they are all directed toward an effort to evaluate what did and what did not occur in a shooting incident.

The various objectives of Shooting Incident Reconstruction are listed below. They are to determine:

- the *range* from which a firearm was discharged
- the *position* of a *firearm* at the moment of discharge
- the *orientation* of a *firearm* at the moment of discharge
- the *position* of a *victim* at the moment of impact
- the *orientation* of a *victim* at the moment of impact
- the *number of shots* in a multiple discharge shooting incident
- the *sequence of shots* in a multiple discharge shooting incident
- the presence and nature of any *intervening material* between the firearm and the victim or struck object
- the *effect* of any intervening material on the subsequent exterior/terminal ballistic performance of projectiles
- the probable *flight path* of a projectile
- the *manner* by which a firearm was discharged
- other *exterior* and/or *terminal ballistic events* that may have special significance in a particular case.

CASE DECISIONS RELATED TO THE ADMISSIBILITY OF SCIENTIFIC EVIDENCE

Frye v. U.S. 293 Fed. 1013 (D.C. Cir. 1923).

Daubert v. Merrell Dow Pharmaceuticals, Inc. 509 U.S. 579, 113 S.Ct. 2786, 125 L.Ed.2d 469 (1993).

Kumho Tire v. Carmichael, 526 U.S. 137 (1999).

REFERENCES AND FURTHER READING

Burrard, G., *The Identification of Firearms and Forensic Ballistics*, A.S. Barnes and Co., NY, 1962.

Davis, J., *Toolmarks, Firearms and the Striagraph*, Charles C. Thomas, Springfield, IL, 1958.

De Forest, P.R., R.E. Gaensslen and H.C. Lee, *Forensic Science—An Introduction to Criminalistics*, McGraw-Hill Publishing Co., NY, 1983.

Faigman, D.L., D.H. Kaye, M.J. Saks and J. Sanders, *Modern Scientific Evidence*, Vol. 2, West Publishing Co., St. Paul, MN, 1997.

Hatcher, J.S., *Hatcher's Notebook*, 3rd edn, The Stackpole Co., Harrisburg, PA, 1966.

Hatcher, J.S., *The Textbook of Pistols and Revolvers*, Wolfe Publishing Co., Prescott, AZ, 1985.

Hatcher, J.S., F.J. Jury and J. Weller, *Firearms Investigation, Identification and Evidence*, The Stackpole Co., Harrisburg, PA, 1957.

Kelly, G.G., *The Gun in the Case*, Whitcombe & Tombs, Ltd, New Zealand, 1963.

Kirk, P.L., "The Ontogeny of Criminalistics," *Journal of Criminal Law, Criminology and Police Science*, Vol. 54, 1963, pp. 235–238.

Kirk, P.L., *Crime Investigation*, 2, ed. J. Thornton, John Wiley & Sons, NY, 1974.

Mathews, J.H., *Firearms Identification*, Vol. I, II, III, Charles C. Thomas Publisher, Springfield, IL, 1962.

Moenssens, A., F.E. Inbau and J.E. Starrs, *Scientific Evidence in Criminal Cases*, 3rd edn, The Foundation Press, Mineola, NY, 1986.

O'Hara, C.E. and J.W. Osterburg, *An Introduction to Criminalistics*, 2nd edn, Indiana University Press, Bloomington, IN, 1972.

Saferstein, R., *Criminalistics—An Introduction to Forensic Science*, Prentice Hall, Englewood Cliffs, NJ, 1977.

Saferstein, R., ed., *Criminalistics—Forensic Science Handbook*, Prentice Hall, Englewood Cliffs, NJ, 1982.

Saferstein, R., ed., *Criminalistics—Forensic Science Handbook*, Vol. III, Regents/Prentice Hall, Englewood Cliffs, NJ, 1993.

Svensson, A., O. Wendel and B.A.J. Fisher, *Techniques of Crime Scene Investigation*, 4th edn, Elsevier Scientific Publishing Co., NY, 1987.

Thorwald, J., *The Century of the Detective*, Harcourt, Brace and World, NY, 1964.

Warlow, T.A., *Firearms, the Law and Forensic Ballistics*, Taylor & Francis Inc., Bristol, PA, 1996.

CHAPTER 2

THE RECONSTRUCTIVE ASPECTS OF CLASS CHARACTERISTICS AND A LIMITED UNIVERSE

BULLET DESIGN AND CONSTRUCTION

Class Characteristics consist of the intended features of an object. The Class Characteristics of bullets would include such obvious things as caliber, weight, method of construction, composition, the design and location of any cannelures, base shape, heel shape, nose shape, and any number of more subtle features. In our normal laboratory efforts they provide a quick and ready sorting process that can quickly pare down the choices of source for a fired bullet. They are not ordinarily thought of as a means of identification but in situations where we are presented with a limited universe, Class Characteristics can provide definitive answers in shooting reconstruction cases. Figure 2.1a,b shows two views of a selection of unfired .38-caliber and 9 mm bullets. From left to right, these consist of a cannelured Winchester aluminum-jacketed bullet, a nickel-plated Winchester jacketed hollow point (JHP) bullet, a Russian full metal-jacketed (FMJ) bullet with a copper-washed finish over a steel jacket, a Remington JHP bullet with a scalloped jacket, a Federal *Hydra-Shok* bullet, a Winchester *Black Talon* bullet with a black copper oxide finish, a CCI-Blount *Gold Dot* JHP bullet, and a Remington *Golden Saber* bullet possessing a brass jacket. Each of these bullets possesses certain distinguishing Class Characteristics. Figure 2.2 shows each of these bullets after discharge and recovery from a tissue simulant. This reveals some additional manufacturing features of potential value for certain bullets such as the central post in the Federal *Hydra-Shok* and the "talons" on the Winchester *Black Talon* (subsequently renamed the *Ranger SXT*).

Figure 2.1

Two Views of a Selection of Unfired .38-caliber and 9 mm Bullets (see Color Plate 1)

(a)

(b)

Profile and base views of eight representative bullets. From left to right:

[1] a cannelured Winchester aluminum-jacketed bullet
[2] a nickel-plated Winchester JHP bullet
[3] a Russian FMJ bullet with a copper-washed finish over a steel jacket
[4] a Remington JHP bullet with a scalloped jacket
[5] a Federal *Hydra-Shok* bullet
[6] a Winchester *Black Talon* bullet with a black copper oxide finish
[7] a CCI-Blount *Gold Dot* JHP bullet and
[8] a Remington *Golden Saber* bullet possessing a brass jacket.

Figure 2.2

The Same Selection of Bullets in Figure 2.1a,b after Discharge into a Tissue Simulant (see Color Plate 2)

The upper row consists of unfired specimens. The middle and bottom rows provide two examples and views of each of these bullets after discharge into a tissue simulant. Note the unique, surviving characteristics of many of these bullets.

HYPOTHETICAL CASE EXAMPLES

The following case examples employ the concept of a *limited universe*. A limited universe represents a situation where there are a finite and limited number of choices for an event. In these cases the analyst is typically presented with two or three types and brands of ammunition whose sources are known or have been established. It is understood and established that these limited choices are the *only* choices for the particular event. An eliminative process for all but one contender and subsequent correspondence in class characteristics of this only remaining choice establishes an identity of source where there is a limited universe of candidates.

CASE 1

Consider a situation where an innocent bystander is killed by an errant shot in a multi-agency police operation. Three law enforcement agencies are involved in the attempted arrest of an armed and highly dangerous subject. One or more members of each agency ultimately fired shots during an exchange of gunfire with the subject. The fatal bullet passed through the victim and was never found. A portion of the bullet's jacket was recovered from the wound track, however. The initial laboratory report describes this item as a fragment of a bullet jacket that lacks any rifling impressions and is therefore *un*suitable for identification purposes.

Agency "A" carried and fired 9 mm Winchester *SilverTip*. Agency "B" carried and fired Federal *Hydra-Shok* and agency "C" Remington *Golden Saber*. The armed subject fired a revolver loaded with plain lead bullets. Given the limited universe for the source of this fatal injury, this case can be solved on the basis of the differing jacket compositions for these three bullets: nickel-plated gilding metal for the *SilverTip*, *plain* gilding metal for the Federal *Hydra-Shok* and brass for the Remington *Golden Saber*.

CASE 2

Let us modify Case 1 to the extent that the innocent bystander lives and has a partially-expanded bullet in her body (visible on X-ray radiographs). This bullet is in an area where the treating doctors conclude that it is safer to leave the bullet in her body rather than remove it. Four agencies fired their handguns in this hypothetical example using the following ammunition: Winchester *SilverTip*, Winchester *Black Talon*, Federal *Hydra-Shok*, and Remington *Golden Saber*. As before, the armed suspect fired a revolver loaded with plain lead bullets. How might the question of responsibility be resolved in this situation?

A possible solution resides in a pair of X-ray films—one in the lateral view and one in the anterior-posterior (A/P) view. It would be quite surprising if such X-ray films did not already exist among the victim's medical records. If this is the case, lateral and A/P films should be requested with a concerted effort to get the clearest possible views of the projectile. If they do not exist, then additional X-ray films should be prepared. If either the barb-like talons of a *Black Talon* bullet or the central post of the *Hydra-Shok* bullet can be seen in one of these films, the question is answered. It would also be answered upon the appearance of the classic profile of an unexpanded, round nose lead bullet of the type contained in the suspect's revolver. Given the differences in the jacket compositions of the law enforcement ammunition, SEM/EDS analysis of the "bullet wipe" around the entry hole in the outermost garment could also result in a resolution of this case.

PROPELLANT MORPHOLOGY

The example of a distance determination based on a powder pattern around a bullet hole in clothing has been previously cited as a simple example of a shooting reconstruction. Figure 2.3 illustrates the conical expulsion of partially burned and unburned powder particles from the muzzle of a handgun at discharge. It is this

Figure 2.3

Gunshot Residue Production during the Discharge of a Semi-Automatic Pistol

In this photograph the bullet is just a few inches beyond the muzzle. Numerous particles of partially burned gun powder have emerged from the muzzle in a conical distribution. A cloud of soot or "smoke" is also visible in the area of the muzzle. A faint plume of sooty material can also be seen escaping upward from the chamber area. The slide of this semi-automatic pistol has just started to move rearward and the fired cartridge is still in the chamber.

predictable and reproducible phenomenon that has served criminalists and firearm examiners as the basis of such distance determinations for decades. These powder particles also possess (and frequently retain) physical attributes that can be exploited to solve certain shooting reconstruction questions.

While it is beyond the scope of this book to describe the various manufacturing methods and the chemistry of classic and modern small arms propellants, the common physical forms are easily illustrated in Figures 2.4a–2.4i. These figures show seven (7) distinct forms of contemporary smokeless gunpowder followed by four granulations of Black Powder and Pyrodex RS (a black powder substitute) on 1/8 in. grids. Since no firearm–ammunition combination is 100% efficient in burning all of the powder in a cartridge, a few to many particles of unburned and partially burned propellant may be left behind in the fired cartridge case, in the chamber in which the cartridge was fired, in the bore of the firearm, and, of course, deposited on objects or surfaces in close proximity to the muzzle. The cylinder gaps of revolvers also represent a source of such deposits that have special reconstructive value as will be pointed out later in this chapter.

The following hypothetical case provides an example of the application of propellant morphology to the matter of shooting reconstruction. A subject known to have been in an altercation with three armed individuals in the parking lot of a bar is shot and killed by a single perforating gunshot wound to the chest. The three subjects were quickly apprehended and found to have the following guns: a 7.65 mm Walther PPK, a Lorcin .32 Automatic, and an Iver Johnson .32 S&W revolver. All three subjects admit to firing a shot but each claim to have discharged a "warning shot" into the air. The fatal bullet was never recovered. Two fired .32 Automatic pistol cartridges were found near the body. Initial laboratory examination established that a "Geco" brand cartridge was fired in the Walther PPK pistol and a Winchester brand cartridge was fired in the Lorcin pistol. The .32 S&W revolver was found to have one expended Remington brand cartridge under the hammer. All of these findings substantiate the admissions of the three suspects insofar as they having discharged their pistols. Live rounds of the corresponding brands were also found in each pistol. The medical examiner's autopsy report describes some powder stippling around the entry wound. The charging bureau at the prosecutor's office wants to know who to charge with murder and who to charge with lesser offenses related to firearms violations.

ANALYTICAL APPROACH

At this point I will expose the reader to a theme that will be repeated many times in this text. *What do we know about the problem?* It is the beginning step in *The Scientific Method*. All three of these firearms are essentially of the same caliber. Given the uncertainty associated with estimating the caliber of the

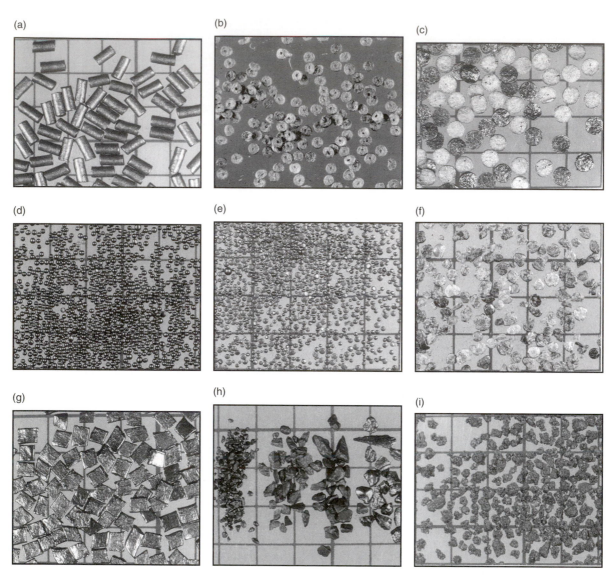

Figure 2.4

Common Physical Forms of Contemporary Smokeless Gunpowder, Black Powder and Pyrodex RS on 1/8-in. Grids

Legend: a = extruded tubular powder, b = perforated disk-flake powder, c = unperforated disk-flake powder, d = spherical ball powder, e = flattened ball powder, f = cracked ball powder, g = lamel powder, h = four granulations of black powder, i = Pyrodex RS.

responsible firearm from bullet hole size in the victim's shirt and the same problem with the diameter of entry wounds in skin, such measurements can *not* lead to a valid resolution of this incident. The mention of powder stippling by the medical examiner offers considerable hope since the intervening clothing stands to have filtered out some of the powder particles. If the fatal wound was sustained in bare skin, the retention of some representative powder particles from the stippled area by the medical examiner is critical to the solution of this case. In this hypothetical example subsequent examination of the victim's shirt reveals numerous particles of spherical ball powder (see Figure 2.4d) around the bullet hole.

Examination of the "Geco" brand ammunition, the fired "Geco" cartridge, and the bore of the Walther pistol all reveal lamel-form powder residues (see Figure 2.4g). The Iver Johnson revolver and its Remington ammunition show unperforated disk-flake powder (see Figure 2.4c). Examination of the fired Winchester cartridge from the Lorcin pistol reveals ball powder residues as does a tight-fitting cleaning patch pushed through the bore of this pistol prior to any test-firing. The disassembly of several of the live Winchester cartridges from the Lorcin's magazine also reveals the propellant to be spherical ball powder. By simple inspection of the Class Characteristics of the propellants and propellant residues, the Geco and Remington shooters are *excluded* and the shooter of the Winchester ammunition is *included*.

Revolvers offer another source and dimension insofar as GSR and powder deposits are concerned. Such residues not only emerge from that muzzle but also emerge in an oval or fan-shaped pattern from the right and left sides of the cylinder gap. These hot and highly energetic gases can blast or burn a characteristic pattern into any surface immediately adjacent to the cylinder gap (Figure 2.5). Figure 2.6a,b illustrate the reconstructive value of muzzle and cylinder gap deposits. Cylinder gap deposits are of special value in possible suicide cases, alleged struggles over a revolver, purported accidental discharges in holsters, or when the revolver in question was placed on or against some surface where it is claimed to have discharged. This subject will be revisited in Chapter 5.

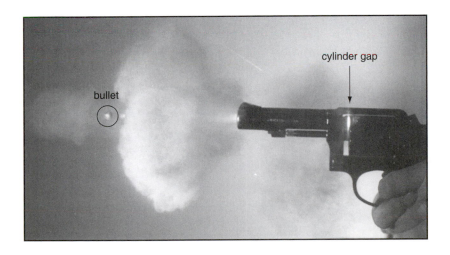

Figure 2.5

Gunshot Residue Production from a Revolver (see Color Plate 3)

Figure 2.6

Diagram 1: Gunshot Residue Deposition from a Revolver

(a)

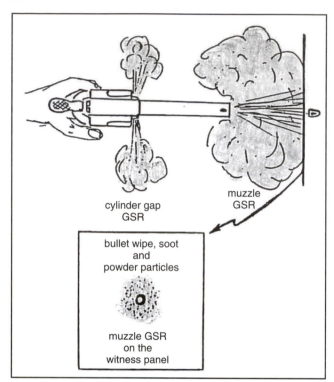

cylinder gap
GSR

muzzle
GSR

bullet wipe, soot
and
powder particles

muzzle GSR
on the
witness panel

(b)

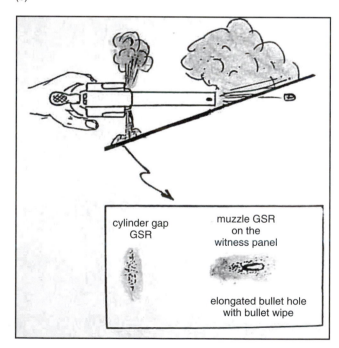

cylinder gap
GSR

muzzle GSR
on the
witness panel

elongated bullet hole
with bullet wipe

THE WORTH OF WEIGHT

To some readers this topic may seem inappropriate to the subject of reconstruction. Others may conclude that it is so elementary that the subject is an insult to their intelligence. But sometimes it is the simplest of things that can solve a case. Something as basic as the weight of a projectile, a bullet core, or a fragment of a projectile can answer a reconstructive question. A number of otherwise very competent examiners have occasionally overlooked the obvious and simple solution to some of the following questions:

1. Is a bullet fragment part of a particular fragmented bullet or some other bullet?
   Answer: If the weights of the two items exceed the weight of the intact bullet, the fragment is from some other projectile.

2. Of what value is the weight of a severely deformed 22 rimfire bullet?
   Answer: A long rifle bullet can be differentiated from a 22 short or 22 long bullet.

3. Is a group of cast bullets from the same mould all of the same alloy?
   Answer: Differences in alloy composition will produce significant differences in bullet weights for bullets cast in the same mould.

4. What was the weight of the unfired bullet based on the weight or either a separated core or a separated bullet jacket?
   Answer: For each manufacturer's bullet, there is a relationship between the total weight of a bullet, its lead or lead alloy core and its jacket (see the table or core and jacket weights in the Appendix).

5. How can the total weight of live rounds of ammunition be useful?
   Answer: Consistency (or inconsistencies) in loading can quickly be detected. Significant differences (such as two different bullet weights, a missing powder charge, or a double powder charge) in otherwise visually indistinguishable ammunition can be detected by weighing the intact cartridges.

6. Of what value is the weight of intact cartridge cases vs. fragments of burst or separated cartridge cases?
   Answer: Weight can serve as a means of ascertaining whether the entire burst cartridge is represented by the fragments presently in the examiner's possession.

7. How can the weight of deformed shot pellets, buckshot, and/or spherical projectiles be useful?
   Answer: The pre-discharge *size* (shot size number or diameter) of shotgun pellets from badly deformed, but otherwise intact, pellets can be determined from their weight.

Table 2.1

Shot and Buckshot Sizes and Average Weights per Pellet

Shot size	T	BBB	BB	1	2	3	4	5	6	7	7½	8	8½	9
Dia. (in.)	.20	.19	.18	.16	.15	.14	.13	.12	.11	.10	.095	.09	.085	.08
Dia. (mm)	5.08	4.83	4.57	4.06	3.81	3.56	3.30	3.05	2.79	2.54	2.41	2.29	2.16	2.03
wt.—Pb (mg)	771	663	561	394	325	265	211	167	128	96	82	71	59	49
wt.—Fe (mg)	541	465	394	276	228	186	148	117	90	68	—	—	—	—
Buckshot														

Shot size	000	00	0	No. 1	No. 2	No. 3	No. 4
Dia. (in.)	.36	.33	.32	.30	.27	.25	.24
Dia. (mm)	9.14	8.38	8.13	7.62	6.86	6.35	6.10
wt.—Pb (g)	4.49	3.46	3.16	2.60	1.90	1.51	1.33

Note: The weights of shot are in milligrams and those of buckshot are in grams.

The last question deserves special attention. Table 2.1 lists the nominal weights in grams and milligrams for American shot sizes. This table also gives the approximate diameter of these shot sizes in English and metric units.

It might also be useful at this point to recall that the diameter of American shot sizes in inches can be derived from the following equation:

$$\text{diameter (inches)} = (17 - \text{shot size \#}) \div 100$$

Example: #6 shot gives 0.11 in. for its diameter from this equation.

The diameter of deformed spherical lead projectiles such as those fired from muzzle-loading rifles and cap and ball revolvers can also be determined from their weight as will be demonstrated. This is especially useful to the battlefield archeologist. Prior to the mid-1800s nearly all firearms fired spherical lead projectiles. Some firearms continued to employ such projectiles during and immediately after the American Civil War. The majority of these were percussion (cap and ball) revolvers. These firearms quickly faded from the scene with the advent of cartridge-firing arms. The renewed interest in these historic firearms, however, has led to the manufacture of numerous, fully functional replicas. On rare occasion such guns have been involved in accidental shootings and even employed by criminals in the commission of a crime. Insofar as modern arms

Table 2.2

*Properties of Interest for
Metals Used in Projectiles*

	Steel/Iron (Fe)	Copper (Cu)	Tungsten (W)	Lead (Pb)	Bismuth (Bi)	Antimony (Sb)
Atomic #	26	29	74	82	83	51
Atomic weight	55.8	63.5	183.8	207	209	121.8
Melting point (°C)	1535	1083	3410	327.5	271.3	630
Specific gravity (at 20 °C)	7.874	8.96	ca. 19	11.35	9.747	6.68
Density (% of Pb)	69.4	78.9	167	—	85.9	58.8
Hardness (Mohs*)	4–5	2.5–3	6.5–7.5	1.5	2–2.5	3.0–3.3

* *Note*: The Mohs hardness scale sets talc as 1 and diamond as 10.

are concerned, spherical lead projectiles are almost exclusively associated with shotgun ammunition in the form of buckshot and the smaller shot sizes primarily used for bird and small game hunting. There are also some pistol and revolver cartridges available that are loaded with small shot. The compositions presently available are lead (both dead soft lead and hardened lead), steel, bismuth, and tungsten-impregnated polymer spheres. Table 2.2 describes some of the physical properties of interest for these metals. Copper has been included because of its use in bulleted ammunition and contemporary frangible projectiles. Antimony is added in relatively small amounts (typically 0.5–5%) to harden lead.

DERIVATION OF SPHERE DIAMETER FROM WEIGHT

In the case of lead spheres, the formulas on the following pages derived from the equation for the volume of a sphere and the density of lead are quite useful in calculating the original diameter of a lead ball. The weight of a sphere composed of any of these metals is directly related to its diameter. This relationship is forensically useful since projectiles, particularly soft ones such as lead, will often deform upon impact. If no metal has been lost during terminal ballistic deceleration, the *weight* of a deformed spherical projectile can be used to derive its original diameter or caliber. Table 2.1 has already revealed how this would be useful for deformed shot from shotguns. Any loss of material can usually be determined by a careful inspection of the deformed lead ball under the stereo-microscope. The diameter of a lead ball is closely related to (but usually not identical to) the caliber of the muzzle-loading firearm from which it was discharged. This concept will be revisited later in this section.

The mathematical derivation for the relationship between the weight of a spherical projectile and its diameter is as follows:

The formula for the volume (V) of a sphere is $4/3\pi r^3$ where r is the radius of the sphere. This formula can be rewritten on the basis of diameter $(d = 2r)$ and simplified to give:

$$V = 0.5236d^3 \tag{1}$$

From the simple relationship between weight, density, and volume $W = D(V)$, a general expression relating the weight (W) of a spherical projectile to its diameter (d) would be:

$$W = 0.5236d^3(D) \tag{2}$$

LEAD SPHERES

The density of pure lead in *grains* per *cubic inch* is 2873.5. These units have been selected since American calibers are usually given in inches and projectile weights are in grains. The metric equivalent for the density of lead in *grams* per mm^3 is 0.011345. This value is useful for projectiles weighed in grams and their diameters measured in millimeters.

For *lead* spheres, equation 2 can be further reduced by inserting these density values to give:

$$W \text{ (in gr.)} = 1504.6d^3 \text{ (} d \text{ in in.) or}$$

$$d = 0.08727 \ W^{1/3} \tag{3}$$

The metric equivalent for d in mm and W in grams is $W = 0.005940d^3$ or

$$d = 5.522 \ W^{1/3} \tag{4}$$

Table 2.3 provides a partial list of commercially produced lead balls for flintlock and percussion rifles, pistols, and revolvers. This table was assembled to illustrate the value of these equations and to provide some insight into the varied sizes and sources of such projectiles as well as a check of the calculated weights (in gr.) vs. the actual weights.

The partial list of lead spheres in Table 2.3 are contemporary, *swaged* balls from various commercial sources. "Bullet" moulds, both contemporary and historical, for casting round balls are readily available. Some of these moulds cast balls in sizes that fall between the values in Table 2.3. Projectiles made in this manner (as opposed to the modern swaging process) are likely to show a

Table 2.3
Commercially Manufactured Spherical Lead Balls

Diameter (in.)	Rifle/pistol (caliber)	Source*	Calculated Wt. (gr.)	Measured Wt. (gr.)
.310	.32	H,C,W	44.8	45
.350	.36	H,S,D,W	64.5	65
.375	.36 rev.	H,S,W	79.3	80
.395	.40	H,W	92.7	93
.440	.45	H,S,D,C,W	128	128
.445	.45	H,S,D,W	133	133
.451	.44 rev.	H,S,W	138	138
.454	.44 rev.	H,S,W	141	141
.457	.44 rev.	H,S,W	144	143
.495	.50	H,S,W	182	182
.530	.54	H,S,D,W	224	225
.535	.54	H,S,W	230	230
.570	.58	H,S,W	279	278
.690	.69/.73	D,W	494	494
.735	.75	W	597	593

* *Sources:* Hornady (H), Speer (S), Denver Bullet Co. (D), CVA (C), and Warren Muzzleloading (W).

casting seam and a sprue mark. Additionally, cast balls may be alloyed with other metals such as tin and antimony both of which will lower the density of the resultant projectile. Lyman's No. 2 bullet metal, for example (a popular lead alloy composed of 90 parts of lead, 5 parts of tin, and 5 parts of antimony), has a density that is 95.7% that of pure lead and a hardness of 15 on the Brinell scale as compared to a Brinell Hardness Number (BHN) of 4 for pure lead. The diameters of spherical lead balls and their relationship to the caliber of the muzzle-loading firearms from which they came can be somewhat confusing. Muzzle-loading single shot pistols and rifles were most often loaded with a *patched* ball—i.e., a swatch of cloth, usually circular and on the order of 0.015 in. thick. Pillow ticking and fine woven linen were common choices for patching material. Very thin deer skin or other types of animal skins were also known to have been used during the era of muzzle-loading firearms. With these firearms the ball was slightly *under*sized and was held in place against the powder charge by the snug-fitting patch. With a properly selected patch and powder charge, the ball never directly contacted the bore of the gun during loading or discharge. At best only faint vestiges of the rifling might print through the patch and on to the ball. The weave of the fabric selected for the patch may be seen embossed on the side or base of the fired ball. The patch itself survives the discharge process and represents important physical evidence. At very close range (a few inches) it will follow the projectile into a wound track. At more distant ranges, the patch will be found within a few yards of the discharge location of the gun.

Percussion revolvers with their front-loading cylinders used a very different approach. A lead ball of slightly *larger* diameter than the cylinder's chambers is mechanically forced down into the opening of each chamber with the ramming arm of the percussion revolver. It should be noted that the front of the ball typically receives a distinct and often identifiable imprint from the face of the ramming arm during the loading process and this mark may survive impact with "soft" targets such as muscle or other tissues. This rammer imprint should *not* be confused with imprints that might be left on a ball by a ball starter or ram rod used with muzzle-loading rifles and single shot pistols. The seated ball retains its position in the chamber of percussion revolvers prior to discharge due to the forced fit it underwent. The bore into which this ball will be driven during discharge is slightly *smaller* than the chamber from which it is expelled. Such a projectile makes *direct* contact with the bore of a percussion revolver (*un*like the patched ball method) and receives land and groove marking around the contacting circumference of the ball and with a diameter (before any impact deformation) equal to that of the bore of the gun. Whether fired from a percussion revolver or from a muzzle-loading pistol or revolver, the exterior ballistic performance of spherical lead projectiles is rather poor* compared to conical projectiles of the same caliber.

*In this context "poor" refers to a sphere's high drag and correspondingly poor ballistic coefficient compared to conical bullets. It does not suggest that spherical projectiles are inherently inaccurate.

STEEL AND BISMUTH SPHERES

The previous equations can be recalculated utilizing the densities of steel and bismuth. For mild steel/iron of 7.87 g/cc (1994 gr./in.^{3}), the relationships are:

$$W_{\text{gr.}} = 1044d^3$$

where d (diameter) is in inches and W (weight) is in grains

$$d = 0.09857W^{1/3}$$

For bismuth with a density of 9.75 g/cc (2469 gr./in.^{3}), the relationships are:

$$W_{\text{gr.}} = 1293d^3$$

where d (diameter) is in inches and W is in grains

$$d = 0.09180W^{1/3}$$

The more useful form of these equations is the final one relating the diameter of out-of-round or deformed spheres of lead, steel, or bismuth to the cube root

Lead	$d = 0.08727W^{1/3}*$
Bismuth	$d = 0.09180W^{1/3}$
Steel	$d = 0.09857W^{1/3}$

Table 2.4

Sphere Diameter in Inches from Weight in Grains for Three Metals

Note: * The cube root of a number can be determined on most contemporary pocket calculators possessing scientific keyboards.

of their weights. These have been restated along with the expressions for lead in Table 2.4.

SUMMARY

The various design and compositional features of projectiles can lead to the absolute exclusion of certain sources of shots and the identification of the specific source of a shot even though such projectiles or projectile fragments are non-identifiable by traditional comparison microscopy. This is possible through the concept of a limited universe.

The propellants used in small arms ammunition are seldom completely consumed during the discharge process and often leave recognizable particles in the bore of the firearm, in the fired cartridge case and on any object or victim in close proximity to the discharge of the firearm. Their varied physical forms and their exterior ballistic properties provide a means of reconstructing certain shooting incidents.

The simple matter of the weight of a bullet fragment, a separated bullet jacket, or deformed spherical projectile can resolve important questions in certain shooting incidents. This is a quick, non-consumptive measurement with reconstructive value that has been overlooked or not fully appreciated in the past.

REFERENCES AND FURTHER READING

Haag, L.C., "Some Forensic Aspects of Spherical Projectiles," *AFTE Journal* 30:1 (Winter 1998).

Haag, L.C., "Physical Forms of Contemporary Small-Arms Propellants and Their Forensic Value," *Am. Jour. of Forensic Med. and Pathology* 26:1 (March 2005).

Watkins, R.L. and L.C. Haag, "Shotgun Evidence," *AFTE Journal* 10:3 (July 1978).

CHAPTER 3

IS IT A BULLET HOLE?

THE QUESTION OF BULLET HOLES IN COMMON MATERIALS

Is a particular mark on, or a hole in, an object caused by a bullet? This can be a relatively common question to crime scene technicians and the forensic laboratory. The answer is easy when a tracking through the hole leads to a projectile. It may not be so easy when one is presented with a defect in some object and no bullet is clearly associated with it. The answer to the question relies in part on some basic properties of projectiles and principles of physics as well as a fundamental concept in forensic science. Last things first. *Locard's Exchange Principle* stands for the proposition that, in theory, there will be a mutual exchange between two objects that come in contact with each other. Pressing your hand against a chalky blackboard (now you have some idea how old the authors is) results in the transference of chalk dust to your hand and the deposition of visible body oils and perspiration on the blackboard. The mutual exchange of material between two objects that come in contact is the guiding principle in trace evidence analysis. This conceptual model is equally important and just as useful in the reconstruction of certain shooting incidents.

The various metals used to manufacture most bullets are all relatively soft (lead, copper, copper–zinc alloys, aluminum). The bearing surface of a fired bullet has also been galled and abraded as a result of its rather violent journey through the gun barrel. This will further promote the transference of bullet metal to a subsequent impact site. The bearing surface of a fired bullet also possesses a coating of GSR that is rich in primer constituents and carbonaceous soot from the propellant. All of these factors combine to produce and promote the transference of material from the bullet to nearly any impacted surface. Traces of these materials will almost always be deposited around the margin of a bullet hole or left in an impact site. This is particularly true in materials such as cloth, leather or wood in which the bullet essentially pushes its way into and through the material. These circumferential deposits are referred to as "bullet wipe." This takes the form of a dark ring around the margin of the bullet hole as depicted in Figure 2.6 in the previous chapter and a number of the photographs in this chapter. Exceptions to the transference of bullet wipe are frangible and brittle surfaces that shatter or flake away as the projectile makes its way into such materials. Sheet metal is another medium that generally does not take up bullet wipe well although metal transfers from the penetrating or perforating bullet may be present. Certain fabrics and garments take up or retain bullet wipe to differing degrees. Cotton takes up and retains bullet wipe well whereas some synthetic fabrics do not. It may be desirable in those situations where no bullet wipe can be seen or detected with optical and simple chemical methods to examine some selected and representative fiber ends from around the margin of the suspected bullet hole under a scanning electron microscope (SEM) equipped with energy dispersive X-ray spectrometer (EDX) attachment. This instrumentation can locate and identify extremely small amounts of adhering GSR *in situ* and without consuming or altering any of the residues.

Beyond the mere transference of trace materials from a bullet during its interaction with a struck surface, bullets also possess considerable kinetic energy in flight that is going to be applied to a relatively small area at the impact site. This not only enhances the transference of trace material from the bullet but also typically leads to characteristic damage to commonly encountered materials (wood, sheet metal, cloth, leather, plastic, rubber, glass, etc.). The case of nylon and polyester fabrics and garments deserve special mention. The brief but intense frictional and crushing action of a projectile forcing its way through either of these fiber compositions produces a unique change at the severed ends of the individual fibers. This takes the form of enlarged or swollen club-like ends around the margin of the hole. These can be seen with a stereozoom microscope. When viewed under the polarizing microscope, the strong birefringence present in the unaltered nylon or polyester fibers will be nearly or totally relieved at the bullet-severed fiber ends as a result of the momentary melting or softening on the fibers during bullet passage. This effect can be easily demonstrated with a few test shots and is quite different that what one will see with otherwise similar holes produced by simply poking a hole in a nylon or polyester garment with some object such as a pencil or even burning a hole in the fabric with a cigarette.

Depending on the nature of the struck object, the responsible bullet will correspondingly suffer damage associated with the surface it struck and frequently acquire trace evidence or characteristic imprints in the bullet from the particular surface or object. This is the other half of Locard's Exchange Principle in action. Bullets that strike the ground, concrete, asphalt and that perforate wood, glass, drywall, and fabrics will all take up adhering traces of these materials. The physical damage that such bullets suffer will also bear a relationship to the nature of the struck surface. Fabric imprints in lead that survive subsequent terminal ballistic events are often so clear that the particular weave and thread type can be seen and compared to any perforated garments or fabrics. Examples for a number of these interactions will be illustrated in Chapters 6 and 7 dealing with bullet penetration and perforation of materials and ricochet respectively.

EMPIRICAL TESTING

The characteristic damage to an impacted surface produced by a bullet should be relatively easy to discriminate from impacts by other objects such as stones, debris, irregular fragments from explosive devices, etc. As will often be the case, the examiner may need to carry out some empirical testing in order to satisfy himself- or herself as to the specific characteristics of bullet damage to the material under evaluation. This may ultimately include one or more test shots into a section or area of the actual evidence material as a definitive means of evaluating the nature of the bullet damage caused by the specific type of bullet, the Locardian transference of trace evidence between the bullet and material as well as any corresponding damage to the bullet. The use of an area in the evidence material to be shot is justifiable on the basis of reducing or eliminating variables that could be present when using other seemingly like materials for such tests. Such a site in a portion of the evidence material for empirical testing should be chosen and prepared with great care to assure that the subsequent tests do not alter or compromise the actual evidence site.

Although some readers may think that empirical testing is too time-consuming and perhaps unnecessary, it is strongly recommended for other reasons. First of all, it can be an integral part of *The Scientific Method*. Second, it can be very useful in persuading a skeptical court that your analysis, your evaluation of the evidence and your subsequent opinions have merit and validity. In designing or selecting a test protocol, start with what is known about the incident under investigation. The following example should be useful.

AN EXAMPLE OF A BULLET HOLE ISSUE—BULLET HOLES IN WOOD

A putative bullet hole has been found in a wooden fence board where a shooting incident took place and it is important to know (1) whether it is indeed a bullet hole, and (2) if it can be associated with one of two subjects known to have fired their guns toward this fence. Shooter A is known to have fired a .38 Special revolver loaded with 158 gr. lead round nosed (LRN) bullets. Shooter B fired a 9 mm semi-automatic pistol loaded with 124 gr. FMJ (gilding metal: 95% copper- and 5% zinc) bullets. The shape and diameter of these bullets are quite similar so the *size* of the hole in the board will *not* allow a resolution of this inquiry. Figure 3.1a,b shows entry and exit bullets holes in a soft pine board due to these two types of bullets.

Figure 3.1

Entry and Exit Bullet Holes in a Soft Pine Board Produced by 9 mm and .38-caliber Bullets

(a)

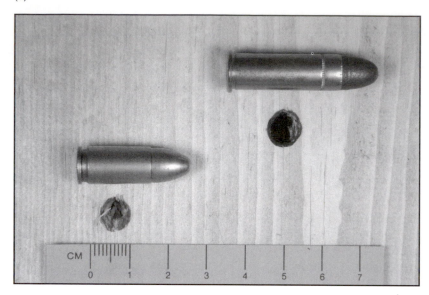

(b)

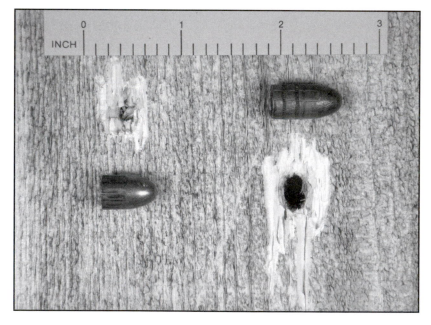

Note how the wood fibers have closed in to a much greater degree in the bullet hole produced by the FMJ 9 mm bullet (left) than with the LRN .38 Special bullet (right). The lead bullet has also deposited a much darker ring of bullet wipe as one would expect. The internal surface of the track produced by the lead bullet is also coated with dark gray lead deposits (not visible in these photographs) whereas the track produced by the FMJ bullet is free of any visible deposits.

CONSIDERATIONS AND SOLUTION

The physical features of a hole caused by bullets perforating wooden boards are straightforward and easy to recognize. The margin of the entry hole will be relatively smooth often with visible bullet wipe whereas the exit hole will typically have chips of wood dislodged from its margin and no bullet wipe. A simple test for lead, the *sodium rhodizonate test*, will show the presence of lead around the margin of this hole *if* it is in fact a bullet hole. The procedures for preparing this reagent and carrying out this chemical test are described in the next chapter. Optical inspection or photography in the infrared spectrum will typically reveal the IR-absorbent carbon in the bullet wipe. This technique is particularly useful when the background or surface is dark and any bullet wipe that might be present cannot be observed under normal lighting. Combined with the physical attributes of the hole in the fence board and the circumferential deposits of lead and carbon residues, question 1 can be answered in the affirmative.

However, the sodium rhodizonate test will be positive for bullet holes in soft wood whether they were produced by a LRN bullet or the FMJ bullet. This is true because the bullet wipe from the FMJ bullet contains lead from the priming mixture (lead styphnate and possibly other lead-containing primer constituents) as well as lead eroded away from the open base of the FMJ bullet and re-deposited on the bearing surface of the bullet (see Chapter 4). Such bullets typically pick up lead residues from a previously fouled bore. Carbonaceous material also stands to be present in bullet wipe from both plain lead and jacketed bullets. Question 2 (relating to discriminating a bullet hole by a jacketed bullet from one produced by a lead bullet of similar caliber) cannot be answered by the sodium rhodizonate test. A test for copper on the other hand will allow for the discrimination of source (given the very limited universe of choices in this example) since only the 9 mm FMJ bullet will have copper residues in the bullet wipe. The sodium rhodizonate test for lead will still be useful in verifying the hole as a bullet hole. The proper protocol for these tests and the preparation of the reagents will be discussed in the next chapter. It should also be pointed out that the plain lead bullet will also leave considerable lead along the interior surface of the bullet's track through the wood whereas the FMJ bullet will leave little or no detectable lead in this area so there may come a point in such an investigation that the interiors of each of these bullet holes may need to be tested for lead with the sodium rhodizonate reagent. *Note*: Plain lead bullets due to their much softer nature will *not* pick up sufficient copper residues from previously discharged jacketed bullets to produce detectable levels of copper in the bullet wipe from lead bullets. The wood fibers in the channel of a bullet's path through wood often close in after the bullet's passage so that it may not be possible to see through such a hole. Any probes passed or forced through such a hole should be chosen carefully so as not to alter the path created by the projectile nor transfer lead or copper deposits to any of the wood.

BULLET HOLES IN SOME TYPICAL MATERIALS

Figure 3.2

Entry Bullet Holes in Typical Materials produced by a Round Nose Bullet and a Hollow Point Bullet: (a) Painted Sheetrock; (b) Cotton Cloth; (c) 22-Gauge Sheet Metal; (d) Suede Leather; (e) Pine Board; (f) Tire Sidewall

Figure 3.2a–3.2f provide a representative sampling of bullet holes produced in some common materials and reproduced to the same scale. All of these bullet holes were produced with the same 9 mm Ruger P-85 pistol. One hundred twenty-four (124) gr. round nose, FMJ bullets and 124 gr. JHP bullets with 0.22 in. diameter hollow points were used throughout (Figure 3.3).

The specific behavior of small arms bullets as they strike, penetrate, and perforate many of these common materials and the response of these materials are discussed in Chapter 6.

(a)

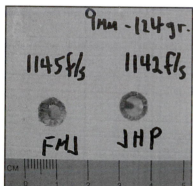

(b)

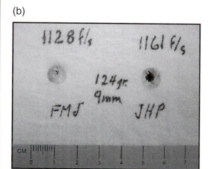

(c)

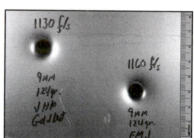

(d)

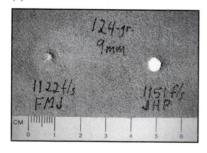

(e)

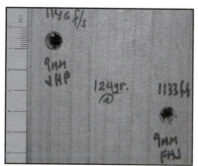

(f)

These materials were all shot using the same Ruger P85 9 mm pistol and the two types of bullets shown in Figure 3.3. Each of the target materials was positioned just beyond a ballistic chronograph located 15 feet beyond the muzzle of the pistol. The line of fire was orthogonal to each target. The velocity values in feet per second have been written on all of the targets. A centimeter scale is included in each photograph and all are printed to the same size so the reader can make direct visual comparisons between bullet holes. Note the "cookie cutter" effect produced by the hollow point bullet in cloth, leather and rubber.

The bullets in these 9 mm cartridges consist of a full metal jacketed round nose design (FMJ-RN) on the left and a JHP design on the right. The diameter of the hollow point cavity is 0.22 in.

BULLET HOLES IN FABRICS AND GARMENTS MADE OF NYLON AND POLYESTER

Representative bullet holes in cotton cloth are illustrated in Figure 3.2b. The general size and shape of a defect in any type of clothing overlying an entry gunshot wound in concert with the presence of bullet wipe around the margins of the hole makes its identification as bullet-caused relatively straightforward. This determination requires a little more caution if there is no gunshot wound that can be aligned or reasonably associated with the defect in the garment. Still, the presence of obvious bullet wipe with its attendant chemistry of carbon, bullet metal, and primer residues effectively establishes causation. However, bullet wipe can be removed in some situations (washing of the garment, long exposure to the weather, burial in certain types of soil, prolonged submersion, etc.) or it may not have been deposited due to the projectile passing through some intervening object. In these situations it may not be possible to identify the source of the defect from the physical attributes alone with two exceptions: nylon and polyester. The frictional forces and crushing action of a projectile passing through garments or fabrics made of or blended with these synthetic fibers undergo a unique and characteristic transformation. The author has given multiple presentations on this phenomenon and has utilized it in a number of past cases but never got around to reducing it to an article in any scientific journal so this chapter would seem the appropriate place. The severed ends of nylon or polyester momentarily soften or melt and take on a swollen, club-like appearance. This can be seen with a stereomicroscope adjusted to the higher powers of magnification such as $30\times - 40\times$ but the ultimate tool is the polarizing microscope. Both of these fibers are highly birefringent when viewed under the polarizing microscope using crossed polars, but the properties that cause this are relieved by this momentary frictional heating so that the enlarged ends of the fibers around the margin of a bullet hole *lack* any birefringence. This phenomenon is best viewed and photographed with normal illumination followed by the insertion of the polarizer and/or the 1-wave plate (Figure 3.4a,b). It is distinctly different from the mere severance of the fibers by other means or even from the burning of a hole in such fabrics with something like a cigarette. Neither will forcing objects like a pencil through the fabric or garment produce this characteristic effect.

The use of nylon and polyester is not just limited to clothing but can be found in duffle bags, baseball caps, sleeping bags, tents, and a host of other materials. The material does not need to be composed entirely of nylon or polyester. The polyester fibers in cotton/polyester blends respond in the same manner. Unlike bullet wipe and GSRs, these thermal/mechanical effects to the projectile-severed ends of nylon and polyester do not wash out or deteriorate with time. They will be present at the ends of severed fibers in entry bullet holes and, if the energy and velocity of exit is sufficient, they may be produced in exit holes as well. While shoring of the nylon or polyester-containing fabric aids in the production of these characteristic fiber ends, it is not required.

There is one final application of this phenomenon that may be overlooked. Fibers snagged by a bullet or punched out by a bullet as it passes through a nylon or polyester-containing material, with a little searching under the microscope, will display this effect. This is useful in differentiating these fibers from fibers that are simply debris or artifacts from the environment in which the bullet was recovered.

Those wishing to study this effect need merely acquire some remnants from a fabric store and carry out some ballistic testing followed by the necessary microscopy.

Figure 3.4

Bullet-Severed Fiber Ends at the Margin of Bullet Holes in Nylon and Polyester Fabric

(a)

(b)

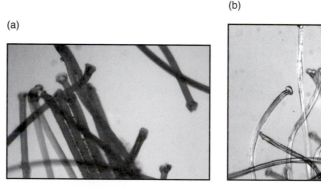

Photomicrographs taken through a polarizing microscope at 100×.
(a) Nylon fibers severed by a lead hollow point 22LR *Stinger* bullet.
(b) Polyester fibers severed by a FMJ 30-Carbine bullet.

SUMMARY

With training and experience, the physical properties of bullet holes and bullet impact sites in most materials are readily distinguishable from defects produced by other objects. The determination is easy when a recognizable projectile is ultimately recovered at the end of a channel in the struck object. In the absence of an embedded bullet, the transference of bullet metal and bullet wipe to the margins of many bullet holes and impact sites provides a means of verification through chemical or instrumental methods.

Empirical testing with comparable ammunition offers a useful and graphic means to illustrate the specific properties of bullet holes or impact sites in the evidence material.

REFERENCES AND FURTHER READING

Cashman, P.J., "Projectile Entry Angle Determination," *JFS* 31:1 (January 1986) pp. 86–91.

Haag, L.C., "Projectile-Induced Mechanical and Thermal Effects in Fibers," *CAC Seminar* (October 1987), *AFTE Training Seminar* (1989), *SWAFS Seminar* (1996).

Laible, R.C., ed., *Ballistic Materials and Penetration Mechanics*, Elsevier Scientific Publishing Co., Amsterdam, Oxford, NY, 1980.

McCrone, W.C., L.B. McCrone and J.G. Delly, *Polarized Light Microscopy*, Ann Arbor Science Publishers, Ann Arbor, MI, 1978.

CHAPTER 4

SOME USEFUL REAGENTS
AND THEIR APPLICATION

As pointed out in the previous chapter, one of the common questions in the investigation of shooting scenes is that of determining whether a hole in or a mark on some object was produced by a bullet. If it can be established as bullet caused, additional questions may arise. For example, what can be said about the nature of the bullet that caused it? Was it a lead bullet? Was it a copper-jacketed bullet? Can directionality be determined? Can anything be said about the velocity or energy associated with the projectile's impact from the nature or pattern of any bullet metal deposits, from the amount of damage, from the degree of penetration or the lack thereof?

Two reagents properly formulated and properly applied can usually answer most of these questions. A third reagent may be necessary in certain situations. These reagents and their use do not require the examiner to be a degreed chemist. Some training and practice along with some procedural controls will allow the examiner to successfully apply these reagents in the field and make reliable assessments concerning the nature of questioned bullet impact or perforation site. The two most common tests are the dithiooxamide (DTO) test for traces of copper and the sodium rhodizonate test for lead. A supplemental reagent for copper detection is 2-nitroso-1-naphthol, hereafter referred to as 2-NN. These tests come out of well-known and long-established microchemical methods for the detection of these metals, and have been adapted to forensic situations. Of these tests, the sodium rhodizonate test is more useful and common but the DTO test for copper can usually resolve issues where is it important to know if the bullet was a copper-jacketed bullet as opposed to some type of plain lead bullet. The structures of these three reagents and their reactions with lead and copper are shown in Figures 4.1–4.3.

As pointed out in the earlier hypothetical example of bullet holes in a wooden fence, lead will be present in nearly all bullet impact marks (including

Figure 4.1

The Dithiooxamide Test for Copper

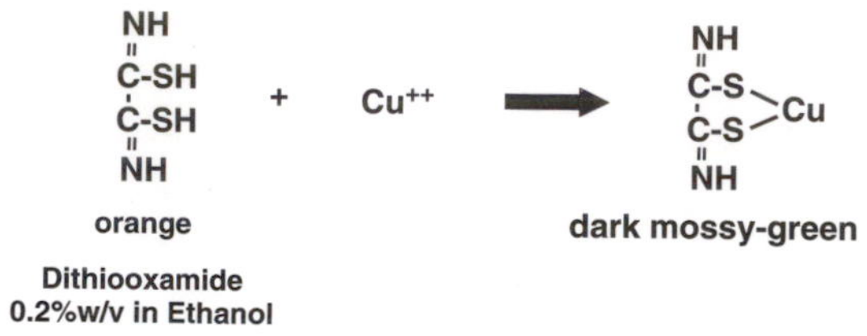

Dithiooxamide
0.2%w/v in Ethanol

Figure 4.2

The 2-Nitroso-1-Naphthol Test for Copper

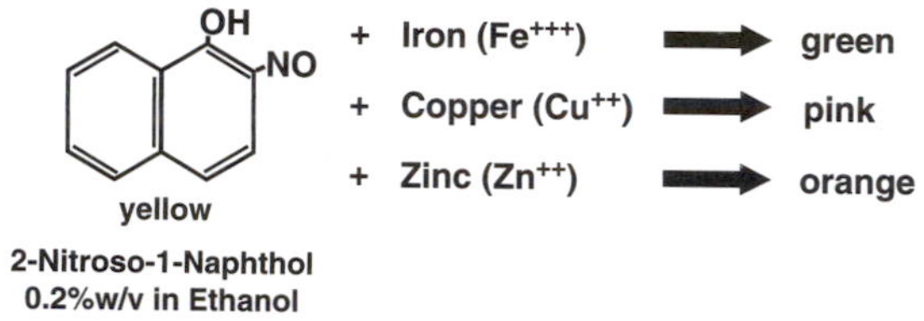

2-Nitroso-1-Naphthol
0.2%w/v in Ethanol

Figure 4.3

The Sodium Rhodizonate Test for Lead

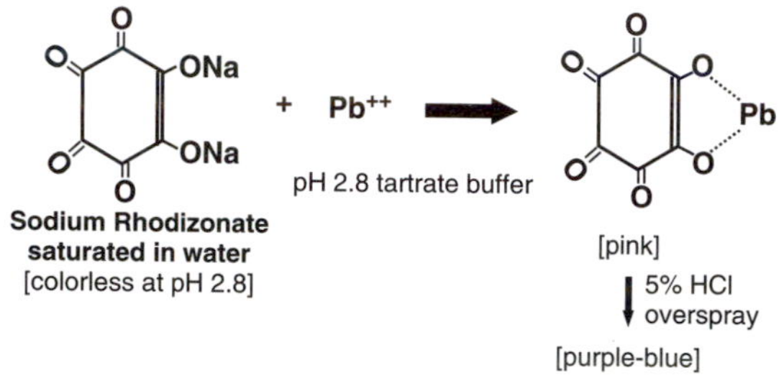

Sodium Rhodizonate
saturated in water
[colorless at pH 2.8]

those from FMJ bullets) and in the wiping around bullet holes. This is because the primer mix of nearly all present-day centerfire ammunition contains lead (most commonly from lead styphnate in the priming mixture). Some of this lead-containing residue from the discharge of the cartridge finds its way onto the bearing surface of the bullet as it makes its way down the bore of the firearm. This is true even for the first shot through a previously cleaned bore with a jacketed bullet. Subsequent shots with jacketed bullets will typically have a higher concentration of lead as a result of "pick-up" from the fouled bore. Full metal-jacketed bullets with their lead cores exposed at the base also generate substantial lead residues during discharge through erosion by the hot powder gases. The temperatures of these gases are on the order of 3000 °C, which is well above the melting and boiling points of lead of 327 °C and 1740 °C respectively. Some of this vaporized lead also becomes a part of the residue on the bullet's bearing surface and will usually transfer to the impacted surface depending upon the nature of the material struck. In summary, the presence of lead around the margins of a hole or in an apparent graze mark may establish it as bullet-caused but *not* necessarily as the consequence of a lead bullet.

Another phenomenon called "lead splash," detectable with the sodium rhodizonate test, quite literally adds another dimension to one's reconstructive efforts. When lead bullets or jacketed bullets with exposed lead tips impact a surface, some of the lead may be partially vaporized followed by condensation on the much cooler adjacent surface. Figure 4.4 shows an example of "lead splash" for a 22 Long Rifle bullet striking a thick aluminum plate at about 1100 fps.

Figure 4.4

An Example of Lead Splash

This photograph has captured a plain lead 22 Long Rifle bullet just after it has impacted a thick aluminum plate. The bullet approached from the lower right corner of the photograph. The impact velocity was approximately 1100 fps and the incident angle was 85° (5° off perpendicular). Gray deposits of partially vaporized lead can be seen on the aluminum plate just above the impact site. Numerous small fragments of lead and the major bullet fragment can also be seen fanning out at low departure angles relative to the surface of the aluminum plate (Digital image by Forensic Photographer Stan Obcamp, Phoenix, AZ).

If the bullet's intercept angle is shallow, the pattern of this "splash" can show the directionality of the responsible bullet (Figures 4.5 and 4.6). Proper application of the sodium rhodizonate and DTO tests using a lifting technique or direct application (depending upon the nature of the surface) can render these deposits and their distribution visible. These results should be photographed and the lift retained if the transfer method has been used. Suggested procedures for the copper and lead tests are given in the following pages. The materials and reagents are readily available from several chemical suppliers. A complete field kit for carrying out copper and lead tests is available from at least one source.

(Author's note: At the time of this writing (2005) lead-free primer mixtures were becoming more and more common but lead contamination of bores previously fouled with lead-containing ammunition and primer mixtures still resulted in lead-positive bullet wipe for many (e.g., 25–50) subsequent shots of the newer lead-free ammunition. This is due to the very tenacious nature of lead residues in the bores of firearms from previous firings with lead-containing ammunition.)

Figure 4.5

Lead Splash as a Result of a Low Incident Angle Impact and Ricochet (see Color Plate 4)

A visible ricochet mark (from the 3 cm to 7 cm mark) present on a brick sidewalk has been "lifted" with tartrate-treated filter paper (shown above the gray ricochet mark) then sprayed with saturated sodium rhodizonate solution. The pink color is due to the presence of lead. The approximate boundary of the ricochet mark has been outlined with a black marking pen on the BenchKote filter paper. Vaporized and minute particulate deposits of lead have fanned out from the actual contact area of the projectile and show the direction of the bullet's travel from left to right.

(a)

Color Plate 1

Two Views of a Selection of Unfired .38-caliber and 9 mm Bullets (see Figure 2.1)

(b)

Profile and base views of eight representative bullets. From left to right:

[1] a cannelured Winchester aluminum-jacketed bullet
[2] a nickel-plated Winchester JHP bullet
[3] a Russian FMJ bullet with a copper-washed finish over a steel jacket
[4] a Remington JHP bullet with a scalloped jacket
[5] a Federal *Hydra-Shok* bullet
[6] a Winchester *Black Talon* bullet with a black copper oxide finish
[7] a CCI-Blount *Gold Dot* JHP bullet and
[8] a Remington *Golden Saber* bullet possessing a brass jacket.

Color Plate 2

The Same Selection of Bullets in Figure 2.1a,b after Discharge into a Tissue Simulant (see Figure 2.2)

The upper row consists of unfired specimens. The middle and bottom rows provide two examples and views of each of these bullets after discharge into a tissue simulant. Note the unique, surviving characteristics of many of these bullets.

Color Plate 10

20-ga. Shotshell and Fired Components (see Figure 14.1)

Color Plate 11

Sectioned 12-ga. Shotshell Containing #8 Shot (see Figure 14.2)

These figures show several common types of wadding in contemporary shotshells. Multiple cardboard or fibrous wads may be in a single shotshell such as the 20-gauge shell depicted in Figure 14.1. The upper wad will take up impressions of the pellets during discharge as will the plastic shot collar unique to certain Winchester brand shells. The sectioned shell in Figure 14.2 contains a one-piece plastic shotcup in which the pellets are nested. Two of the four petals of this shotcup can be seen in this figure.

DITHIOOXAMIDE (DTO) TEST FOR COPPER RESIDUES

Important Note: If *both* the lead and copper tests are to be carried out, the DTO test for copper *must* be done *first*.

The reason for this is the fact that the mild acidic solution used to transfer lead residues will *also* transfer copper residues. If the lead test were carried out first and it was deemed important to carry out the copper test, any copper in the bullet impact site would likely have been previously removed by the lead test.

The mildly alkaline (basic) ammonium hydroxide solution used with either test for copper (DTO or 2-NN) will *not* remove lead residues since they are insoluble in the alkaline solution employed in the tests for copper. Copper will be selectively removed or rendered reactive by the ammonium hydroxide solution leaving the lead behind for subsequent detection by the sodium rhodizonate test.

The DTO and/or the 2-NN reagent and materials described herein allow copper-containing residues in bullet impact sites to be detected and made visible through a simple color-complexing reaction and a lifting technique.

By way of a background, DTO (also known as rubeanic acid) is a specific colorimetric reagent for copper. Several chemical reactions have been proposed for the coupling of copper ions with the DTO reagent. The reaction proposed by Jungreis is shown in Figure 4.1. The more important matter is this reagent's specificity for copper. The reaction produces a color that has been variously described as a mossy gray-green color to a charcoal-green color in the presence of trace amounts of copper. Because copper is much harder than lead, there is no such thing as "copper splash" with common small-arms projectiles and any positive response will only occur in locations where the copper-jacketed bullet made direct contact with the surface tested with one and possibly two exceptions. In close range discharges (inches to several feet) small particles of copper jacketing material may be stripped from the bullet during discharge and become a part of the overall gunshot residue deposited around a bullet hole. The second exception relates to the appearance of certain brands of frangible ammunition. This special purpose ammunition is intended for indoor ranges and training situations. A number of the bullets used in this type of ammunition either contain or are composed of powdered copper in a plastic matrix. The firing of such bullets generates numerous particles of copper that often appear and behave like partially burned powder particles expelled from the muzzle of the gun. The DTO reagent would, of course, react with such particles raising the possibility of using this test in conjunction with the sodium rhodizonate test and the Modified Griess Test in a distance determination procedure. As previously mentioned, lead bullets and jacketed bullets with exposed lead noses can produce "lead splash" upon impact and leave much greater quantities of lead on the impacted surface. An example of this type of bullet is shown in

Chapter 2, Figure 2.1a (the fourth bullet from the left). Even FMJ bullets have been known to produce lead splash where the impact energy is sufficiently high to tear the jacket and expose the inner lead core of these bullets. Fiegl describes the DTO reagent as about 15 times more sensitive than the sodium rhodizonate test for lead but several competing factors in shooting investigations tend to offset its sensitivity. These include the greater hardness and higher boiling point of copper over lead and the ability of bullets with exposed lead to "splash" upon impact and overwrite the underlying copper deposits. Additionally, the color produced from the complexing of DTO with copper ions is not a very exciting or conspicuous color and it can, at times, be difficult to see against anything but a clean, white background. As little as 0.1 μg of copper can be detected in a 1 cm spot on white filter paper with this reagent. This is also the case for the 2-NN reagent.

PRE-TEST CONSIDERATIONS

Before *any* testing is carried out, the examiner should give some thought to the case, the nature of the surface to be tested, what can be seen at and in the questioned impact site and what is known and not known about the incident under investigation.

Testing for both copper and lead is *not* required to verify a hole or impact site as bullet-caused. It may be desirable, however, to use both tests in certain cases where the bullet types are known and/or the presence of copper would be useful in reconstructing certain ballistic events in a shooting incident (such as the bullet-hole-in-the-fence hypothetical given in Chapter 3).

Materials and Reagents:
- Small Sprayer Unit (two recommended)
- Whatman BenchKote®
  Note: Sheets of any smooth surface filter paper can also be used in lieu of BenchKote.
- DTO in ethanol (0.2%w/v solution) having a light orange color
  Note: DTO is a stable compound both as the dry powder and as a 0.2%w/v reagent in ethanol and can be stored at room temperature.
- Ammonium Hydroxide Solution (2:5 dilution of concentrated ammonium hydroxide solution).

This too, is stable at room temperature for weeks to months when kept in airtight containers. If this solution does not have a distinct ammonia odor prior to use, a new solution should be prepared. *Note*: The ammonium hydroxide concentration is not particularly critical but the 2:5 dilution of concentrated NH_4OH solution (28–30% NH_3) is recommended. Solutions as strong as a 1:1

dilution have been used by others but the strong ammonia odor is objectionable to many.

Sections (squares) of BenchKote® (a plastic-backed form of filter paper manufactured by the Whatman, Inc. of Clifton, New Jersey) constitute the third consumable item. These are employed as a lifting medium in concert with the ammonium hydroxide solution. BenchKote is not mandatory but it does offer several advantages over plain filter paper. It can be cut to various sizes and shapes as needed for the particular surface and one can write or draw on the plastic backing. This backing adds strength to the paper side and serves as a moisture barrier during the lifting process.

THEORY

Copper residues are soluble in both acidic and ammoniacal solutions. Lead residues are *in*soluble in ammoniacal solutions (indeed, if otherwise water soluble, they will be *precipitated* in the presence of OH^- ions) but will be subsequently solubilized by acetic acid or tartrate buffer solutions used with the Modified Griess Test for nitrites and the sodium rhodizonate reagent. If copper and lead residues are both present in bullet wipe or impact transfers, contact with the 2:5 ammonium hydroxide solution will preferentially transfer some of the copper residues and leave the lead residues in place. Once the residual NH_4OH solution has dried (evaporated) from the object or surface being tested, the application of the pH 2.8 tartrate buffer solution used with the sodium rhodizonate test will solubilize some of the lead in the same residue and allow it to react with this reagent. From the foregoing it should be apparent that if one wishes to test for *both* lead and copper, the DTO test for copper *must* precede the sodium rhodizonate test for lead. If done in the reverse, the acidic nature of the lifting reagent for the sodium rhodizonate test will lift *both* copper *and* lead. It would only be a stroke of good luck that sufficient copper were left behind for subsequent detection with the DTO or 2-NN reagents if the lead test were carried out first.

DITHIOOXAMIDE PROCEDURE FOR COPPER RESIDUES

BULLET HOLES IN CLOTHING

1. The section of clothing containing the possible copper-containing bullet wipe should be placed over a water-repellant substrate such as waxed paper or the plastic side of a separate piece of BenchKote.
2. The filter paper side of a suitably sized section of BenchKote is moistened to a glossy sheen with 2:5 ammonium hydroxide solution from a small sprayer (allow adequate ventilation).

3. The moistened side of the BenchKote paper is pressed firmly against the putative bullet hole and adjacent area around the hole. Maintain firm contact for about 30 sec. but do *not* cause the filter paper to move or slide across the surfaces being tested. (The hole should be partially visible or detectable by feel through the translucent BenchKote paper. This will allow the location of the hole and any other "landmarks" to be delineated on the plastic surface with a black marking pen.) These marks are for subsequent orientation purposes after the processing of the test paper has been completed. The author typically places a small dot on the plastic backing at the center of the hole being tested along with the tracing of one or more landmarks such as a seam, a button hole, or the edge of a sleeve or collar.

4. The BenchKote paper is inverted and visually inspected prior to any further treatment. Photography is highly recommended at this point for the following reasons. As previously stated, the color complex between DTO and copper ions, while specific for copper, is not particularly exciting and can look like the mere transference of dirt or grime. Verification that no transference has occurred that might later be confused with a DTO-copper reaction is very important before proceeding to the next step. In the event the transference of some material that has a similar color to the DTO-copper response, the examiner should consider using the 2-NN reagent instead of the DTO reagent. Another option with or without the presence of any potentially confusing color transference is to simply allow the ammonium solution to dry, carefully protect and package the "lift" for later processing in the laboratory. If copper residues have been transferred to the filter paper, they will still be there days, weeks, months even years later and can be rendered visible with the DTO reagent.

5. Following satisfactory completion of the previous step, the filter paper side is sprayed *lightly* with the 0.2% alcoholic DTO reagent after verifying that the reagents are working with a known copper transfer or deposit in one or more of the corners of the BenchKote filter paper. A dark greenish-gray ring corresponding to the margin of the hole constitutes a positive test for copper-containing bullet wipe. Although this chemical complex between copper and the DTO reagent is typically stable over long periods of time, color photography of the results is strongly advised. *Note*: If the filter paper side is still quite wet, it may be desirable to let it dry somewhat before overspraying with the alcoholic DTO reagent. There is often a tendency to apply more of the DTO reagent than necessary. The examiner should realize that there is vastly more reagent in each drop of this solution than the amount of copper likely to be on the transfer paper so drenching it with reagent is clearly counterproductive. Partial

drying prior to the application of the DTO reagent will improve sensitivity and contrast.

BULLET GRAZE OR RICOCHET MARKS

Go directly to Steps (2), (3), (4), and (5) above.

Note: Prior microscopical examination of impact marks will often reveal flecks of copper jacketing material lapped or piled up on raised or abrasive particles on the struck surface. This simple non-consumptive examination, if available, should not be overlooked and, if such flecks of metal are present, macrophotography is strongly encouraged.

Additional Note: As pointed out previously, if the ammonia solution "lift" shows a transference of material that could be confused with the color of the DTO-copper complex (or stands to obscure it), the author recommends the alternate 2-nitroso-1-naphthol (2-NN) reagent. This reagent is also made up as a 0.2%w/v solution in ethanol in which it has a light yellow color. It is not as stable as DTO and should be stored in a refrigerator when not in use. As with the DTO reagent, a known copper transfer should be tested first to verify the viability of the reagent. This reagent can also detect as little as 0.1 μg of copper in a 1 cm diameter spot on smooth white filter paper.

2-NN PROCEDURE FOR COPPER RESIDUES

This reagent should be considered where the ammonium hydroxide "lift" has a color to it that might be confused with the DTO–copper reaction or might obscure the DTO–copper reaction. The same technique as described in Step (5) of the DTO procedure is employed. The appearance of a pink color against the light yellow background of the reagent indicates the presence of copper. This color is much easier to see against any dingy background color on the BenchKote or filter paper "lift" and should be photographed. It may be desirable to outline any positive pink color reaction with a pencil because there is more to do in this procedure. A few false-positives have been observed with this reagent (the formation of a pink color even though copper is not present) and consequently a kfollow-up treatment should be carried out. This involves allowing the 2-NN-treated "lift" to reach near-dryness followed by a light over-spraying it with the DTO reagent. If the pink color developed with 2-NN is indeed due to the presence of copper, it will disappear and be replaced by the DTO–copper color reaction.

Figure 4.6 shows a BenchKote "lift" (using the 2:5 ammonium hydroxide solution) of a ricochet mark produced by a copper-jacketed bullet. The copper-positive area has been cut lengthwise after which one half has been sprayed with

Figure 4.6

*A Comparison of the
2-NN Test and the DTO
Reagent on a Ricochet
Mark "Lifted" with
BenchKote and
Ammonium Hydroxide
Solution (see Color
Plate 5)*

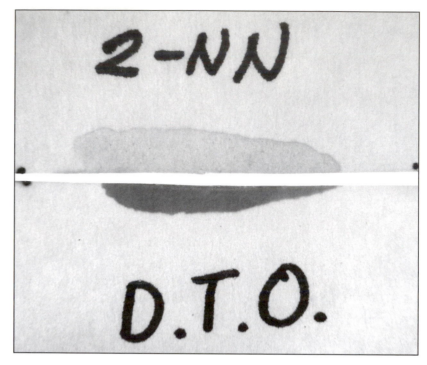

A "lift" of a ricochet mark produced by a copper-jacketed bullet. The lift has been sectioned lengthwise then the halves treated separately with the 2-NN and DTO reagents for copper.

Figure 4.7

*The Results of
Over-Spraying a 2-NN
Test with the DTO
Reagent (see Color
Plate 6)*

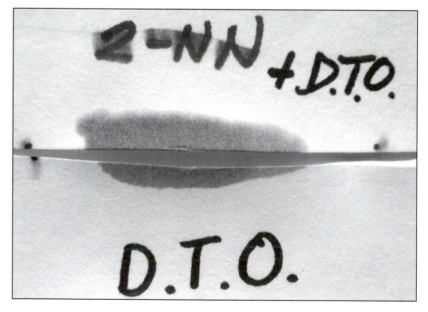

The copper-positive 2-NN response shown in the previous figure after spraying with the DTO reagent. The DTO reagent replaces the 2-NN and gives the dark gray-green color for copper.

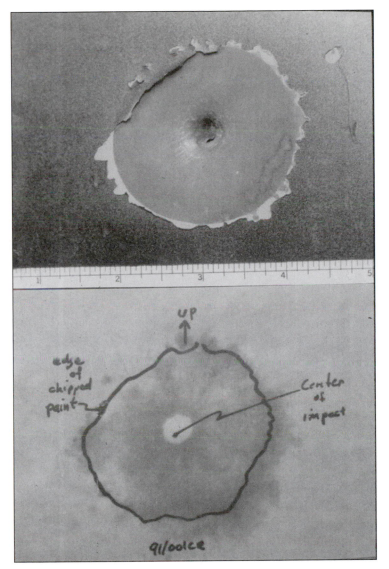

Figure 4.8

A Suspected Bullet Impact Site in a Car Door (see Color Plate 7)

An area of deformed metal and missing paint is shown in the upper half of this photograph. The responsible object failed to perforate the sheet metal and was never found but the driver of the vehicle told investigators that a person in another car shot at him with a small caliber pistol. The lower half of the photograph shows the *BenchKote*—tartrate buffer "lift" shortly after being treated with sodium rhodizonate solution and mounted on a light box. The outline of the missing paint has been marked during the lifting process. This impact site was ultimately determined to be the consequence of a 40 gr., .22-caliber lead bullet that failed to perforate the car door because of its low velocity. Note that the lead splash developed covers the area of missing paint and even extends out onto the surviving paint. The small, unresponsive area in the center of the lift was due to the inability of the lifting paper to make contact at the deepest point of the impact site. The pattern of lead splash and the deformation of the sheet metal show this to have been a near-orthogonal strike.

DTO and the other half with the 2-NN reagent. Figure 4.7 shows the effect of overspraying the 2-NN-treated portion with the DTO reagent.

Note: The DTO and 2-NN tests are also capable of giving positive responses with impact marks and bullet holes produced by copper-plated, steel-jacketed bullets.

SODIUM RHODIZONATE TEST FOR LEAD RESIDUES

INTRODUCTION

These reagents and materials allow lead-containing residues in "bullet wipe" and/or projectile impact sites to be made visible by a simple color-complexing reaction. This reaction is shown in Figure 4.3.

Depending on the nature of the object to be tested, such lead-containing traces may be visualized *in situ* (on the actual object) or "*lifted*" onto BenchKote or filter paper as was done with the tests for copper. The criterion for deciding to apply the reagents directly to an object or through a lifting technique will be described in the Procedure section.

MATERIALS AND REAGENTS FOR THE SODIUM RHODIZONATE TEST

- Small Sprayer Unit (two recommended)
- pH 2.8 aqueous Tartrate Buffer, 3%w/v (1.9 g of sodium bitartrate + 1.5 g of tartaric acid per 100 ml of distilled or deionized water*)

  *For those with pH testing capabilities, a 3%w/v solution of tartaric acid in water can be adjusted to pH 2.8 with the addition of 3–4 reagent grade sodium hydroxide pellets.

  Note: Add a small amount of preservative such as benzalkonium chloride and/or refrigerate if this reagent is to be stored for any length of time.

  Saturated aqueous Sodium Rhodizonate (Rhodizonic Acid, Disodium derivative) solution.

  Note: This reagent is unstable once it is mixed with water and must be prepared just prior to use. Control tests (described later) are used to verify its reactivity after standing for any length of time such as 15 minutes or more after preparation.

  The concentration of the sodium rhodizonate reagent is *not* particularly critical. It can be prepared in the field by simply adding small amounts of the dark, powdery reagent to the chosen volume of water until a moderately strong orange-brown solution is formed (comparable to strong tea). A slight excess of undissolved reagent is acceptable and represents a saturated solution.
- Whatman BenchKote®

  Note: Sheets of smooth filter paper can also be used in place of BenchKote but this product offers the same distinct advantages as described for the DTO test.

• Dilute Hydrochloric Acid Solution (5 ml. of 37% HCl + 95 ml. of distilled water)

Note: This reagent may be optional in field use and can be used later as a final step if deemed necessary. It is stable over time and at room temperature unless the cap is left loose.

This solution is used as either an overspray or a spot treatment after obtaining the positive pink color response with the tartrate buffer and sodium rhodizonate solution. Its use can be deferred until a later time if deemed appropriate by the examiner.

THEORY

This procedure is based on a well-known colorimetric test for lead which is sensitive to microgram quantities of this metal either in the form of particulate deposits from primer residues, vaporized bullet metal, bullet fragments, lead-containing "bullet wipe," or other impactive transfers by lead-containing projectiles.

The pH 2.8 tartrate buffer solution solubilizes a portion of the lead in the direct application method and allows it to react with the sodium rhodizonate reagent. Similarly, the tartrate buffer causes a transference ("lifting") of some of the lead deposits onto a more suitable medium and background (smooth white filter paper or BenchKote paper), in those situations where direct application is not desirable or practical. The same technique of firm pressure without slippage or movement, marking landmarks, and placing orientation marks on the "lift" as described for the DTO test applies. The tartrate buffer also decolorizes the orange-brown color of the sodium rhodizonate reagent and leaves the pink lead–rhodizonate color complex according to the chemical reaction previously shown in Figure 4.3.

The 5%HCl reagent is optional and is used if there is any doubt or question that the pink color developed with the sodium rhodizonate reagent is due to lead. The addition or overspraying with 5%HCl solution will turn the pink color to a blue-purple color with lead. The reader is forewarned that the color intensity will be reduced and the blue color may fade with time. A color photograph of the test results prior to 5%HCl addition is strongly recommended (Barium and strontium may give vaguely similar colors with the rhodizonate/tartrate reagents but they are decolorized after overspraying with 5% HCl).

It should also be pointed out that the *entire* area of the pink response need *not* be sprayed with the 5%HCl solution. The author frequently masks off the majority of the positive pink area and then sprays a small, representative site with the 5%HCl solution. Placing a drop of the HCl solution on a small, selected area of the pink response is an alternative method for carrying out the confirmation test.

SODIUM RHODIZONATE TEST PROCEDURE FOR LEAD RESIDUES

VERIFICATION OF REAGENTS

Verification that the reagents are working correctly is accomplished by lightly marking a corner of a piece of BenchKote or filter paper with the nose of a lead bullet or placing a drop of a known solution of soluble lead salt such as lead acetate containing approximately 0.02% soluble lead by weight. This area is then misted with a *light* spray of the tartrate buffer solution then oversprayed with freshly prepared sodium rhodizonate solution. An immediate pink color should form where the known lead sample was used to mark the test paper and the orange-brown color of the rhodizonate reagent should decolorize in a few minutes. If the optional hydrochloric acid treatment is to be used, a subsequent overspray with this reagent should cause the previous pink color to turn to a blue-purple color often with some loss in intensity if the lead concentration was low. A preferable method for hydrochloric acid step is to simply place one or two drops of this solution in selected areas and then observe and document the immediate pink-to-purple color change for lead.

DIRECT APPLICATION METHODS

Where the substrate is white or light-colored and *absorbent*, one may elect to treat the surface directly. Examples of such surfaces would be lightly colored cotton garments and a lightly colored pine board with a questionable bullet hole or graze mark in it. In such instances the same approach as described for the verification of reagents is employed but before applying *any* of the reagents described in this chapter, some careful thought and possible pre-testing need to be considered. If, for example, it is deemed important to test for the presence of copper, there are multiple factors that must be considered and evaluated. Although the DTO reagent is more sensitive than 2-NN, the development of the dark gray-green copper–DTO complex is not likely to be seen or distinguished in a ring of bullet wipe or sooty area. Furthermore, the object must be lightly moistened with the ammonium hydroxide solution to yield a good reaction with the DTO reagent. The treated object must be thoroughly dried (to remove the ammonium hydroxide) before the application of the tartrate buffer and sodium rhodizonate solution otherwise the test for lead will fail. If the 2-NN reagent appears to be a better choice (due to background color problems for the DTO test), you are now confronted with the possibility of the development of a pink color (due to copper) to be followed by the sodium rhodizonate test which also gives a pink color. There are several choices for a solution to this quandary. The first is to return to the lifting technique with filter paper or BenchKote treated with the ammonium hydroxide solution. An alternative is to divide or partition the object or material to be treated along some line of symmetry and only treat half of the object or material with the other half left for treatment with the tartrate buffer and sodium rhodizonate reagent. Pre- and post-treatment steps should be photographed with and without a scale and under the same conditions of lighting. Some form of orientation mark(s) along with an item number or site identifier in the field of view are recommended. If a traditional film camera is used, the use of a gray card to calculate the proper exposure is very useful where bright or white objects are involved. Digital cameras have the advantage of immediate playback allowing the adequacy of the exposure to be observed. An additional photograph would follow any treatment of a positive sodium rhodizonate response with the 5%HCl solution.

"LIFTING" OR TRANSFER METHOD

This method is useful with dark colored objects, immovable objects, non-porous objects, mildly bloodstained objects, and/or difficult-to-reach objects or surfaces. The success of this method does require some skill and sound judgment on the part of the examiner or investigator. Over- or under-wetting with the tartrate buffer solution, incomplete contact or movement of the lifting paper can cause problems. Heavily bloodstained objects may require special processing by the forensic laboratory prior to testing for lead residues. Hydrophobic (water-repellant) substrates also present difficulties that may require some thought and evaluation prior to proceeding with the testing of the questioned area or site. Such things as nylon garments, oil-based enamel finish, and plastics fall in this category. The author has found it useful to add a *small* amount of liquid detergent to the tartrate buffer solution or about 10 ml of reagent alcohol per 100 ml of buffer solution to act as a wetting agent. Some experimentation on an area known to have *no* evidentiary value is highly desirable prior to proceeding on the questioned area of item or object.

If the substrate to be tested can be wetted without the tartrate buffer beading or running, proceed as follows:

- Prepare a section of BenchKote or filter paper of sufficient size to cover the questioned site *and* some of the surrounding unaffected area to serve as a "blank."
- Evenly moisten the evidence item to the extent that the tartrate buffer is *not* running off or forming puddles but to the degree that the subsequent pressing of the BenchKote or filter paper against the item will cause a blotting

action to occur. (Very porous surfaces such as cinder block, bricks, concrete surfaces may also necessitate a light pre-moistening of the transfer paper.) These materials are especially troublesome because they are very alkaline and often neutralize the tartrate buffer thereby preventing the transfer of any lead residues. The 15% acetic acid solution normally used with the Modified Griess Test may be a preferable lifting agent. It may even be necessary to lightly mist such surfaces with 15% acetic acid and carry out a second "lift" with these very alkaline materials.

- The transfer paper must be thoroughly pushed and pressed into the surface without allowing it to slip or slide.
- Orientation marks should be placed on the backside of the transfer paper (the side toward the examiner) *before* it is lifted from the surface.
- If dry areas are seen, careful lifting of one side of the transfer paper and re-spraying of the substrate is appropriate.
- Once it is certain that the limits of the transfer paper's contact with the substrate have been defined and documented, the transfer paper may be turned over, placed on a suitable surface and sprayed with fresh sodium rhodizonate solution to produce an even yellow-brown color. Any transferred lead residues will immediately appear pink (*Note*: right–left reversal will be present when one views or photographs the contact side of the transfer paper).

If the surface to be tested is hydrophobic, difficult to work with, or deemed undesirable to spray the object itself, proceed as follows:

- Prepare a section of the transfer paper as described above.
- Evenly moisten the transfer paper until it is shiny wet and translucent but not runny.
- Promptly press the transfer paper firmly against the substrate as previously described. If properly wetted, you should be able to partially see through the BenchKote.
- Make appropriate orientation marks on the transfer paper *before* lifting it from the surface.
  Note: As previously mentioned, right–left reversal occurs with the lifting technique; therefore, one may wish to photograph the fresh, translucent lift on a lightbox or while taped to a window exposed to daylight. Alternatively, the photo lab can be instructed to reverse the negative when making a print. Digital photographs are easily reversed with the computer. These techniques will make it easier for non-technical people to understand the spatial relationships for any lead deposits lifted from the evidence item.
- If the hydrochloric acid confirmatory step is deemed necessary, the lift (or a selected portion of the lift) can be oversprayed with the 5% HCl to give

the blue-purple color with lead. This color change should be promptly photographed.

- After the transfer paper has dried, it should be stored inside plastic protective sheets like those used for photographs.

DIMETHYLGLYOXIME (DMG) TEST FOR NICKEL RESIDUES

Nickel is a silver-white metal with an atomic weight of 58.71. Its melting point is 1555 °C and it has a calculated boiling point of 2837 °C. Nickel is a hard metal (Moh's hardness of 3.8 compared to lead, copper, and steel at 1.5, 2.5, and 4 respectively) belonging to the iron–cobalt group of metals. Nickel easily takes on a polish. This is readily seen in ammunition where bullets, buckshot, or cartridge cases plated with nickel have a shiny, almost mirror-like sheen to them. Nickel resists oxidation, and nickel-plated cartridge cases left at a scene for months and years look as shiny as they did when deposited there.

A number of brands and types of bullets have a shiny nickel coating over their gilding metal or mild steel jackets. Certain pistol bullets among the Winchester *SilverTip* line have possessed a nickel-plating over a copper alloy jacket. Pistol bullets and a few rifle bullets of foreign manufacture likewise come with a nickel-plating as do certain loadings of buckshot.

After an evaluation of the circumstances of a shooting incident, it may become apparent that one or more shooters discharged ammunition with nickel-plated bullets. In such cases it may be appropriate to consider the use of the DMG test. Just as with the DTO test for copper, a DMG test for nickel could resolve some important question in a shooting investigation. Another advantage of this reagent is that it is a clear colorless solution making the color development much more visible and more easily interpreted. The testing process is carried out by the transfer method using the same 2:5 dilution of concentrated ammonium hydroxide solution employed with the DTO and 2-NN reagents for copper. The DMG reagent is available as either the pure white to off-white powder or as a 1%w/v solution in ethyl alcohol, or the pure reagent can be purchased and the appropriate strength solution can be prepared in or by the laboratory.

CHEMISTRY OF THE NICKEL–DMG REACTION

According to Fiegl, DMG forms a stable bright red insoluble salt with nickel salts in neutral, acetic acid, or ammoniacal solutions. When put in the same environment, two DMG molecules form a ring around a single metal nickel ion and binds to them in a chelating process. [Figure 4.9] The resultant compound has been used as a sun-fast pigment in paints, lacquers, cellulose compounds and cosmetics.

Figure 4.9

The Dimethylglyoxime Test for Nickel

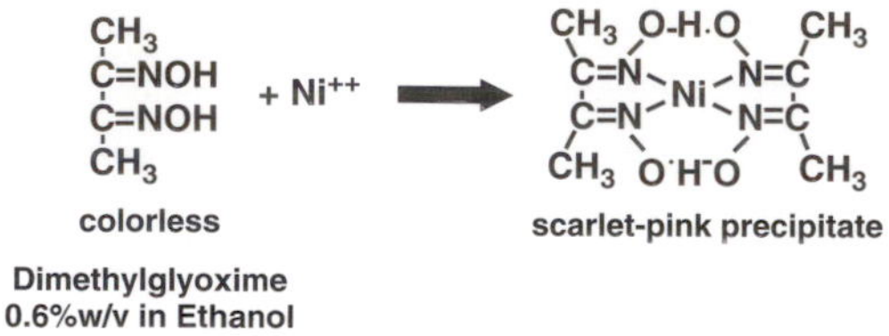

$$\begin{array}{l} CH_3 \\ C=NOH \\ C=NOH \\ CH_3 \end{array} + Ni^{++} \longrightarrow \begin{array}{l} CH_3 \ O\text{-}H\cdot O \ \ CH_3 \\ C=N \diagdown Ni \diagup N=C \\ C=N \diagup \ \ \diagdown N=C \\ CH_3 \ O^-H^- O \ \ CH_3 \end{array}$$

colorless   scarlet-pink precipitate

Dimethylglyoxime
0.6%w/v in Ethanol

DMG TEST PROCEDURE

Some thought must be given to the incident under investigation. Just as one needed to determine the reconstructive value of a test for copper, so it is when considering the DMG test for nickel. In fact, it can even be more complicated if the examiner is faced with discrimination of bullet holes or impact marks produced by nickel-plated projectiles, copper-jacketed bullets, or plain lead bullets. At least two techniques are available to the examiner.

[A] Sections of BenchKote paper or filter paper that have been pretreated with an alcoholic solution of DMG and dried are used in this procedure. The author has found no detectable difference in the performance of such pretreated papers using solutions of 0.2%w/v or 1%w/v. These pretreated test sheets are stable over long periods of time and can be kept in a manila envelope or folder. A corner of the pretreated test panel previously cut to the appropriate size is lightly moistened with the same 2:5 ammonium hydroxide solution used with the DTO and 2-NN reagents. A drop of a standard nickel solution on the order of 0.005% soluble nickel is placed on the moistened area. A common U.S. 5-cent coin (the one with Jefferson's bust on one side and Monticello on the other) can be used as an alternate method for reagent verification by pressing one of these coins firmly against the moistened area for about 30 sec. These coins are composed of a 75% copper—25% nickel alloy. Once the positive response for nickel is noted, the remaining area of the transfer panel is lightly sprayed with the ammonium hydroxide solution, then firmly pressed against the suspected bullet hole or impact site along with an appropriate surrounding area. The "lift" is then removed, inverted, and inspected for the scarlet-pink response for nickel. If this response is the consequence of a nickel-plated bullet (as opposed to nickel-plated shot), copper is also likely to be present and a subsequent overspray with the DTO reagent should reveal this since the DMG reagent does not interfere with the DTO–Cu reaction. If the nickel-positive response is due to nickel-plated shot, there should be no copper present. Instead, large amount of lead should be present when this same site is later tested using the tartrate buffer transfer technique and sodium rhodizonate reagent.

The pre-impregnation of the test paper provides the advantage that the DMG reagent is evenly distributed across the test medium. Moreover, the color change that occurs will be more representative of the substance at its original location when using a pretreated transfer paper. Finally, as previously pointed out, the DTO test for copper can be carried out by lightly overspraying this same test paper.

There is at least one disadvantage to the pre-impregnated transfer papers, however. If the detection of copper is also important and the lift pulls up the same sort of dingy color that was described during the DTO procedure, the use of 2-NN is severely, if not totally, compromised when a positive response for nickel has occurred. This is for the obvious reason that one is now trying to see a pink color against and already strong pink background.

The previous example of a positive nickel response from nickel-plated shot compounds this potential problem in that the large amount of lead residue that stands to be present might result in a dingy area of transfer against which the DTO test (to show the *absence* of copper) is obscured. This sort of quandary should be thought out in advance and if it is a clear possibility, then technique B should be used.

[B] This technique goes back to a lift of the area with the plain filter paper or straight BenchKote paper moistened with the 2:5 ammonium hydroxide solution. As before, judgments should be made about any need for a copper test and the presence or absence of a dingy transferred color that would mask the DTO test. If a copper test is deemed useful, then carefully cut the test paper in half by cutting through the area of special interest. After verifying the performance of the reagents with a known nickel and copper source as described in technique A, lightly spray one half of the lift with the appropriate copper reagent and the other half with the DMG reagent. Document the results as before, dry and retain the test papers.

BARREL RESIDUES

Since the use of nickel-plated bullets is relatively uncommon and the presence or absence of nickel residues in the bore of a gun may have important reconstructive value, it is appropriate to add a method for the testing of gun barrels. One round of a nickel-plated bullet will typically leave nickel deposits that will produce an unequivocal DMG response whereas subsequent shots with common copper-jacketed bullets or lead bullets with greatly reduce or even negate any positive response for nickel.

To test the bore of a firearm for nickel deposits, soft cotton gun cleaning patches pre-impregnated with the DMG are recommended. Alternatively, a 2:5 ammonium hydroxide moistened cleaning patch can be used. A patch from the same supply should be moistened and tested for a positive response with a known nickel source before processing the bore of the submitted firearm. When the examiner is ready to test the bore of the firearm, the pretreated DMG test patch is lightly moistened with the ammonium hydroxide solution and pushed through the bore just as would be done in a normal cleaning procedure making certain that the patch fits tightly in the bore. If an ammonia-only patch has been used, it is then oversprayed with the DMG in alcohol reagent. This patch should also be chosen to fit tightly in the bore. It is further suggested that this patch be worked back and forth in the bore to enhance the removal of any nickel residues.

Note: Nickel-plated firearms should not be tested for the obvious reason. Likewise, it is usually pointless to test the bore of a shotgun since nickel-plated shot is usually nested in a plastic shotcup and does not come in contact with the bore of the gun.

SUMMARY AND CONCLUDING COMMENTS

Two reagents, DTO and 2-NN, have been described for the detection of traces of copper in bullet wipe, bullet impact sites, and for particular residues generated during the discharge of copper-containing frangible ammunition. One or both of these tests need only be carried out when the detection of copper stands to be of importance in the case at hand or until such time that totally lead-free ammunition is common.

The sodium rhodizonate test for lead will reveal both the presence and the pattern of lead deposits around and in bullet holes, bullet impact sites, and in the overall gunshot residue deposits associated with close proximity discharges. Such lead deposits can confirm a hole or damage site as bullet-caused. Lead-containing "lead-in" marks associated with low incident angle projectile strikes and/or the location of "lead splash" at a bullet impact can also establish the directionality of the impacting bullet. These characteristic marks will be discussed further in the chapter dealing with ricochet.

Whether one employs the lifting technique or the direct application technique, areas that extend beyond the specific site in question should also be treated with all these reagents. This serves as a reagent blank (insuring both that the reagents themselves are not contaminated with the particular metal and that the surface being tested does not contain detectable levels of lead, copper, or nickel). The use of cotton swabs or commercial test sticks to test suspected bullet impact sites is discouraged for the reason that it is difficult to completely rule out the presence of lead or copper in the material or surface being tested. Moreover, *no* pattern information is provided when using cotton swaps to test selected sites.

With rare exception the DMG–nickel response, the DTO–copper response, and rhodizonate-developed lead deposits are stable over time when stored at room temperature and out of strong light. However, the subsequent hydrochloric acid treatment with the sodium rhodizonate test reduces the sensitivity of the test (approximately tenfold) and may result in a gradual fading of the lead-specific blue-purple color. Color photography would be appropriate for the documentation of any color reactions that the examiner develops.

A test for nickel used in some ammunition has also been described, and may, after a careful evaluation of what is known about the case, serve as a useful adjunct to the usual copper and lead tests.

It should be understood that any copper, lead, and nickel that are lifted by the transfer technique have *not* been destroyed or consumed. They have simply been rendered visible by a color-complexing reaction. This means that alternative procedures (e.g., instrumental methods such as SEM-EDX) could be employed to further test these color complexes if the chemical identities of these complexes come to be in serious question. This can be done on small, representative areas that are both color-positive and color-negative by excising a small square or rectangle out of the lift with a scalpel. This is mounted on a stub designed for SEM–EDX spectrometry, appropriately carbon-coated then analyzed. In those instances where the lifting method has been used for lead or copper, it is very unlikely that all of the lead or copper has been transferred to the lifting paper. Typically there will be additional lead and copper left behind on the substrate so that either the test can be repeated or it can be tested by some other means and some later time.

It must also be understood that traces of these metals may not be transferred or solubilized in a sufficient amount to respond to the particular reagent. This is especially true in the case of nickel residues since this metal resists oxidation and solubilization by either ammonium hydroxide or tartrate buffer. Stated another way, positive responses for any of the tests described in this chapter are useful and potentially meaningful. The failure to detect lead, copper, or nickel in bullet wipe or a bullet impact site does not necessarily rule out the presence of the particular metal in the bullet's composition or construction.

If at any time the examiner has any uncertainty about the effectiveness of the planned protocol, a control test should be carried out on a non-evidentiary area of the substrate.

Finally, as with all chemicals, special precautions should *always* be exercised to avoid absorption or inhalation of these reagents. The use of rubber gloves and a fume hood are appropriate when working in the laboratory. If the examiner is carrying out such tests in the field, rubber gloves are still a requirement along with an open area free of bystanders and with the person doing the test located upwind of the object being treated.

REFERENCES AND FURTHER READING

Bashinski, J.S., J.E. Davis and C. Young, "Detection of Lead in Gunshot Residues on Targets Using the Sodium Rhodizonate Test," *AFTE Journal* 6:4 (1974) pp. 5–6.

Fiegl, F., *Spot Tests in Inorganic Analysis*, 5th edn, American Elsevier, New York 1958.

Gunsolley, C.R., "Dimethylglyoxime: A Spot Test for the Presence of Nickel," presentation at the 2002 BATFE National Firearms Academy and the 2004 Shooting Incident Reconstruction Course at the Gunsite Training Academy, Paulden, AZ.

Haag, L.C., "A Microchemical Test for Copper-Containing Bullet Wipings," *AFTE Journal* 13:3 (1981).

Haag, L.C., "A Method for Improving the Griess and Sodium Rhodizonate Tests for GSR on Bloody Garments," *SWAFS Journal* (April 1991) also *AFTE Journal* 23:3 (July 1991).

Haag, L.C., "Phenyltrihydroxyfluorone: A 'New' Reagent for Use in Gunshot Residue Testing" *AFTE Journal* 28:1 (January 1996).

Haag, M.G., "2-Nitroso-1-Naphthol vs. Dithiooxamide in Trace Copper Detection at Bullet Impact Sites" *AFTE Journal* 29:2 (Spring 1997), pp. 204–209.

Jungries, E., *Spot Test Analysis—Clinical, Environmental, Forensic and Geochemical Applications*, John Wiley & Sons, NY, 1985.

Kokocinski, C.W., D.J. Brundage and J.D. Nicol, "A Study of the Uses of 2-Nitroso-1-Naphthol as a Trace Metal Detection Reagent," *JFS* 25:4 (October 1980).

Lekstrom, J.A. and R.D. Koons, "Copper and Nickel Detection on Gunshot Targets by Dithiooxamide Test," *JFS* 31:4 (October 1986) pp. 1283–1291.

Shem, R.J., "The Vaporization of Bullet Lead by Impact," *AFTE Journal* 25:2 (April 1993).

CHAPTER 5

DISTANCE AND ORIENTATION DERIVED FROM GUNSHOT RESIDUE PATTERNS

THE PRODUCTION OF GUNSHOT RESIDUES

The term "gunshot residues" includes unconsumed powder particles, carbonaceous material from the incomplete combustion of the propellant, primer constituents, and ablated bullet metal. In certain situations, this term also includes vaporized lead and/or bullet lubricants. These materials are expelled from the muzzle of a firearm during the discharge process and at close range will be deposited on nearly any surface. The dimensions of the pattern and the density of certain discharge products provide a means for estimating the distance between the muzzle of a gun and the surface bearing such discharge products. Probably the most useful of these discharge products are partially burned and unburned propellant particles, sooty material, and primer constituents. They may also include vaporized lead and materials present in the bore of the responsible firearm from previous firings. Vaporized lead normally arises from the discharge of lead bullets or FMJ bullets with lead cores and exposed lead bases. The sooty material (sometimes called "smoke") typically consists of carbonaceous material, primer constituents as either vaporous materials or very fine particulates. It may also include vaporous lead from the previously described sources. The firing of a jacketed bullet through a previously leaded bore (from shooting lead bullets) will also produce large amounts of vaporous lead. This will diminish with the discharge of each subsequent round of jacketed ammunition. All of these discharge products provide varying degrees of useful information relating to the range of fire when they are deposited on any surface including the skin of gunshot victims.

Forensic pathologists often provide range-of-fire estimates in their reports based on soot deposits and/or powder stippling or tattooing patterns around an entry wound. Their opinions are usually based on past experience and general considerations. Any critical assessment regarding range of fire will require some thoughtful empirical testing whether these patterns are on skin or some inanimate object. Powder particles expelled from the muzzle of a firearm have velocities comparable to the projectile, and since they are relatively hard, they may produce physical damage (stippling) to any surface they strike. Such surfaces include wood, painted metal, plastic, leather, and wallboard. Stippling of skin is well known and arises from the same mechanism, namely the impact of very energetic particles of unconsumed and partially consumed gunpowder. In this situation the powder particles produce small, hemorrhagic injuries in a living individual. When the mass and energy of these particles is sufficient that they enter and embed themselves in the skin, the term "tattooing" may be applied by some medical examiners. Other forensic pathologists make no distinction between "stippling" and "tattooing" and often use these terms interchangeably. However, the term "tattooing" would *not* be applied to powder patterns in inanimate objects. Figures 5.1 and 5.2 show powder stippling in painted wallboard and automotive sheet metal respectively. Figure 5.3 shows a powder stippling pattern in wood with an added feature of special reconstructive value.

Figure 5.1

Powder Stippling in Painted Wallboard

The path of this close range shot from a .38-caliber revolver was from the lower right of the photograph. Multiple particles of gun powder can be seen adhering to, and embedded in the paint. The pattern formed by these powder particles is elongated because of the oblique angle of fire.

Once documented as to location, and photographed, the path of this bullet should be determined.

Following these efforts the entire area of powder-stippled wallboard should be cut out and impounded for any later comparisons of propellant morphology and/or muzzle standoff distance determinations.

Figure 5.2

*Powder Stippling in
Painted Sheet Metal*

This shot was fired directly into a panel of painted sheet metal from a standoff distance of about 6 In. using a .357 Magnum revolver. The energy of the partially burned and unburned particles of gunpowder was sufficient to stipple and even remove small areas of paint at the individual impact sites.

Figure 5.3

*Powder Stippling
in Wood*

This shot was sufficiently close that soot and unburned powder particles have been deposited around the entry bullet hole in this wooden gate. It was also discharged after a shot was fired through this board from the opposite side. The sequence for these two shots was determined by the presence of powder particles embedded in the area of blown out wood around the exit bullet hole on the right.

A portion of this powder pattern involves an area of missing wood particles around an exit bullet hole. This figure is a re-creation of a case where two armed individuals were on opposite sides of a gate. According to the surviving shooter he was standing very near the gate when the decedent fired a shot through the gate barely missing him. He immediately returned fire striking and killing the subject standing outside the gate. This account is supported by the presence of powder embedded both in the interior surface of the gate ("I was standing very near the gate") and in the blown-out areas of wood around the exit bullet hole ("I returned fire"). In this example it was the mere presence and location of embedded powder particles that answered the critical question of shot sequence. It is the spatial distribution, composition, and density of GSR and the patterns they possess that often allows distance and/or orientation of the firearm to be determined. The expulsion of powder residues from the muzzle of a firearm follows a conical distribution with distance much like a shotgun discharge in miniature. Depending on their size, density, and shape, these particles can easily produce powder patterns out to several feet. Spherical ball powder residues from centerfire ammunition will travel the farthest (powder patterns as far as 4 feet) due to its superior exterior ballistic properties. Flattened ball powder comes in second and flake powder residues will travel the shortest distance producing powder patterns at distances on the order of 18–24 in. (See Chapter 2 for a review of the various physical forms of small arms propellants.) With any and all physical forms of gunpowder, a distance will be reached with the particular gun-ammunition combination where *no* discernable powder pattern is recognizable although a few scattered powder particles may be found adhering to the surface of the "target." These distances are on the order of 4 to as much as 15 feet. Figure 5.4 summarizes the effects of standoff distance and GSR deposition.

In addition to the physical form of the propellant, its burning rate along with the weight of the powder charge, the efficiency of the particular load, and the barrel length of the gun all have a bearing on any powder pattern that might be produced at some selected standoff distance from a recording medium. A long barrel will generally result in a *reduction* in the amount of unconsumed powder emerging from the muzzle but it can *increase* the amount of vaporous lead eroded from the bases of FMJ bullets containing exposed lead cores. Keeping all other factors the same, a shorter barrel produces more unconsumed powder at the muzzle just as one would expect but it also results in a greater dispersion of these particles because of the higher pressures at the muzzle. This in turn increases the diameter of the powder pattern and reduces the density of the powder pattern. *Density* here refers to the number of powder particles per unit area at some standard distance from the bullet hole. Figure 5.5a,b shows powder patterns on white filter paper (BenchKote) at the same standoff distance of 6 in. with the same FMJ .357 Magnum ammunition fired from a 4 in. barrel and an 18.5 in. barrel.

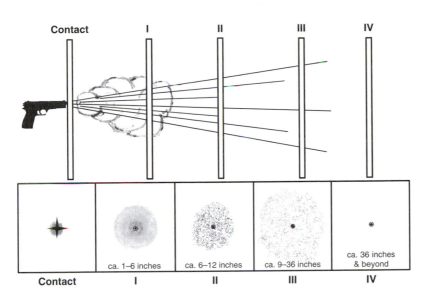

Figure 5.4

The General Characteristics and Behavior of Gunshot Residues with Range

Typical characteristics:

Contact   Blast destruction, stellate tearing of skin or clothing, very intense soot possible around the edges of the entry site but mostly on the *inside* of the garment or driven into the wound. The outline of certain contacting parts of the firearm (e.g., front sight, barrel bushing, etc.) may be printed in the skin adjacent to the entry hole. This latter phenomenon is referred to as a *muzzle imprint*

Zone I   Intense, dark sooting with dense deposits of unburned and partially burned powder particles around the bullet hole. Blast destruction is still possible in clothing. Powder tattooing and stippling of the skin and stippling of certain inanimate objects such as wood, drywall, painted surfaces, and plastics.

Zone II   No visible soot to some faint sooting may be discernable. A circular deposit of powder particles will be present around the bullet hole. Powder tattooing and stippling are likely particularly with ball powder and poorly burning propellants.

Zone III   No visible soot. A roughly circular deposit of widely dispersed powder particles present around the bullet hole. Powder particles are often loosely adhering at the greater distances. The modified Griess test may raise nitrite-positive sites where powder particles struck but later became dislodged. Powder stippling of skin is still possible, particularly at the closer distances.

Zone IV   No discernable pattern of firearms discharge products is present. A few scattered and loosely adhering powder particles may be found but lack any pattern. Bullet wipe will be present around the margin of the entry hole regardless of the distance from which the shot was fired.

Finally, there are some propellant–bullet combinations that are more efficient than others. This means that one can encounter cartridges of a particular brand, caliber, and bullet weight that are loaded to the same muzzle velocity and peak pressure but that produce very different amounts of gunshot and unconsumed propellant residues. Figure 5.6a,b represents powder patterns produced with the same revolver at the same standoff distance with two different lots of 125 gr. JHP Remington .38 Spl. ammunition – one containing 18 gr. charges of spherical ball powder and the other containing 5.5 gr. of unperforated disk-flake powder.

Figure 5.5

Powder Pattern at a 6 in. Standoff Distance: (a) .357 Magnum Cartridge Fired from a 4 in. S&W Revolver; (b) .357 Magnum Cartridge Fired from a 18.5 in. Carbine

(a)

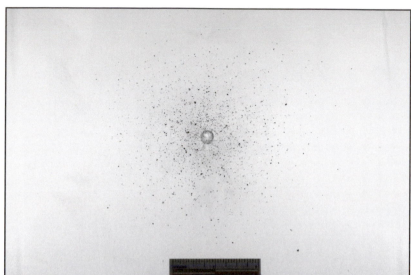

(b)

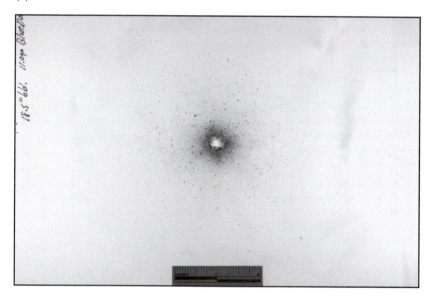

Both of the .357 Magnum cartridges used to prepare these powder patterns were loaded with 11.0 gr. charges of a medium-burning, unperforated disk-flake powder and 170 gr.-FMJ bullets. The much greater barrel length resulted in more of the propellant being consumed (reduced powder pattern in the lower figure) but it also allowed more time for the hot powder gases to vaporize some lead from the base of the FMJ bullet (dark gray deposits around the bullet hole).

(a)

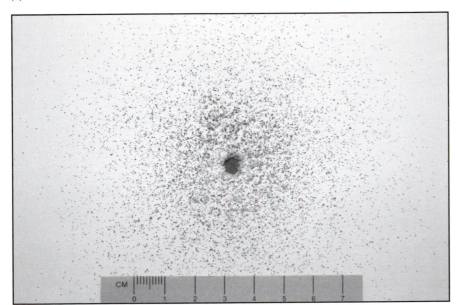

(b)

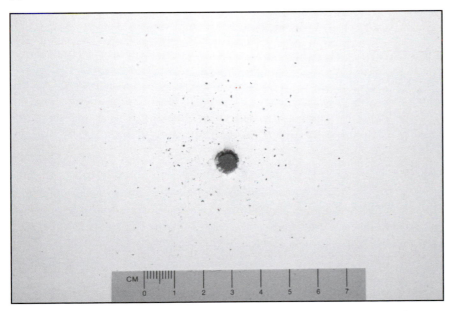

These powder patterns were produced with the same Colt revolver at the same standoff distance of 6 in. with two different lots of 125 gr. JHP Remington .38 Spl. ammunition—one containing 18 gr. charges of spherical ball powder and the other containing 5.5 gr. of unperforated disk-flake powder. These cartridges produced comparable muzzle velocities but, as can be seen in these figures, they produced *very* different powder patterns.

These cartridges produced comparable muzzle velocities but, as can be seen in these figures, they produced *very* different powder patterns. The lesson to be learned here is to insure that any test patterns are produced with ammunition comparable to the ammunition discharged in the actual incident. Reliable numerical distance determinations can only be accomplished by empirical evaluation of the type of gun and ammunition involved in the specific case. Numerical range-of-fire estimates based on past experience but without such testing leaves firearm examiners, crime scene investigators, and pathologists on tenuous ground, and open to legitimate attack.

The soot or "smoke" cloud generated during the discharge of a firearm rapidly expands and dissipates in a generally spherical form. These vaporous to fine aerosol particles travel much shorter distances than partially burned and unburned powder particles. Visible soot/smoke deposits seldom extend more than 6–10 in. beyond the muzzle of the gun with modern ammunition. At very close standoff distances these deposits are very dark and localized around the bullet hole. As the muzzle-to-surface distance increases, a gradient often appears in the soot/smoke deposits. With some gun-ammunition combinations reproducible patterns or ring-like deposits may occur. Finally, there will come a distance at which the soot/smoke deposits are barely discernable although chemical methods may render them visible.

TARGET MATERIALS

A variety of target materials have been used for the preparation of exemplar powder and GSR patterns with a submitted firearm. These materials include heavy white blotter paper, card stock, jean twill cloth, fresh pig skin, foam board, and BenchKote®. Initial test shots at selected distances are typically carried out and compared to the evidence pattern. When patterns are obtained that are close to the evidence pattern in diameter and density, some examiners carry out the final test shots with samples of the actual evidence material (garment, wallboard, etc.) taken from an area that does not affect or compromise the evidence pattern. The author supports this approach because it can refine the examiner's estimate of range of fire by removing the variable created by the use of a target material other than the actual surface upon which the powder-GSR pattern exists. Patterns on human skin necessitate the use of some form of target material. As a result of multiple tests carried out by the author and others, the author generally prefers BenchKote for this purpose. Fresh, white pig skin has been used, but besides being difficult and messy to work with, it offers little advantage over BenchKote, heavy blotter paper, or jean twill cloth. Its only appeal is the fact that it *is* skin but it still does not provide the vital reaction to stippling and tattooing that takes place with living human skin.

INTERPRETATION AND REPORTING OF RESULTS

It has been said that all measurements are estimates and so it is with range-of-fire determinations based on powder-GSR patterns on both inanimate objects and gunshot victims. Other than contact shots, the examiner must carefully assess the factors of pattern diameter, the presence or absence of soot, the intensity and diameter of any soot deposits, the presence or absence of stippling (or tattooing with skin), and the density of any powder pattern. Each of these factors relate to range of fire for a particular gun-ammunition combination. Close range shots into curved or irregular surfaces and/or shots fired at non-orthogonal angles must also be dealt with in a thoughtful manner and allowances made in estimating the range of fire. In some cases range-of-fire estimates with uncertainty limits of ±2 in. are possible. In more typical to extreme cases the examiner may set uncertainty limits of ±12 in. The issue is usually not whether the gun was 4 in., 6 in., or 8 in. from the victim's shirt but whether the shot was fired at a close range or from several feet away. In other cases where the issue is self-inflicted vs. inflicted by another, the critical issue may be whether the standoff distance is within or without arms reach of the victim. Take, for example, a powder pattern on a victim's shirt for which the examiner, after multiple test-firings, reports the range of fire as "12 to 36 in. with the 24 in. pattern most closely representing the evidence pattern on the victim's shirt." If the responsible firearm is a rifle with a 26 in. barrel, the trigger is beyond the decedent's reach for all of these standoff distances. In the absence of special devices such as a yardstick to depress the trigger, the use of the foot to fire the rifle or an impactive discharge, a self-inflicted injury can be excluded. The total absence of any powder or GSR deposits on a surface, on a gunshot victim, or the victim's clothing presents a special problem. Some examiners are unwilling to say anything about range of fire other than it is not a contact shot. Others have made statements to the effect that "The firearm was fired from a distance greater than——," where the fill-in value is the distance beyond which no powder particles could be found on the target material. This is a perilous statement for several reasons. First, the test-firings are typically carried out under ideal conditions and into a near-ideal target material (some form of white, retentive material such as jean twill cloth). The evidence surface, whether it be the clothing of a gunshot victim, the victim's skin, or some inanimate object at a shooting scene, is likely to experience some loss or reduction of powder particles and/or GSR deposits through handling, bleeding, medical intervention, or exposure to the elements before collection so that the examiner is often looking at an understatement of the original pattern. At relatively close ranges (a few inches) this is not a problem, but if the shot was fired from several feet away where only a few powder particles arrived at the evidence surface, the loss of these few particles could result in no evidence being found on the victim or submitted object. Second, the presence of an intervening object may be difficult to exclude. Most any intervening object, such as a pillow used as a silencer, a window, or curtain through which a shot was fired, will filter out the powder particles and other GSRs. For those that wish to make some interpretive statement when no powder or GSR deposits are found on an evidence item or gunshot victim, the following is offered: "No powder particles or gunshot residue was detected around the bullet hole in the——. Test-firing of the submitted gun and ammunition deposited identifiable powder residues out to a distance of 6 feet. In the absence of an any intervening object(s) and loss of any adhering powder particles, these findings would indicate that the shot was fired from a distance greater than 6 feet." The foregoing has not been presented as a recommendation but for the reason that negative findings invariably prompt one litigant or the other to request an interpretation of the negative finding.

REVOLVERS

Revolvers offer special reconstructive opportunities as a result of their design. The presence of the gap between the front of the cylinder and the barrel results in the escape of very hot, high-pressure gases containing all of the same previously described GSR materials. This narrow gap between the face of the cylinder and the back of the barrel is typically on the order of 0.004–0.006 in. These energetic gases emerge from each side of a revolver in a narrow, elliptical pattern with the top strap and bottom of a revolver's frame effectively blocking these gases in the upward and downward directions. Any surface immediately adjacent to the cylinder gap or within a few inches of the side of the revolver will receive GSR deposits and may even suffer physical damage or sustain very intense soot deposits. Such deposits are often found on the inside surface of one of the hands of suicide victims as a result of grasping and supporting the revolver around the cylinder gap. This type of deposit may also occur on the hand or hands of a gunshot victim who has attempted to deflect a revolver fired by someone else. Revolver discharges occurring with the gun essentially tangential to any surface (clothing, a tabletop, a folded pillow being used as a silencer, the interior of a holster) all produce strong cylinder gap deposits as well as muzzle blast effects and GSR residues. These deposits not only provide positional information but also provide a close estimate of the revolver's barrel length – a useful parameter in a "no-gun" case. All of these concepts are illustrated in Figure 5.7.

The front face of the cylinder will typically possess a visible deposit of GSR around any chambers in which cartridges have been discharged. These circular deposits are called "flares" or "halos" and can be both conspicuous and diagnostic (see Figure 5.8). Flares are somewhat fragile and easily disturbed so

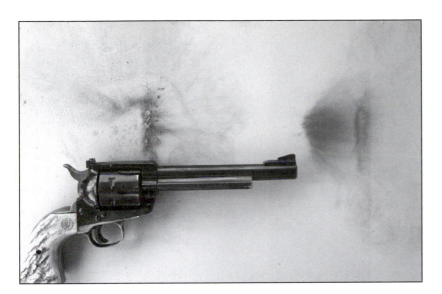

Figure 5.7

Muzzle and Cylinder Gap
Deposits

This .357 Magnum revolver was fired while held parallel to the wall and at a standoff distance of about 2 in. The revolver has been positioned just below the GSR patterns from the cylinder gap and the muzzle to illustrate the relationship between the barrel length of the responsible gun and the GSR deposits. It should also be noted that the cylinder gap discharge was so energetic that it removed some of the paint on this wall.

Figure 5.8

"Flares" on the Face of a
Cylinder from a Revolver

The face of the cylinder removed from the revolver shown in Figure 5.7 possesses two "flares" at the 11 o'clock and 1 o'clock positions. A careful inspection will also show that the 1 o'clock flare overlaps the 11 o'clock flare. This is a consequence of the sequence of these shots.

the presence of one or more flares on the face of the cylinder should be noted and documented before any processing of the firearm for fingerprints takes place and certainly before any test-firing. The presence of "fresh" (undisturbed, powdery-gray) flares on the face of a revolver's cylinder allows one to state that "The revolver has been fired at least (*number of flares*) times since its last thorough cleaning." This language is recommended because multiple shots could conceivably be fired in one chamber leaving the front margins of the remaining chambers free of any flares. The position of any flares at the time of recovery of a revolver can also be of critical importance. With a single shot from a revolver and no manipulation of the gun's mechanism afterward, the single "fresh" flare will be on the face of the chamber under the hammer. If it is *not* under the hammer, it can be reasonably concluded that the gun's mechanism, particularly the cylinder, has been manipulated or rotated in some way after the shot was fired. This can be a critical piece of information in suicide vs. homicide determinations for obvious reasons when the fatal wound is immediately incapacitating.

THE MODIFIED GRIESS TEST FOR NITRITE RESIDUES

The primary constituent in all smokeless propellants is nitrated cellulose. During the discharge process particles of partially consumed and even unconsumed propellant are violently expelled from the muzzle that contain nitrite and nitrate compounds. The reconstructive value of the presence and pattern of these particles on skin, clothing, and other surfaces has already been discussed. But there are situations where the nitrite-bearing particles are masked by a dark background, the particles have been dislodged, or their surviving morphology is so altered that they cannot be recognized as powder particles. These particles and many of the sites where they impacted a surface contain traces of nitrites ($-NO_2$) and nitrates ($-NO_3$). Nitrates are relatively common in nature and can be found in a number of materials *not* associated with small arms propellants. Nitr*ites* on the other hand are neither common nor particularly stable in the environment and *are* present in readily detectable amounts in nearly all smokeless and black powder residues following discharge. It is these nitrite residues and the pattern they form around a bullet hole that are detected and rendered visible by the modified Griess test. The only exceptions encountered by the author have been several instances involving spherical ball powder that did not degrade sufficiently during the discharge process to produce a positive response for nitrites even though numerous unconsumed particles of the propellant could be seen on the garment.

The current Griess (pronounced like the word grease) test has evolved over the last 75 years. The reagents for nitrites in the earlier formulations were found to be serious health hazards and have been replaced with less dangerous chemicals in the modified Griess test.

The basic chemistry involved is the formation of an orange azo-dye between alpha-naphthol and a diazonium compound of sulfanilic acid. In cases involving clothing this is accomplished by steaming (heating) the evidence garment with acetic acid vapors which in turn convert nitrites into nitrous acid (HNO_2), which is a volatile compound. A specially treated panel (desensitized photographic paper) containing sulfanilic acid and α-naphthol in the emulsion layer is placed in direct contact with the surface of the evidence item. As with the previous transfer techniques used for lead and copper tests, multiple reference marks should be made on the transfer paper prior to its removal from the garment. A layer of cheese cloth soaked in 15% acetic acid is placed on the opposite side of the garment and heat is applied with a common steam iron using the "cotton" setting. This drives the hot acetic acid vapors through the garment, converting any nitrites to nitrous acid where they immediately react with the reagents in the emulsion layer of the desensitized photographic paper. An alternative technique involves placing the acetic acid solution in the reservoir of the steam iron and omitting the cheese cloth. By either technique, nitrite-containing spots and particles will produce bright orange spots on the transfer paper. The transfer technique described here requires the object being tested to be relatively thin and porous so that it can be steamed from the backside with the reactive panel on the opposite side containing any nitrite residues. If this "sandwich" arrangement is not possible due to the nature of the object, alternatives will have to be considered such as direct application of the reagents or a moistening of the emulsion side of the desensitized and treated photographic paper followed by firm and intimate contact between the transfer paper and the evidence item for several minutes. So long as adequate time has been allowed for the acetic acid fumes to liberate any nitrites as nitrous acid, the author has found no need for using a steam iron on object other than garments. Readers who believe that the heat supplied by a steam iron is necessary for an improvement in detecting nitrites should experiment with a test pattern by processing half of the powder pattern with prolonged physical contact vs. the steam iron technique. Smooth filter paper previously treated in the same fashion with the sulfanilic acid/α-naphthol solution as desensitized photographic paper and then moistened with a light spray of 15% acetic acid can be substituted for the photographic paper as this latter product is becoming less common in your average photo shop. All of these procedures require some skill, consequently it is recommended that any examiner who does not perform the modified Griess test on a regular basis practice on some test powder patterns produced on comparable substrates prior to processing the evidence item(s).

If the sodium rhodizonate test for lead is contemplated or desired, it can be carried out *after* the modified Griess test has been completed and documented. Color photography of any pattern or positive response both with and without a scale is recommended since these colors may fade or a background discoloration may develop over time. The reader should be reminded that acetic acid will solubilize and transfer some portion of any lead particulates or vaporous lead deposits on the object being tested. It is also important to note that the Griess test, *unlike* the bullet metal tests described earlier, is a one-time test. Any nitrites present are converted into nitrous acid and rapidly react with the chemicals in the Griess test paper to form colored spots. Prompt documentation of the results of this test is therefore of critical importance.

Materials—Modified Griess Test

Acetic acid (15% v/v aqueous solution)
Alpha-naphthol (0.3% w/v in methanol)
Sulfanilic acid (0.5% w/v in distilled water)
Sodium nitrite (0.5% w/v aqueous solution)
Distilled water
Methanol
Large sheets of photographic paper* (* smooth, quantitative grade filter paper or selected brands of Inkjet photographic paper may be substituted)
Photographic "hypo" (fixer) solution (aqueous sodium thiosulfate solution)
Cheese cloth or filter paper
Steam iron

THE PREPARATION OF REAGENTS AND MATERIALS

In the original method, desensitized photographic paper is used as the test paper. More recently, it has been discovered that certain brands of Inkjet photographic paper can be substituted for desensitized photographic paper and gives comparable, in some instances, superior results. If traditional photographic paper is to be used, the silver salts must first be removed from the emulsion layer by soaking and rinsing large sheets of the photographic print paper in a tray containing fixer solution (obtained from a photo shop). This step must be carried out in darkness. Once processed with the "hypo" solution, these sheets are then allowed to dry before treatment with solutions of sulfanilic acid and α-naphthol. Those having access to a police photo-lab have an advantage in that the photo-lab should be able to prepare these sheets thereby sparing the examiner this somewhat awkward and unfamiliar step. No pretreatment is necessary if a suitable Inkjet photographic paper is used.

Equal volumes of the sulfanilic acid solution and α-naphthol solution (e.g., 100 ml + 100 ml) are mixed and placed in a clean photographic tray. Each sheet of dry, desensitized photographic paper (or Inkjet photo-paper) is momentarily dipped in this solution then allowed to dry on a clean, flat surface. Once dry, these sheets should be placed in a sealable envelope, dated, and stored in a refrigerator.

Cotton-tipped swabs are moistened with an aqueous solution of sodium nitrite, allowed to dry, then stored in an air-tight container. These will be used to verify the efficacy of the test panels prior to application to the evidence item(s). This is done by moistening a swab with a drop of acetic acid then touching it to an edge or corner of the test paper.

Visual and microscopical examination with the stereomicroscope should first be carried out on the evidence items before any chemical testing. If the item or garment is dark or blood-stained, infrared viewers or photography can often render carbonaceous soot visible. Radiographic films exposed with soft X-rays can often reveal powder particles underneath a coating of blood. Once these preliminary steps have been completed, the modified Griess test can be carried out with one of the previously described techniques depending on the nature of the object to be tested.

Although the modification of the Griess test amounted to substituting α-naphthol for N-(1-naphthyl)ethylenediamine dihydrochloride (a known carcinogen) in order to reduce the health risks associated with carcinogens, the user must consider *all* of these reagents as potential health hazards. The use of rubber gloves, a fume hood, complete avoidance of inhalation of vapors, and subsequent contact with these materials is to be scrupulously avoided.

GUNSHOT AND PRIMER RESIDUES ON THE HANDS

In the absence of visible GSR on the hands of a suspected shooter, instrumental methods may be used to detect and identify these very low levels of GSR. Nearly all of these tests are directed toward the *inorganic* elements associated with firearms discharge products although considerable research is being devoted to the numerous organic constituents present in GSR. These organic constituents include unconsumed nitrocellulose, stabilizers, plasticizers, flash inhibitors, and propellant modifiers. This research shows that these compounds provide very useful information to the forensic analyst and investigator but their presence on the hands and clothing of shooters is not yet sought in average casework simply because a standard collection and analysis procedure has yet to be worked out and adopted. Such procedures have been established for the inorganic constituents. The collection of the inorganic residues is carried out by one of the following two methods:

1. Mild acid swabbing of selected areas of the hands followed by flameless atomic absorption spectroscopy analysis (FAAS) of the extracts of these swabs for elevated levels of metals (lead, barium, and antimony) associated with common primer formulations.

2. Sticky tape lifts of the hands that are subsequently analyzed by SEM/EDX analysis. This technique has special advantages in that it provides high resolution images of very small particles and allows their elemental composition to be determined without consuming or altering the particles. This is important because the very high temperatures and pressures associated with firearms discharges generate spheroid particles on the order of 1–10 microns (μm) in size and composed largely of lead, barium, and antimony when derived from common primer formulations. For these reasons the SEM/EDX technique has become the predominant method for GSR/primer residue detection and identification on samples collected from the hands.

Micro-vacuuming and particle collection on special filter discs have also been developed for the processing of selected items of clothing from suspected shooters. The use of SEM/EDX only stands to increase as new and varied primer formulations continue to be introduced by nearly every primer manufacturer. These "environmentally friendly" primers contain elements not previously associated with firearms discharge residues. These include zinc, titanium, potassium, boron, strontium, silicon, calcium, zirconium, magnesium, aluminum, sulfur, and manganese. The morphology of these particles is only ascertainable with a scanning electron microscope and since a number of these elements are relatively common in the environment, the combined analytical power of the SEM with the EDX accessory is, and will continue to be, mandatory for their identification.

Some discussion regarding the current status and usefulness of GSR testing of the hands of suspected shooters is appropriate. The value of such testing has fallen short of the original hope of identifying a recent shooter and excluding non-shooters. This is not to say that such tests are without value or that collecting samples is a futile exercise. The typical reporting language used in American crime laboratories where a positive result is obtained reads something like: "The subject either fired a gun, handled a gun or was in close proximity to a firearm when it was discharged." Given these choices for a positive result, many readers may regard the value of such evidence as very low and not worth the effort involved in taking samples from suspects' hands and the subsequent expense of analyzing them. But this negative reaction is not well thought out. The *circumstances* of each case must be considered when the sampling of one or more individual's hands for GSR/primer residue is contemplated. Consider a case in which three individuals admit to having been in a car where a gun was discharged but all deny firing a gun. Testing these subjects is probably a futile exercise because they are all likely to test positive for GSR/primer residue due to the relatively confined space in which a gun was discharged. Testing the hands of a suspected suicide victim with a loose contact gunshot wound to the chest is also likely to give a positive result simply because the hands would have been in close proximity to the gun at the moment of discharge. This is also true if the victim was murdered, so a positive finding of GSR/primer residue does not distinguish a murder from a suicide. Conversely, a negative finding, particularly in a living individual, does not exclude a subject as having fired a gun. The microscopic particulate residues associated with GSR can be removed by the washing of the hands and through normal activities with the passage of a few hours. This latter fact is the reason that most, if not all, GSR collection procedures set a cutoff time after which no samples will be taken. Moreover, some firearm-ammunition combinations and/or the skin of some individuals do not consistently leave or retain detectable levels of GSR/primer residue on the hands. So after all of the foregoing negativism, when is the collection and analysis of such samples useful? These tests stand to be useful when the time interval between the incident and the collection is short (minutes to an hour or two) *and* in situations where the individual denies owning a gun, shooting a gun, handling a gun, or being anywhere near the discharge of a firearm or the immediate scene of a shooting. In this situation a positive finding would be very incriminating. There are a number of other considerations that cannot be addressed within the limits of this chapter, but suffice it to say that there are instances where the collection and analysis of hand swabs or sticky tape lifts from individuals can provide useful and incriminating evidence. The desirability of sample collection simply needs to be well thought out as opposed to collecting such samples because the materials are available. Properly collected samples can be retained indefinitely and analyzed when such analysis is deemed useful but it should be remembered that a negative finding does not preclude the tested subject as having fired a gun. The absence of evidence is not necessarily evidence of absence.

SUMMARY

The various constituents of GSR have been described along with their short-range exterior ballistic properties. The reconstructive value of visible and chemically detectable GSR deposited on various surfaces has been presented. The reagents and general procedures for the application of the modified Griess test for nitrites have been described. While a detailed protocol is desirable for the routine use of this reagent with clothing, the author urges examiners to consider a more thoughtful approach to non-standard surfaces by carrying out some preliminary empirical testing to refine and select the best technique for the ultimate application of the modified Griess test to the evidence at hand.

The same philosophy has been presented with regard to the testing of suspected shooters' hands for trace amount of GSR/primer residue, namely a thoughtful assessment of the suitability of the subject for sampling and the probative value of any positive result.

REFERENCES AND FURTHER READING

Barns, F.C. and R.A. Helson, "An Empirical Study of Gunpowder Residue Patterns," *JFS* 19:3 (July 1974) pp. 448–462.

Bashinski, J.S., J.E. Davis and C. Young, "Detection of Lead in Gunshot Residues on Targets Using the Sodium Rhodizonate Test," *AFTE Journal* 6:4 (1974) pp. 5–6.

Davis, T.L., *The Chemistry of Powder and Explosives*, Angriff Press, Hollywood, CA, 1943.

Dillon, J.H., "The Modified Griess Test: A Chemically Specific Chromophoric Test for Nitrite Compounds in Gunshot Residues," *AFTE Journal* 22:3 (July 1990) pp. 49–56.

Dillon, J.H., "A Protocol for Gunshot Residue Examinations in Muzzle to Target Distance Determinations," *AFTE Journal* 22:3 (July 1990) pp. 257–274.

DiMaio, V.J.M., C. Petty and I.C. Stone, "An Experimental Study of Powder Tattooing of the Skin," *JFS* 21:2 (April 1976) pp. 367–372.

Dodson, R.V. and R.F. Stengel, "Recognizing Vaporized Lead from Gunshot Residue," *AFTE Journal* 27:1 (January 1995).

Fiegl, F., *Spot Tests in Inorganic Analysis*, 5th edn, American Elsevier, New York, 1958.

Haag, L.C., "A Method for Improving the Griess and Sodium Rhodizonate Tests for GSR on Bloody Garments," *SWAFS Journal* (April 1991) also *AFTE Journal* 23:3 (July 1991).

Haag, L.C., "American Lead-Free 9MM-P Cartridges," *AFTE Journal* 27:2 (April 1995).

Haag, L.C., "Phenyltrihydroxyfluorone: A 'New' Reagent for Use in Gunshot Residue Testing," *AFTE Journal* 28:1 (January 1996).

Haag, L.C., "Reference Ammunition for Gunshot Residue Testing," *CACNews* (Second Quarter 2000) also *AFTE Journal* 32:4 (Fall 2000) and *SWAFS Journal* 23:1 (February 2001).

Haag, L.C., "Sources of Lead in Gunshot Residue," *AFTE Journal* 33:3 (Summer 2001).

Haag, L.C., "Skin Perforation and Skin Simulants," *AFTE Journal* 34:3 (Summer 2002).

Haag, L.C. and R. Bates, "Preliminary Study to Evaluate the Deposition of GSR on Unfired Cartridges in the Adjacent Chambers of a Revolver," *AFTE Journal* 32:4 (Fall 2000).

Haag, M.G. and E.J. Wolberg, "The Scientific Examination and Comparison of Skin Simulants for Distance Determinations," *AFTE Journal* 32:2 (Spring 2000) pp. 136–142.

Jalanti, T., P. Henchoz, A. Gallusser, and M.S. Bonfanti, "The Persistence of Gunshot Residue on Shooters' Hands," *Science & Justice* 39:1 (March 1999) pp. 48–52.

Jungries, E., *Spot Test Analysis—Clinical, Environmental, Forensic and Geochemical Applications*, John Wiley & Sons, NY, 1985.

Malikowski, Shawn G., "Alternative Modified Griess Test Paper," *AFTE Journal* 35:2 (Spring 2003) p. 243.

Meng, H. and B. Caddy, "Gunshot Residue Analysis—A Review," *JFS* 42:4 (July 1997) pp. 553–570.

Nichols, R.G., "Expectations Regarding Gunpowder Depositions," *AFTE Journal* 30:1 (Winter 1998).

Nichols, R.G., "Gunshot Proximity Testing—A Comprehensive Primer in the Background, Variables and Examination of Issues Regarding Muzzle-to-Target Distance Determinations," *AFTE Journal* 36:3 (Summer 2004).

Rathman, G., "Gunpowder/Gunshot Residue Deposition: Barrel Length vs. Powder Type," *AFTE Journal* 22:3 (July 1990) pp. 318–327.

Stone, I.C., L. Fletcher, J. Jones, and G. Huang, "Investigation into Examinations and Analysis of Gunshot Residues," *AFTE Journal* 16:3 (July 1984) pp. 63–73.

Veitch, G., "An Examination of the Variables that May be Encountered in Gun Shot Residue Patterns," *AFTE Journal* 13:2 (April 1981) pp. 35–54.

CHAPTER 6

PROJECTILE PENETRATION
AND PERFORATION

OVERVIEW

Common materials struck by projectiles include sheetrock (wallboard), wood, glass (3 types), sheet metal (filing cabinets/vehicles/road signs), asphalt, concrete, construction block/bricks, rubber (tires), plastic (truck bedliners, patio furniture), clothing, and other fabrics (upholstery-furniture). Bodies will be discussed in Chapter 10.

There are essentially three possible outcomes of orthogonal and near-orthogonal (perpendicular in both planes) impacts to most of these materials.

1. The projectile will be stopped without penetrating the material.
2. The projectile will penetrate the material where it may become lodged or it may disintegrate and the fragments rebound from the material. A lead bullet fired into a wooden fence post would be an example of the former and the same type of bullet fired into a cinder block wall would exemplify the latter.
3. The projectile will perforate the material.

Changes will take place in the projectile and the impacted material in a predictable and characteristic manner once the dynamics and properties of each are understood.

With low incident angle strikes to materials other than clothing, the projectile will ricochet from the object. Low incident angle impacts will be discussed in detail in Chapter 7, where the important matter of ricochet is presented.

The Locardian view of the likely exchange between projectile and impacted material is a good starting point for all of these encounters followed by some thoughts about the relative hardness of projectile and substrate, as well as the nature of the yielding or failing process for the impacted material. A further concept that will be useful is to divide target materials into two main groups: (1) *non*-yielding materials such as concrete and heavy steel and (2) *yielding* materials such as sheet metal, wood, and glass. Yielding materials can be further divided into the smaller groups of malleable (sheet metal) and frangible (glass/ sheetrock) materials. Some of the more common characteristics of specific materials are discussed in the following sections.

SHEETROCK/WALLBOARD

This common material is used in home and office construction. It is composed of gypsum (calcium sulfate) with a heavy paper coating on both sides. The surface normally seen (and usually first struck by gunfire) is frequently painted or covered with wallpaper. Popular thicknesses in the United States are $\frac{1}{2}$ in. and $\frac{5}{8}$ in. Sheetrock is rather easily defeated and perforated by most common small arms projectiles. By way of example, the approximate threshold velocity (V_T) for .38 caliber/9 mm bullets to perforate $\frac{1}{2}$ in. sheetrock is about 100–150 fps (30–46 m/s), depending on the weight of the bullet and its orientation at the moment of impact. These velocities are on the order of what one can achieve with a common slingshot. The reader that has a ballistic chronograph and wishes to conduct some empirical testing is urged to practice his or her marksmanship with a slingshot and some fired bullets of a caliber and weight of interest, then launch them into a panel of sheetrock positioned just beyond the chronograph. A projectile failing to perforate sheetrock will often leave a clear impression of itself, including its orientation upon impact. This impression may also contain the outline of the rifling impressions on the responsible bullet (see Figure 6.1). The location of such an impact site should be measured and documented after which the area containing the impact site should be excised and impounded as evidence. This can be accomplished with a common utility knife available in most any hardware store or with a small reciprocating saw. Nonperforating bullet imprints have occurred in past cases where a bullet has passed through a gunshot victim and exits in a destabilized manner and with a velocity on the order of 100–150 fps as it strikes an interior wall.

Once the threshold velocity necessary to perforate the particular thickness of sheetrock is substantially exceeded, velocity *losses* for near-orthogonal impacts are on the order of 30 fps (9 m/s) for such bullets. This significant difference between the impact velocity necessary to perforate a panel of sheetrock and the velocity loss experienced during the perforation process is a recurring phenomenon for all thin materials struck by projectiles and will be discussed in detail in the section on sheet metal.

If there is an air space inside the wall free of insulation, the dislodged gypsum from the bullet's path will often be impactively deposited on any opposing nearby and downrange surface such as the sheetrock on the opposite side of an

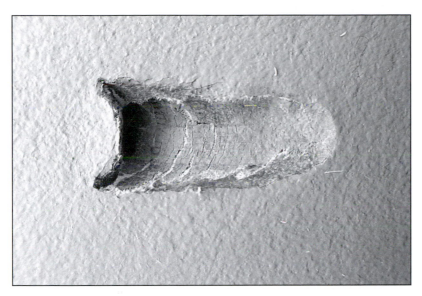

Figure 6.1

A Bullet Impression in Painted Sheetrock

A decelerated and destabilized .38-caliber, 158 gr. LRN bullet struck the painted surface of this wallboard leaving a 3-dimensional outline of itself to include faint rifling marks on the bullet and the outline of two knurled cannelures. The appearance of the type of bullet that produced this impression can be seen in the insert in Figure 6.8.

interior wall or any other supporting surface such as an exterior wall. In non-orthogonal strikes, the location of such a deposit has an unusual and somewhat counter-intuitive relationship relative to the path of the projectile. As the bullet approaches its future exit site during the perforation process in the sheetrock, a spall of gypsum will be propelled away from this site at an angle that is essentially orthogonal to the exit surface. This is important to recognize and understand since one may mistakenly believe that these ejected deposits of dislodged gypsum on the opposing surface represent a point of reference for the projectile's flight path. This is incorrect except where the projectile entered and perforated the sheetrock at an orthogonal angle. An example of this is shown in Figure 6.2. We will also see this interesting behavior in other brittle or frangible materials such as panels of glass when struck by bullets at non-orthogonal angles.

Here again the author urges any reader who actually processes shooting scenes to take the time to construct a mock wall out of sheetrock and a few 2 × 4 boards and then fire a few shots through this wall at orthogonal and non-orthogonal angles. Such an exercise also provides other opportunities to practice making the angular measurements described in Chapter 8, as well as an opportunity to evaluate deflection issues.

The frangible nature of many paints on the entry surface of sheetrock and the underlying sheetrock itself may greatly reduce or even obviate the transference of

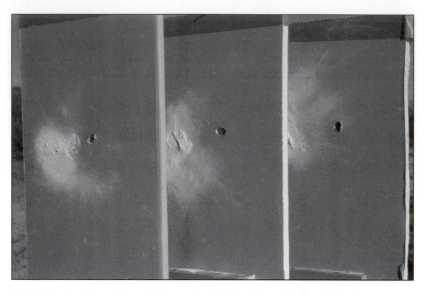

This illustration shows three of six panels of 5/8 in. sheetrock set at a 45° intercept angle to the flight of a 9 mm FMJ-RN bullet. The bullet perforated a previous panel to the left and out of the field of view in this photograph. The flight of this bullet was from left to right. The bullet holes can be seen to right of each of the spattered deposits of powdered gypsum. The deposits were propelled away from the exit surfaces at right angles to each of these surfaces. They then traversed the 6 in. space between the panels and were impactively deposited at the location visible in this photograph.

Although the bullet was noticeably destabilized as evidenced by the out-of-round bullet holes, *no* detectable deflection occurred.

Note: Figure 6.4 shows all six panels with a trajectory rod passed through the bullet holes.

bullet wipe. Differentiating entry from exit is easy, however. The entry or impact side of sheetrock will faithfully record the orientation and shape of the bullet that struck it. In other words, the entrance hole will depict the profile of a destabilized or tumbling bullet, the extended "talons" of a *Black Talon* bullet that has previously mushroomed as a result of perforating a victim (see Figure 6.3), or the normal round or ovoid hole from a direct strike by a stable, nose-first bullet. Close proximity discharges can produce stippling of the sheetrock and, of course, deposit GSR on its surface. An example of this was shown in Figure 5.1 in Chapter 5.

Deflection of bullets that perforate common sheetrock is essentially nil. Figure 6.4 shows a trajectory rod passed through a series of bullet holes from a 124 gr. FMJ 9 mm bullet that perforated 6 panels of ⅝ in. sheetrock mounted at 45° angles to the bullet's flight path. This non-deflecting behavior makes the back-extrapolation of bullet holes through sheetrock walls very reliable insofar as the values of the vertical and azimuth angles are concerned. Bullets themselves are little affected by passage through sheetrock except to say that they will be destabilized in their subsequent flight and, if of the hollow point design, their hollow point cavities will be plugged with gypsum as a consequence of a direct strike.

This .45-caliber *Black Talon* bullet was fired through 3 in. of tissue simulant mounted approximately 3 feet in front of this panel of painted wallboard. Passage through the tissue simulant caused the bullet to fully mushroom. It then struck and perforated the sheetrock in a nose-forward orientation and with its "talons" properly extended leaving their characteristic outline around the margin of this bullet hole.

Figure 6.4

Bullet Perforation of Multiple Panels of Sheetrock

The 5/8 in. thick panels of sheetrock in this simple holder are approximately 6 in. apart and oriented at a 45° angle to the bullet's flight path. The bullet was a 124 gr. 9 mm FMJ-RN bullet fired from a distance of approximately 4 feet with a Beretta Model 92FS pistol. Although the bullet was clearly destabilized by the second panel, the trajectory rod passed through all six bullet holes shows that there was no measurable deflection. *Note*: Figure 6.2 provides a close-up view of panels 2, 3, and 4 before the trajectory rod was passed through the bullet holes.

WOOD

The appearance of entry and exit bullet holes in wood follows a common sense model with the wood fibers forced inward around the margin of the entry hole and chips of wood frequently expelled or turned outward at the exit site. Most types of wood acquire detectable, often visible, deposits of bullet wipe around the margin of the entry hole. Both lead and copper (in the case of copper alloy jacketed bullets) can usually be detected in such bullet wipe by the lifting technique many months after the bullet hole was created. Lead bullets and jacketed bullets with exposed lead noses typically leave strong deposits of lead throughout the bullet's track in wood—a phenomenon that can be exploited in bullet holes that are years old and no longer possess detectable copper or lead around the exterior margin of the entry hole. A jacketed bullet on the other hand will usually leave traces of lead at the entry point but not along the interior channel. The wood fibers along the channel of a bullet hole relax to varying degrees after the bullet's passage so that the resultant bullet hole is usually smaller than the bullet that produced it. Care must be taken in choosing an appropriate probe, both in composition and diameter for insertion through a bullet's path through any wooden object. Brass or copper rods should be avoided, and if tests for lead are contemplated, a probe free of lead residues on its surface is in order. Projectile nose shape, projectile hardness, impact velocity, and, of course, the nature of the particular wood all play a significant role in the properties of the final bullet hole and channel diameter.

Bullet deflection as a result of perforation of relatively thin specimens of wood (wooden fence boards, small tree branches, gun stocks) is typically small, e.g., 1°–2°. Bullet destabilization, however, is common, as is the plugging of hollow point cavities with wood particles. Soft point and hollow point bullets seldom expand as a consequence of wood perforation but often acquire embedded wood particles in their noses and hollow point cavities. Non-orthogonal impacts and penetration/perforation of wood will produce an elliptical entry hole. The arcsine of the ratio of the minimum diameter divided by the maximum diameter of the best ellipse representing the outline of the entry hole can often provide a reasonable *estimate* of the incident angle of this strike. The best ellipse formed by the margin of the entry hole, or the bullet wipe around it, can most easily be constructed from a good, straight-on photograph of the bullet hole and the drawing tool on a computer. Once such an ellipse is constructed, it can be enlarged proportionally to facilitate a measurement of the minimum and maximum diameters. The actual dimensions of the bullet hole are unimportant. It is the *ratio* of any faithful representation of an elliptical (non-orthogonal) bullet hole. Figure 6.5 illustrates this technique for one of three bullet holes in a thin board. The arcsine ($\sin^{-1}$) function on a pocket calculator with scientific functions is used to derive the equivalent incident angle. As with many techniques

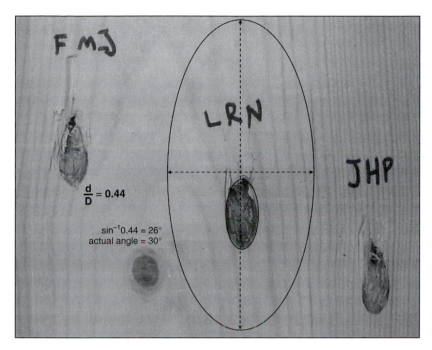

Figure 6.5

Non-Orthogonal Bullet
Holes in Wood

These bullet holes were produced by three types of .38-caliber bullets fired into this thin board at the same nominal incident angle of 30°. This photograph was taken from a position orthogonal to the three bullet holes. A computer drawing tool was used to draw the best-fitting ellipse around the margin of the entry hole produced by the LRN bullet. This ellipse has been copied and enlarged after locking the aspect ratio. The arcsine function was used to derive the approximate intercept angle from the ratio of the width to the length of the ellipse. The calculated value and the true value are shown above.

described in this book, the reader is urged to carry out some empirical tests with the type of ammunition and wood involved in a specific case so as to establish some accuracy and confidence limits for this technique. This technique is a useful adjunct to the traditional probe method for bullet path determination particularly where the perforated board is relatively thin ($\frac{1}{2}$ in. or less) since the uncertainty in path measurements *increases* with *decreasing* thickness.

SHEET METAL

A lengthy discussion on bullet perforation of sheet metal has been given by Nennstiel in the *AFTE Journal*. The essential parameters from his work have been integrated with a number of observations and measurements by this author.

The most common form of sheet metal encountered in shooting reconstruction cases is that found in motor vehicles. This is customarily 22-gauge steel measuring about 0.031–0.032 in. (0.79–0.82 mm) in thickness. Other commonly encountered sources of sheet metal in shooting incidents are office furniture (metal filing cabinets), certain home appliances (washing machines, refrigerators, ovens, dish washers), and road signs. Sheet metal can be perforated by virtually all small arms projectiles given sufficient impact velocity. For any particular bullet, there is a *threshold velocity* that must be exceeded for the bullet to perforate the particular thickness of sheet metal. Obviously, the angle of incidence also enters into the determination of *threshold velocity* (V_T) but V_T values are ordinarily measured for orthogonal impacts. At any velocity *less than* the threshold velocity for a given bullet–sheet metal combination, the metal will undergo some amount of deformation due to its malleable nature. This can be related to the impact velocity of the projectile. Most projectiles will likewise suffer some deformation that can also be related to impact velocity through subsequent empirical testing. Figure 6.6 shows a lineup of LRN bullets that struck sheet metal with ever increasing impact velocities nearly all of which were less than the threshold velocity for this 125 gr. 9 mm bullet–0.032 in. sheet metal combination. It is important to note that the demonstrative aid shown in Figure 6.6 is for this bullet and metal combination only, and cannot be applied universally. This holds true for many of the illustrations in this text.

Figure 6.6

Bullets Deformed by Impact with Sheet Metal

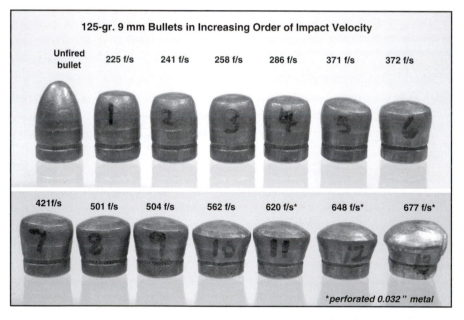

This figure shows a lineup of LRN bullets that struck automotive sheet metal with ever increasing impact velocities. An unfired bullet is also shown at the far upper left. The regular progression of the flattening of these bullets with the attendant thickening of their diameters with increasing impact velocity has obvious forensic implications and reconstructive value. *Note*: The listed impact velocities are in feet per second (fps) and can easily be converted to meters per second by dividing each value by 3.2808.

Bullet deformation and sheet metal deformation are two interrelated phenomena that have obvious reconstructive implications. Consider a bullet like #6 in Figure 6.6 that is found below a lead-positive indentation in a car door. Inspection of the interior of the car door reveals that there were no supporting structures at the impact site. The fact that the bullet did not perforate the sheet metal is an immediate indication that this 9 mm, .38-caliber bullet was traveling at a relatively low velocity. Some subsequent empirical testing in the laboratory shows comparable bullet deformation when shot into automotive sheet metal at impact velocities of 350–400 fps (107–122 m/s). Given the normal muzzle velocity for bullets of this type of approximately 1000 fps (305 m/s), thoughts of a long-range shot or deceleration by an intervening object should immediately come to mind. The fact that the bullet struck nose-first would offer greater support to the long-range shot than passage through an intervening object but this should not be ruled out simply because of the nose-first impact. The reader should also revisit Figure 3.2 in Chapter 3 depicting a strike by a .22 LRN bullet fired from an adjacent car that failed to perforate the driver's door. Its failure to perforate is a clear statement about impact velocity.

The next concept in this discussion of thin metal "targets" is that: *a striking projectile will either be defeated (stopped) or it will perforate the target and have considerable remaining velocity.* This also applies to other thin "targets" such as skin, rubber, glass, thin boards, and clothing.

SHEET METAL PLUGS AND TABS

The perforation process will also frequently involve the production of a small metal *plug* punched out by the bullet. This sheet metal *plug* may become "welded" to the nose of the bullet, particularly if the metal is unpainted and the bullet is jacketed. More often, it will be ejected in front of the bullet, and then quickly overtaken and passed by the bullet. Figure 6.7 gives a profile view of the ejection of a circular plug of sheet metal immediately in front of the 9 mm FMJ bullet that produced it.

If a perforating bullet strikes sheet metal at some angle other than orthogonal, the sheet metal plug will be ovoid in shape (an *ovoid plug*). In the case of a tumbling or destabilized bullet striking the sheet metal in yaw, a *tab* of sheet metal will be punched out. At close range (a few feet) these plugs and tabs of sheet metal are injurious missiles in their own right, and may be found in close proximity to, or within, the gunshot wound produced by the responsible bullet. The circular and ovoid shape of a recovered *plug* tells much about the intercept angle of the bullet that produced it, just as a *tab* is the consequence of a destabilized bullet. Occasionally the characteristic outline of the margins of a hollow point cavity in a pistol bullet can be seen in the concave side of a *plug* and the image of a knurled cannelure in a *tab* produced by a cannelured bullet. Transfers of bullet metal,

Figure 6.7

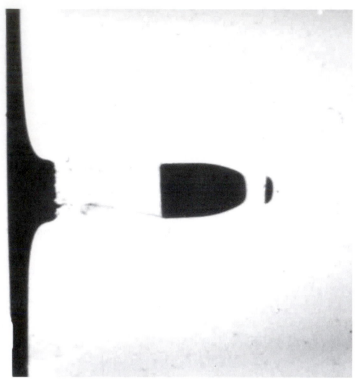

This high speed photograph shows the ejection of a circular plug of sheet metal immediately in front of a 124 gr. 9 mm FMJ-RN bullet that has just perforated a panel of 0.030 in. sheet metal loosing approximately 70 fps of its impact velocity in the process.

At very close range this sheet metal plug can produce a satellite injury to a gunshot victim or be found in the wound track. This same plug can produce small defects or impact sites in inanimate objects when the distances from the bullet hole are short (inches to a few feet).

Within a few more inches of flight, the bullet will overtake and pass the sheet metal plug as it decelerates much more rapidly than the bullet that produced it. Sheet metal plugs may, in certain circumstances, remain attached to the nose of the causative bullet. (Photograph from the work of R. Nennstiel.)

including copper in the case of jacketed bullets, will be present on the concave side of plugs and tabs. In one high-profile case, the author was able to demonstrate a physical match between a sheet metal *tab* and the fatal bullet that produced it. These two items struck the victim within an inch of each other and allowed the critical issue of the entry path of the fatal bullet into the vehicle to be established.

The next concept of importance is that once *the threshold velocity necessary to achieve perforation is exceeded, the velocity loss (V_{Loss}) experienced by the perforating bullet is much less than the threshold velocity (V_T), e.g., $V_{Loss} << V_T$. Additionally, the velocity loss becomes essentially constant regardless of impact velocity once the impact velocity (V_{IMP}) becomes well in excess of the threshold velocity necessary for perforation.* This can be more easily understood by studying Figure 6.8 taken from the extensive work of Ruprecht Nennstiel at the *Bundeskriminalamt* (BKA) Laboratory in Wiesbaden Germany.

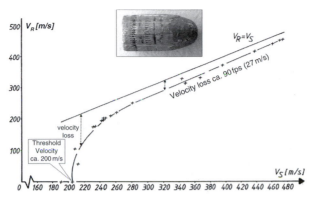

Figure 6.8

Threshold Velocity and Velocity Loss for a .38-Caliber, 158 gr. Lead Round Nose Bullets Striking 0.32 in. Thick Sheet Metal

The horizontal "X" axis represents the striking velocity (V_S) before perforation of the sheet metal. The vertical "Y" axis represents remaining velocity (V_R) after perforation of the sheet metal. The straight, diagonal line in this graph represents the no-velocity loss line, i.e., the absence of any intervening material. The curved line was constructed from exit velocity values (+) for these bullets at various impact velocities. At the threshold velocity of about 660 fps (200 m/s) and below, these bullets fail to perforate the sheet metal. A further inspection of the exit velocity line shows that as the impact velocities exceed approximately 900 fps, the velocity loss becomes nearly constant at about 90 fps. It should also be noted that *velocity loss* is not the same as the *threshold velocity* and that *velocity loss* is substantially less than the *threshold velocity* necessary to perforate the material.

The threshold velocity for the standard 158 gr., .38 Spl. bullet shown above was approximately 660 fps (ca. 200 m/s) and the nominal velocity *loss* about 90 fps (27 m/s) after producing a hole of about 0.45 in. diameter. V_T values for orthogonal strikes to 0.032 in. sheet metal for a few other common bullets are: 540 fps (165 m/s) for the 50 gr. FMJ 25Auto bullet, 425 fps (130 m/s) for the 73 gr. FMJ 32Auto bullet and 360 fps (110 m/s) for the 124 gr. FMJ 9 mm Luger bullet. The velocity *loss* experienced by the FMJ 9 mm bullet, for example, was about 65 fps (20 m/s). Table 6.1 provides some supplemental data regarding velocity losses for some representative bullets perforating sheet metal. The phenomenon of velocity *loss* being much less than the *threshold velocity for*

32 Auto 73 gr. FMJ-RN	116 fps/35 m/s
9 mm L 115 gr. FMJ-RN	74 fps/23 m/s
9 mm L 115 gr. Win. S.T.-JHP	85 fps/26 m/s
9 mm L 124 gr. FMJ-RN	66 fps/20 m/s
9 mm L 124 gr. LRN	74 fps/23 m/s
9 mm L 147 gr. FMJ-TC	52 fps/16 m/s
38 Spl. 158 gr. LRN	94 fps/29 m/s
40 S&W 180 gr. FMJ-TC	55 fps/17 m/s
45 Auto 230 gr. FMJ-RN	65 fps/20 m/s
7.62 × 39 mm 123 gr. M43 FMJ	56 fps/17 m/s
5.56 × 45 mm 55 gr. M193 FMJ	72 fps/22 m/s
5.56 × 45 mm 62 gr. M855 FMJ	60 fps/18 m/s

Table 6.1

Some Average Velocity Loss (V_{Loss}) Values for Some Common Bullets after Perforating Standard 22 Gauge (~0.032 in./0.8 mm) Sheet Metal *perforation* was also noted without special discussion in the previous section on sheetrock. It also holds true for all "thin" targets whether they are brittle, frangible, elastic, such as skin, or malleable, like sheet metal.

The importance of the near-constancy of velocity *loss* once the threshold velocity is substantially exceeded comes in several forms. For example, if we can make some reasonable estimates of the bullet's impact velocity in an object or victim downrange of a perforated panel of sheet metal, we can simply add the nominal value of velocity *loss* due to sheet metal perforation to estimate the impact velocity of the bullet when it struck the sheet metal. This, in turn, can often be related to range-of-fire estimates where the gun and ammunition are known and the initial strike to the sheet metal is a direct strike.

Deflection as a consequence of sheet metal perforation is typically small, e.g., 0.5°–1.5° for various 9 mm bullets fired through 0.032 in. (0.82 mm) thick sheet metal at an incident angle of 45° and impact velocities on the order of 1000–1200 fps (330–370 m/s).

BULLET HOLE SIZE

The *size*, or diameter of the bullet hole in sheet metal relative to the bullet that caused it makes an interesting study. Most FMJ bullets, for example, will typically leave a hole slightly *smaller* than the bullet that produced it at "low" impact velocities. With higher velocities the hole diameter will increase and ultimately become slightly larger than the bullet's diameter. Just what constitutes "low" and "high" velocity associated with bullet hole diameter will have to be worked out by actual testing for the particular bullet and sheet metal combination. This velocity-related phenomenon is believed due to the slightly elastic nature of the perforated sheet metal. At the lower impact velocities the perforated sheet metal relaxes slightly after the bullet's passage resulting in a slight reduction of the hole diameter. At higher impact velocities the acceleration of the sheet metal away from the margins of the bullet hole is such that it exceeds and overcomes the slight relaxation of the metal so that the final bullet hole is slightly larger than the bullet that produced it.

Additionally, the amount of final inward deformation of the sheet metal surrounding the bullet hole also bears a relationship to impact velocity, with less and less deformation as impact velocity increases. At relatively low impact velocities, the metal surrounding the impact site has more time to yield and deform as the bullet acts on the metal. Much of this deformation is retained after the bullet breaches and perforates the sheet metal. At much higher impact velocities, the surrounding sheet metal has much less time to stretch and deform during the perforation process. The diameter limits of this deformation can be measured by placing a straight edge across the deformed area and over the center of the bullet hole followed by a careful marking of the two points where the straight edge loses contact with the surface of the sheet

Impact Velocity (fps)	Bullet Hole Diameter (inches)	Width of Sheet Metal Deformation (inches)
1693	0.215	1.10
1850	0.216	1.05
2227	0.221	1.00
2279	0.223	0.93
2425	0.228	0.96
2692	0.228	0.95
2819	0.233	0.99
3051	0.228	0.75
3220	0.230	0.75

Table 6.2

Bullet Hole Diameter and Metal Deformation for .223-caliber 55 gr. M193 Bullets Fired into 0.028 in. Thick Sheet Metal from an AR15 Rifle

metal. Table 6.2 provides a comparison of bullet hole diameter and static metal deformation for a series of orthogonal shots into 0.028 in. (0.7 mm) sheet metal with an AR15 rifle and 55 gr. M193 FMJ-BT jacketed bullets.

Lead and lead alloy bullets typically produce holes that are larger than the bullet that caused them. This is due to some flattening of these bullets and increase in their diameters during impact and prior to perforation of the sheet metal. For example, the diameters of bullet holes in 22-gauge sheet metal produced by common .38 caliber (.358 in. diameter), 158 gr. LRN bullets at impact velocities on the order of 1000 fps measured 0.43–0.46 in.

GLASS

There are three forms of glass that are commonly encountered in shooting incidents. Since bullets that perforate any type of glass will frequently sustain characteristic damage and embedded particles of the struck glass, a brief review of the basic chemistry of glass and its special properties make a useful starting point for this unit. The optical, physical, and chemical properties of glass particles embedded in recovered bullets allow it to be discriminated from other silica-containing minerals such as sand and quartz.

The common glass used to manufacture windows and most containers is known as soda-lime glass because substantial amounts of sodium carbonate and calcium oxide are mixed with pure silica sand (SiO_2) often along with smaller amounts of other inorganic materials. The materials are then heated until a relatively viscous molten mass is created. Colors are produced by the addition of small amounts of metallic ions. For example, Fe^{++} is added to produce green glass, and Fe^{+++} for brown glass. This molten mass may be injected into forms or molds to make containers such as jars and beverage bottles or headlight lenses, or it can be formed into sheets of uniformed thickness to make windows and other forms of plate glass. The most common, contemporary means of manufacturing sheets of plate glass is to float the molten glass on a bed of molten tin. The "float" method of glass manufacture produces a very smooth panel of glass of uniform thickness that lacks "ream" lines. Ream lines are associated with much older methods of glass manufacture and are of no particular importance to the criminalist or firearms examiner involved in the evaluation of bullet holes in glass.

The tin float method of manufacture does result in a unique and useful property for glass made by this method. The side that was in contact with the tin will fluoresce a dull yellow-green color when examined in the dark under a shortwave ultraviolet (SW-UV) light whereas the internal composition and opposite surface will not. This seeming bit of trivia can be useful in certain cases for the reconstruction of shattered windows and ultimately determine the direction of bullet impact. Sheets of such glass may be as thin a several millimeters and would typically be found in small picture frames. Thicker forms would be used in common windows in homes and some commercial buildings. Such glass may be as thick as 6 mm. The names *plate glass* or *single strength glass* has been used for this basic form of glass.

Two sheets of such glass may be joined together with a plastic layer between them to form a sort of sandwich. This is the so-called *safety glass* or *laminated glass* and is the standard for windshields in automobiles. These panels of glass with the thin polyvinyl plastic layer between them are typically molded to have curvature due to their use in modern automobiles. This laminate layer holds the glass in place during traffic accidents to minimize additional injuries to vehicle occupants.

A third form of glass called *tempered glass* or *double strength glass* is used in many applications because of its greater resistance to breakage. Additionally, because shards of this type of glass have relatively dull edges, there is a reduced likelihood of serious injury when it is broken. The use of the term "safety glass" is discouraged because it has been used to describe *both* laminated windshield glass and tempered glass.

PROJECTILE IMPACT AND PERFORATION OF GLASS

In shooting incidents, *plate* glass is most frequently encountered in windows in residences and many businesses. *Tempered* glass is used in the side and rear windows of motor vehicles, and *laminated* glass in vehicle windshields. Each of these displays some unique properties and effects when struck by projectiles consequently they will be treated separately.

PLATE GLASS

Cone Fractures and Direction of Fire in Plate Glass

The first property of interest insofar as the effect of projectile impact is the *cone fracture*. The classic example of this is the displaced cone of glass in a plate

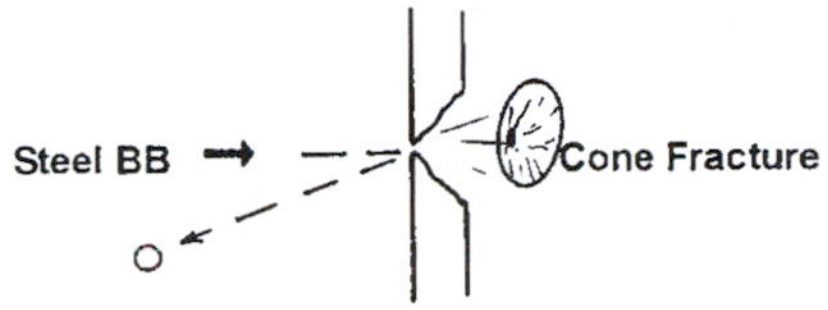

Steel BB ➡   Cone Fracture

Figure 6.9

A Steel BB Strike to Plate Glass with Cone Fracturing

Most BBs fired from BB guns and air rifles do not perforate the glass in windows and doors even though there is often a hole at the center of the impact site. This is because of the relatively low velocity of these projectiles. With or without perforation by a projectile, a cone-shaped area of glass will be ejected from the side opposite to that of the impacting projectile. Cone fracturing occurs in plate and laminated glass. It also occurs in tempered glass for the first bullet or projectile strike that fails such glass.

glass window struck by a steel BB. The cone-shaped area of missing and ejected glass is on the side *opposite* the side against which the impact occurred. A small hole is frequently present that is usually smaller than the steel 0.175 caliber BB that produced it. Given their generally low velocity and the thickness of typical plate glass windows in stores and commercial buildings, BBs seldom perforate such glass, and if a careful search is made, they can usually be found on the ground near the impact site. Figure 6.9 illustrates the classic BB gun cone fracturing of a panel of plate glass.

The real value of the cone fracture exemplified in its simplest form in the BB impact site is that it establishes the *direction of fire*. This is one of the primary questions put to investigators and laboratory analysts. Beyond the simple BB strike, cone fracturing also occurs with much more substantial and energetic projectiles that perforate panels of plate glass. Careful examination of the margin of the bullet hole will reveal a somewhat stair-stepped or tapered flaking away of the glass around the *exit* side of the hole. The glass will be flat and smooth right up to the edge of the bullet hole on the *entrance* side of the hole. It is interesting to note that this same effect is found in other brittle and/or ceramic materials. This includes bone, and in fact forensic pathologists often use this same phenomenon in determining direction of fire in perforating gunshot wounds to the skull. The glass that is flaked away from the margin of the exit bullet hole becomes a part of that glass that is propelled downrange and, combined with the pulverized glass that was immediately in front of the bullet, becomes a source of injury-producing evidence with reconstructive value. Pseudo-stippling of the skin of gunshot victims located near the exit side of a window through which a bullet has passed is a useful example.

Radial and Concentric Fractures, Direction of Fire in Plate Glass

During the initial interaction between a projectile and a panel of single strength glass, the bullet undergoes some amount of flattening at the contact point. The glass will begin to yield somewhat without any breakage or fracturing. This is illustrated in Figure 6.10a. The amount of projectile flattening will depend on a number of factors to include the hardness and design of the bullet (LRN vs. steel-jacketed spitzer point), the impact velocity, the intercept angle, the thickness of the glass, and how large an area the panel of glass occupies before reaching support structures. A quarter-inch thick panel of glass 4 in.[2] and firmly mounted in a four-sided frame will yield less to a bullet's initial impact than a 3-foot square panel firmly mounted in a comparable frame. Lead "splash" may occur with lead bullets and jacketed bullets with exposed lead noses as these bullets encounter the glass with increasing likelihood as impact velocity increases.

As the projectile continues to deform and bend the pane of glass, the glass will begin to fail and open up on the backside by means of crack formations. These cracks radiate out or away from the point of the application of force like the spokes of a wheel (see Figure 6.10b). This is because glass can undergo compressive forces, which occur on the impact side, more than it can endure stretching forces, which occur on the exit side. These fractures are called *radial fractures*. As they propagate outward from the developing bullet hole they are

Figure 6.10

(a) The Pre-Failure Yielding of Plate Glass to a Bullet's Advance; (b) The Initiation of Radial Fractures on the Backside of the Glass; (c) Rib Marks on the Edges of Radial Fractures

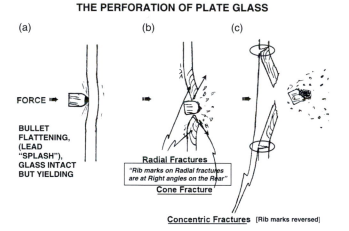

THE PERFORATION OF PLATE GLASS

The three phases of glass breakage and perforation by a projectile are illustrated in these drawings. Prior to failure, the glass actually yields slightly to the bullet's advance. Flattening of the nose of the bullet also occurs during this first phase (a). If the projectile is made of lead or has an exposed lead nose, lead splash begins with some of this vaporized lead ejected back toward the source of the projectile. (b). Radial cracks open up on the backside of the glass and propagate outward. The rib marks on these radial fractures are formed during this process. Pulverized glass and cone fracturing also occur on the backside. The continued bending of the glass causes concentric cracks and fractures to open up on the entry side of the glass (c). The damaged and destabilized bullet continues its flight amid a shower of pulverized glass particles.

creating pie-shaped pieces of glass. Because of the momentum transfer to these developing pie-shaped pieces, stretching of the glass now occurs on the impact side of the glass and in a concentric pattern resulting in *concentric fractures* with the glass opening up on the impact side. Concentric fractures form rings with the bullet hole at or near the center. An appreciation of these events is important because the examination of the edges of radial fractures can allow the direction of fire to be established independently from the cone fracture method. This is useful and important because the area of cone fracturing is relatively small and it may not survive and be identifiable in the static aftermath. Pieces of glass with surviving radial fractures are frequently present and allow for the determination of direction of fire based on *rib marks*. These features form as the glass opens up on the backside (side opposite the application of force) and as the radial fractures propagate outward from the bullet impact point. These rib marks start at right angles to the backside surface of the glass and turn toward the source of breaking force. These are shown in Figure 6.10c. Conversely, rib marks that occur on concentric fractures will be at right angles to the *entry* side rather than the exit side. Unlike rib marks on radial fractures, the direction to which rib marks on concentric fractures turn has no diagnostic value. This does not present a problem so long as the examiner can locate the radial fractures.

Figure 6.11 provides an example of a pie-shaped piece of glass with rib marks on all three edges. The 4-R memory aid (Figure 6.10b) should be useful in sorting out which edges are concentric and radial fractures, and at which corner the breaking force was applied. The more difficult matter may be knowing which side of a piece of submitted glass was on the exterior and interior of the building. If such glass is going to be collected by someone other than the laboratory analyst, it is of critical importance that field investigators or crime scene technicians mark the exterior or interior side of any pieces of glass removed from a projectile-struck window. In those situations where submitted pieces of glass have not been labeled as to interior or exterior surface, it may be possible to resolve this through careful inspection of the two surfaces of the pieces for such things as adhering putty from the mounted edges of the glass, rain spots, weathering effects, paint overspray, etc. This may require the further collection and submission of an additional piece of glass from the struck window whose exterior or interior surface *has* been marked.

In the absence of a physical match between this documented piece and one of the evidence pieces, the previously described features on this reference piece can be compared to the evidence pieces. One final method offers an additional means for resolving the dilemma and that is an examination under an ultraviolet light. If the glass was manufactured by the tin float method, one side of the glass will fluoresce a dull greenish-yellow color and the other side will not. Compare the fluorescent properties of the evidence piece(s) with a piece of known orientation in the bullet-struck window.

Figure 6.11

*The Arrangement and
Interpretation of Rib
Marks on Pie-Sections
of Plate Glass*

Sketch of a Pie-Shaped Piece of Glass with the Edges Opened Out

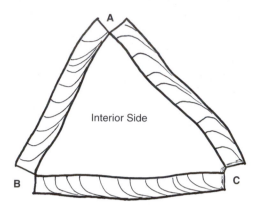

This illustration represents a typical case situation where a section of glass from a shattered window is submitted to the examiner. The scene investigator has marked the surface that was toward the interior of the residence from which the sample was collected. No obvious cone fractures were present on this specimen from which the direction of force could be determined. Two of these edges must be radial fractures and the third a concentric fracture. Applying the "4-R" test (Rib marks on Radial fractures are at Right angles on the Rear) leads to only one solution. Edges AC and BC are radial fractures. The rib marks turn and point to C where the breaking force was applied. The "4-R" test also shows that the force came from the exterior side of this piece of glass. The selection of corner A or B as possibilities for the focal point of the breaking force is quickly excluded by the application of this test.

A preferable alternative to all of the foregoing would be to thoroughly tape the remaining glass in the window and submit the entire window; however, this is not always possible or feasible. Properly marking the pieces collected at the scene will save much time in trying to distinguish interior from exterior surface.

In summary, the rib marks on radial fractures can be used to determine the direction of fire. The examiner may find the following "4-R" memory aid useful: **R**ib marks on **R**adial fractures are at **R**ight angles on the **R**ear. The location of the point of initial breakage (or application of force) is shown by the *direction* of the rib marks on the radial fractures in that they turn and point to this site. (In Figure 6.11 edges AC and BC are radial fractures, the force was applied at corner C and the projectile came from outside to inside.)

As a general statement, the length and density of the radial fractures can provide an indication of the impact velocity of the projectile. For a high-velocity bullet (e.g., 2000 fps+), the radial fractures will typically be more numerous and much shorter than those produced by a low velocity bullet (<1000 fps) perforating the same thickness of single strength glass. Figure 6.12 illustrates these effects in a panel of common window glass

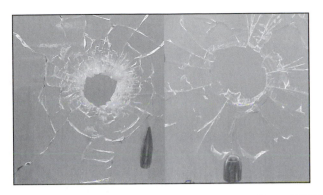

Figure 6.12

High and Low Velocity Bullet Holes in a Pane of Plate Glass

Left: A bullet hole in common window glass produced by a 147 gr. 7.62 mm NATO bullet with an impact velocity of approximately 2800 fps.
Right: A bullet hole in the same pane of single strength glass created by a 230 gr. .45 Automatic bullet with an impact velocity of approximately 850 fps.
Examples of each bullet serve as a scale.

perforated by a high velocity rifle bullet and a low velocity pistol bullet. It should also be apparent from an inspection of this figure and examples of the bullets fired through this panel of glass that hole size bears little or no relevance to the caliber of the bullet that produced the hole other than to say that the responsible bullet is likely to be smaller than the minimum diameter of the bullet hole in glass.

Rib marks can be somewhat difficult to photograph but careful manipulation of the light source or fogging of the broken edges with burning magnesium ribbon will allow these marks to be recorded. Others may simply elect to make a sketch similar to Figure 6.11. Tracing an outline of the actual piece of glass can also be useful in this regard.

Sequence of Shots in Plate Glass—the "Crack" Rule or "T" Test

In those cases where two or more shots strike plate glass and a *radial* fracture meets a preexisting fracture, it will not cross through the preexisting fracture. This phenomenon is not valid for drawing conclusions regarding sequence when dealing with concentric fractures. As with cone fracturing, this phenom-enon also occurs in various plastic substitutes for window glass and other brittle materials such as skull bone. A search for these radial crack "T" intersections and their location will permit the examiner to determine the *sequence* of the shots. Figure 6.13 provides a stylized example of sequence determination based on intersecting radial fractures. The rib marks and/or cone fractures will estab-lish the *direction* of fire. Combined with the directional determination, one can ascertain who fired first and from which side of the panel of glass the first and subsequent shots came.

Figure 6.13

Sequence of Shots in Plate Glass: The "Crack" Rule

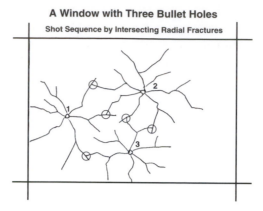

A Window with Three Bullet Holes

Shot Sequence by Intersecting Radial Fractures

This sketch depicts three shots through a glass window. The determination of shot sequence is made on the basis of the circled "T" intersections. Radial fractures are stopped when they reach a preexisting crack in the glass. In this idealized example, the sequence of all three shots can be established. Cone fracturing around the bullet holes or subsequent examination of the rib marks on the radial fractures will allow the direction of fire to be determined.

TEMPERED GLASS

This product starts out as a hot panel of single strength glass, the flat surfaces of which are quickly cooled by blasts of air. This "tempering" process results in a piece of glass that is generally much more resistant to breakage than its untempered counterpart, and is consequently often called double strength glass. It is employed in side and rear windows of vehicles and is also used in a number of other applications such as commercial store and doorway windows, glass-enclosed shower stalls, and some Arcadia door windows. Although it is much stronger than the same thickness of plate glass, when it does fail, it almost instantaneously breaks into many small pieces that are generally cubic to rectangular in shape. This effect has been called "dicing" by glass manufacturers, and is a desirable feature from an enhanced safety standpoint. This, however, creates some serious problems from a shooting reconstruction standpoint. Foremost is the fact that once broken, the pane of tempered glass is *very* fragile and, in fact, often falls out of the framework that was supporting it. Pressure differentials between the inside and the outside of a vehicle with the windows up and doors closed may facilitate the complete failure and removal of projectile-struck tempered glass from its framework. Close range blast effects from the responsible firearm may also result in the removal of the glass from the window. Even when the failed tempered glass in a vehicle survives the shot and remains in place it may fall out shortly thereafter due to vibration, hitting bumps, curbs, potholes in the roadway, high "G" loadings in turns or spin-outs, subsequent impacts with other vehicles or other objects. Such post-shooting events may give the false impression that the shot that broke such a window occurred at the location where the shattered glass is located. From the foregoing it should be apparent that this is not necessarily the case. This fragile evidence is also at risk of being lost during the towing or loading of the vehicle onto a flatbed truck. Careless post-event manipulation of the shattered window (such as opening and closing a door containing such a window) is another potentially disruptive event. All investigative endeavors on any part of the vehicle should only be undertaken after the projectile-struck window is adequately photographed, documented, and reinforced with clear tape or plastic film. This latter effort *may* prevent the glass from falling out but taping the shattered glass provides no guarantee for its survival. The best case situation involving tempered glass is one in which the window possesses a layer of tinted film. This material usually acts very well to preserve most, if not all, of the shattered pane.

Sequence of Shots in Tempered Glass

When a panel of tempered glass *does* survive sufficiently to locate the bullet hole(s), the *first* strike will have intense but relatively short radial fractures around it. After a few inches these radial fractures will begin to wander and quickly result in the usual mini-jigsaw puzzle of square and rectangular pieces of glass. Any subsequent shots through this "diced" glass will simply dislodge areas of previously broken glass leaving somewhat irregular holes with *no* radial fractures. These facts allow the *first* shot into the glass to be identified (see Figure 6.14).

Cone fracturing for the first shot through tempered glass is much more subtle than with normal plate glass and, in fact, may not be identifiable in some cases. Rib marks are also lacking on the edges of the radial fractures in tempered glass.

It is interesting to note that the velocity *loss* experienced by a bullet of particular weight and design perforating an unbroken panel of tempered glass is only slightly greater than the velocity loss suffered by subsequent bullets of the same type fired through the failed or "diced" glass. This may, at first, seem counterintuitive but the reason for this is that the kinetic energy used to produce the initial failure of the intact tempered glass is relatively small. The bulk of the velocity loss suffered by the bullet is the result of the momentum transfer associated with the downrange acceleration and ejection of the dislodged glass. An inspection of the bullet holes in tempered glass will reveal that the amount of dislodged glass in subsequent shots is similar in quantity to the first shot, consequently the velocity loss experienced by subsequent bullets

Figure 6.14

*Shot Sequence in
Tempered Glass*

This remarkable photograph shows a tempered glass rear window of a vehicle that has been struck by a group of 00-buckshot pellets at an incident angle of about 45° (left to right and slightly upward). The first pellet to break the glass created the left-most hole from which numerous radial fractures extend in all directions. The other pellets, arriving fractions of a second later, created holes that *lack* radial fractures because they struck failed glass.

The first projectile to break a pane of tempered glass will possess radial fractures. All subsequent projectile strikes will merely knock out previous shattered glass.

is quite similar to the bullet associated with the first strike. This is made clear by way of some examples:

A panel of 0.19 in. (4.8 mm) tempered glass struck orthogonally by 230 gr. FMJ-RN .45 Automatic bullets with impact velocities on the order of 825 fps experienced a velocity loss of 52 fps for the first strike. The second and third shots through the shattered pane of tempered glass produced velocity loss values of 55 and 50 fps.

Another panel of the same 0.19 in. thickness when shot with two rounds of 115 gr. FMJ-RN 9 mm ammunition having nominal impact velocities of 1100 fps yielded velocity loss values of 184 and 136 fps respectively.

Tempered glass of 0.13 in. (3.3 mm) thickness shot orthogonally with two, 147 gr. JHP 9 mm bullets having impact velocities 949 and 923 fps experienced velocity losses of 74 fps (first strike) and 70 fps (second strike).

From these data it should be apparent that there is no substantial difference in velocity loss for shots into intact vs. shattered tempered glass.

LAMINATED GLASS

As pointed out previously, this special purpose glass is simply two preshaped pieces of normal plate glass, each typically about 0.10–0.12 in. thick, with a thin layer of clear plastic resin between them. This unique arrangement provides a significant measure of occupant protection from inward-ejected fragments of glass when a windshield is struck by a hard object, such as stone thrown up by another vehicle. This arrangement also reduces the seriousness of injuries to front-seat occupants who strike and fracture such glass during the high decelerative forces of a frontal collision.

These desirable attributes and the increased thickness of laminated glass have some very *un*desirable effects on bullet behavior and subsequent reconstructive efforts in shooting incidents where projectiles have struck and broken (or perforated) vehicle windshields. These include frequent separation of bullet jackets from their lead cores (particularly with JHP or soft point pistol bullets), *un*reliability of the "T" test as a means of determining shot sequence, and substantial deflection of perforating bullets.*

*Bullet deflection as the result of windshield perforation can be substantial, e.g., 10–15° in some extreme cases. With shots fired from close range from the front of the vehicle, this deflection often occurs in a direction that is counterintuitive, and empirical testing may be necessary to properly assess the nature and degree of deflection for a specific combination of projectile, glass, incident angle, and impact velocity.

Common pistol bullets fired into typical automotive windshields from positions in front of the vehicle frequently show a consistent *downward* deflection of 1°–5°. Failure to recognize and consider this phenomenon when present will result in back-extrapolations that place the shooter closer to the vehicle than he really was as a result of a falsely elevated pre-impact flight path.

At reduced impact velocities the direction of bullet deflection may reverse itself. Bullet deflection as a result of glass perforation is discussed further in Chapter 7.

Although the glass in laminated windshields is plate glass, the bonding of two pieces of glass together and mounting them in a framework frequently results in subsequent crack propagation with the passage of time. This is much like an annoying crack in one's windshield that continues to grow over days and weeks. Stress and/or thermal changes as well as subsequent movement of the vehicle may result in the formation of one or more "T" points between the cracks emanating from two or more shots; however, unlike a single sheet of plate glass, the configuration of this junction is *not* necessarily the consequence of shot sequence. Rather it may merely be the result of continued crack growth. In an actual shooting incident, it would seem *very* unlikely that someone would have the opportunity to examine the arrangement of the radial and concentric fracture lines in both layers of glass in a windshield *immediately* after the shots were fired. Even here there is no guarantee of reliability of the "T" test for sequence determination. The author's advice is "Don't try it on laminated glass windshields." All is not lost or hopeless however. The direction of fire in laminated windshields can usually be determined on the basis of cone fracturing just as with single panes of plate glass. The clock orientation of the customary oval holes in common windshields provides some indication of the *azimuth angle* of the shot. An oval hole oriented in the 12 o'clock by 6 o'clock orientation was fired from an essentially straight-on position (from in front of the vehicle), whereas an 8 o'clock by 2 o'clock orientation came from a decided left to right direction. Conversely, a 4 o'clock to 10 o'clock orientation of the long axis of a bullet hole in a windshield indicates a right to left flight path as viewed from in front of the vehicle. The approximate azimuth angle relative to the windshield glass at the point of impact can be refined somewhat by marking or establishing the vertical axis line through the bullet hole followed by the line through the best estimate of the long axis of the bullet hole. It is important to note that the view of this angle must be from directly above the bullet hole. Measurements taken from any other view will result in an erroneous measurement. An example of the proper view is shown in Figure 6.15. The best accuracy that can be expected from this technique is on the order of ±5° of the true value where a symmetrical, oval bullet hole is present. A fixed downrange impact point within the vehicle is of considerable assistance so long as some evaluation or consideration of deflection is included in the back-extrapolation of the responsible bullet's path.

EVIDENCE OF GLASS IMPACT ON BULLETS

As mentioned previously, the initial engagement between a typical pistol bullet and an unbroken panel of glass will create a flat spot on most bullets, the size and orientation of which is related to a number of factors. One of these is the intercept angle between the bullet and the glass. This smooth, flat spot may not survive subsequent impactive events, but when it does, it can be useful in answering certain questions such as *Did the bullet that perforated the driver's side window then strike and kill the driver come from a rooftop shooter or someone standing on the street adjacent to the victim's vehicle?* At this particular crime scene, the rooftop in question and the location of the victim's vehicle form a vertical angle of approximately 45°. If the fatal bullet shows an essentially orthogonal flat spot from its impact with the tempered glass side window, the rooftop hypothesis can be effectively excluded. Figure 6.16 shows a series of 9 mm FMJ bullets that

Two techniques to identify the vertical axis for this bullet hole in a laminated windshield have been employed in this photograph. A plumb bob and line have been passed through the bullet hole after the vehicle was leveled in the examination bay. Two rectangles of white paper were placed above and below the bullet hole after which a projecting laser level was used to project a vertical line across the center of the hole. The black lines were then drawn across these pieces of paper and out onto the surface of the windshield. The right-to-left path of the bullet is represented by the long axis of this relatively symmetrical bullet hole.

Figure 6.15

Azimuth Angle Estimates for Shots into Laminated Windshields

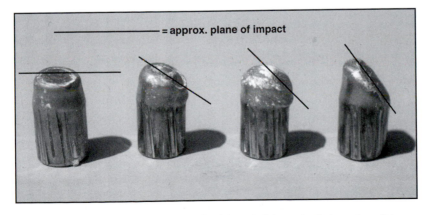

These four, FMJ-RN 9 mm bullets were fired through flat panels of glass at varying intercept angles. The approximate relationship between each bullet and surface of the struck glass is represented by the black line.

Figure 6.16

Intercept Angle Estimates from Bullet Deformation during Impact with Glass struck panels of glass at the approximate angles indicated by the lines drawn across the flattened areas.

The passage of any bullet through any of the three types of glass will also result in abrasive pitting and checking of the nose and ogive portion of the bullet (see Figure 6.17). Even the bearing surface of the bullet may acquire some of this abrasive damage from the bullet's violent journey through the pulverized cloud of glass it has created. Areas showing this abrasive damage will typically survive subsequent impactive events and the reader should become familiar with the appearance of such damage. This is most noticeable when such bullets are examined under the stereomicroscope with a high intensity light source. It is relatively easy to prepare some "training" specimens. Simply fire some representative bullets through typical panels of glass placed in a cardboard box positioned over or against the entry port of an appropriate bullet recovery device (water recovery tank, etc.). A panel or layer of rubber cut from old inner tubes on the backside of the glass will prevent the majority of glass fragments from going into the water tank.

Not only will the abrasive effects be apparent, but microscopical examination will typically reveal sparkling inclusions of powdered glass. These glittering inclusions may be so numerous and obvious in bullets from actual casework that a few particles can be dislodged and mounted on a microscope slide where inspection under the polarizing microscope will reveal that they are

Figure 6.17

The Effects of Glass on a Perforating Bullet

A fired, but undamaged, 7.62NATO rifle bullet appears on the left of this figure. On the right is the same type of bullet after perforation of a 0.087 in. (2.2 mm) panel of plate glass at 2372 fps.

Note the sand-blasted appearance of the ogive of this bullet due to the abrasive effects of the pulverized glass through which this bullet passed.

sharp, angular isotropic particles. This is quite different from soil and sand grains most of which show weathering effects and are anisotropic. Glass residue present in the nose of such bullets may look like plain white powder due to its high level of pulverization. This may be mistaken for gypsum residue by the inexperienced observer, but glass can be easily differentiated from gypsum with a little thought and effort.

An alternate and more sophisticated analytical approach to characterizing the inclusions in the surface of recovered bullets is SEM/EDX spectroscopy. This non-destructive technique does *not* require the removal of any particles and allows their morphology, method of deposition, and elemental composition to be ascertained *in situ*. As with the polarizing microscope, glass particles will have sharp angles and edges whereas soil grains will typically show evidence of weathering. The elemental compositions of the two types of particles will also be revealing. A typical soda-lime glass spectrum, in addition to silicon and oxygen, will show the presence of sodium, magnesium, and calcium, whereas a silica or quartz sand grain will usually appear rounded and yield a spectrum showing only silicon and oxygen. This is due to its chemical composition, SiO_2. Quartz sand is the closest contender in composition to manufactured glass. Other soil minerals are radically different in elemental composition.

Sequence of Shots through Tempered Glass from an Examination of the Recovered Bullet

The *first* bullet to impact an intact panel of tempered glass will acquire a smooth, flat spot at the contact point. The previously described dicing behavior of tempered glass can produce an interesting and useful effect on the nose of *subsequent bullets* that strike it. Although a diced panel of tempered glass may seem fragile and easily displaced when one pushes against it with a finger, to a bullet traveling 800 to several thousand feet per second, a collision with such glass is still a major event. This, in part, is evidenced by the comparable velocity losses for the same type and weight of bullet fired through unbroken and broken tempered glass. As a typical pistol bullet encounters a pane of diced tempered glass, the individual small squares and rectangles of glass struck and shattered by the nose of the impacting bullet produce "facets" in the bullet rather than a single, smooth flat spot. *If* the nose survives subsequent terminal ballistic events, such "facets" *identify* the bullet as a shot through a previously failed and diced section of tempered glass. Examples of such bullets are shown in Figure 6.18a,b.

The presence of pulverized and embedded glass in a recovered bullet is usually apparent when such a bullet is examined under the stereomicroscope. On occasion it may be desirable or necessary to examine bullets suspected of

(a) (b)

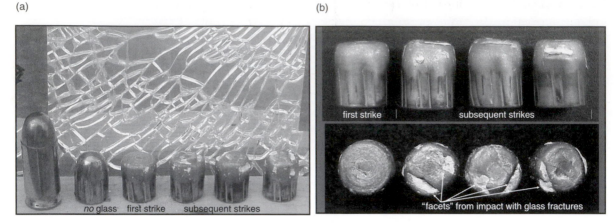

Figure 6.18

Shot Sequence and Facet Formation on Bullets through Tempered Glass (See Color Plate 8)

These figures show three views of four .45 Automatic bullets that have been fired through tempered glass. A live round and an unfired bullet are shown at the far left of Figure 6.18 along with a section of failed tempered glass.

Note the smooth, even impact damage to the "first strike" bullet and the "facets" on the bullets that struck the failed and diced tempered glass. The white powder is pulverized glass.

perforating glass under the scanning electron microscope. When coupled with an EDX system, an SEM allows the *direction of deposition* and the basic elemental composition of the embedded particles to be determined. Glass particles have a characteristic sharp, angular appearance under the SEM (see Figure 6.19).

Figure 6.19

Scanning Electron Microscope View of Glass Particles in a Jacketed Bullet

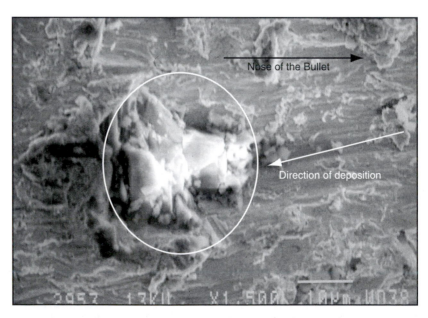

SEM View of characteristic glass particles (circled) in a bullet jacket at 1500× magnification. Note the very angular appearance of these shattered particles of glass.

Their elemental composition typically shows a very large silicon peak with sodium, magnesium, and calcium present (soda-lime glass), whereas sand, silica, and/or quartz grains are nearly pure silicon dioxide and have weather-worn, rounded shapes.

RUBBER

This elastic material behaves much like skin in that something similar to an abrasion rim can often be seen around the margin of a bullet hole. This "pseudo-abrasion ring" in rubber roughly approximates the caliber of the projectile, but the actual bullet hole will be smaller, and in some situations *much* smaller than the bullet that produced it. Automobile tires are the most common example of rubber struck by gunfire (see Figure 3.2f in Chapter 3). Their black color makes locating the pseudo-abrasion rim very difficult. The holes produced by FMJ bullets and lead buckshot pellets are so small that they could easily be confused with a nail puncture. The DTO and/or sodium rhodizonate tests may resolve the question but the author has encountered situations where these tests failed to reveal traces of copper or lead around the margin of known bullet holes. Hollow point pistol bullets such as those commonly used by the American law enforcement produce more easily recognized bullet hole as a result of a "cookie-cutter" effect. These bullets punch out a plug of rubber that, in some instances, may be recovered from the hollow point cavity of the responsible bullet. The result is a relatively conspicuous hole in the tire. In some instances it is possible to physically fit this plug or rubber from the hollow point cavity of a bullet back into the area of missing rubber in the struck tire.

Empirical testing has revealed that there is no discernable difference in bullet holes produced in the sidewalls or tread area of demounted tires vs. the same tire inflated and mounted on a wheel. This is good news if one wishes to carry out some ballistic tests on tire rubber to demonstrate the appearance of bullet holes fired from different angles or produced by different types of bullets.

Deflation tests carried out on typical tubeless automobile tires perforated by FMJ and hollow point pistol bullets of 9 mm to .45 caliber resulted in very slow deflation (up to several minutes) with the FMJ bullets. Deflation times on the order of 20–30 sec. were obtained for the hollow point bullets. Bullets that have been ricocheted into tires will usually produce an irregular entry hole due to the deformed shape of the ricocheted bullet and the likelihood that such bullets will strike the tire in a yawed orientation.

Projectiles striking and perforating a tire will either hit some interior location on the metal wheel or will perforate and exit the opposite sidewall of the tire. These exit sites can be just as difficult if not more difficult to locate as entrance holes. Since tires rotate and can turn as well in the case of front tires, the position and location of the wheels of a shot vehicle should be marked and documented *before* the vehicle is moved. This can be done with a brightly colored spray paint by making a stripe on the wheel and sidewall of the tire that extends onto the surface upon which the tire is resting. Documentation should include one or more close-up photographs that allow one to see the various markings on the tire and their relationship to the ground and the immediate features of the vehicle itself. It is also important to mark the relationship between the tire and the wheel upon which it is mounted. If any suspected bullet holes can be seen prior to moving the vehicle, these should be marked in some way and documented through photography. Bullets that perforate a tire but then strike the metal wheel often breakup or ricochet around inside the tire without exiting. This means that both the recovery of any bullet or bullet fragments and the reconstruction of the bullet's flight path will necessitate the demounting of the tire. Without the orientation marks described above, the examiner will not be able to properly align the entry hole in the rubber with the subsequent impact mark on the metal wheel. The fact that tires revolve when a vehicle is in motion presents some complication and some interesting reconstructive opportunities. In some cases it may be possible to determine whether the wheel was rotating or stationary when one or more bullet struck and perforated a tire. Downrange fixed features of the vehicle such as the frame, shock absorbers, or springs that were either struck or missed by the exiting bullet provide reference points for evaluating the issue of a moving or stationary vehicle at the time of the shooting. Such endeavors must integrate the path of the bullet or bullets through the tire, with any downrange impact points in the vehicle itself, as well as the location of the shooter at the time the shots were fired. Finally, there may be some positions where a struck area on a tire is not ballistically accessible, i.e., a bullet cannot be fired into the tire along the path taken through the tire without first perforating some portion of the vehicle such as the fender. This is especially important with the front tires since the turning of the steering wheel to various positions will significantly alter the ballistic accessibility of certain areas on these tires. This concept of ruling out certain bullet routes to the struck tire should be considered early on. It is one of the fundamental concepts of the *Scientific Method*. Alternatively, a bullet path through a fender that extends to the tire creates another means of assessing whether the tire was in motion or not. The ideal case is one in which a bullet perforates a fender, then the exterior sidewall of the tire followed by an exit from the interior sidewall and comes to rest in the frame or wheel well of the vehicle.

PLASTICS

The term "plastics" includes a rather wide variety of materials ranging from relatively hard and brittle materials, such as Plexiglas, polycarbonate windows, and patio covers, to polyvinyl shower curtains, bright orange hunting vests, and polypropylene bedliners for pickup trucks. The former respond to gunfire much like glass, bone, and ceramic materials. This behavior includes cone fracturing, the propagation of radial fractures, and adherence to the crack rule regarding shot sequence. For these reasons no further discussion would seem necessary for this category of plastics.

"Soft" plastics, on the other hand, display several properties and responses to bullet impact that differ from other materials. The thick polypropylene bedliners popularly used in pickup trucks are particularly interesting and deserve some discussion. Bullet holes produced in this material shrink somewhat after the bullet's passage so that the inside diameter of such bullet holes is an understatement of the responsible bullet's caliber but the interior of this channel will often possess a negative image of the rifling characteristics present on the responsible bullet. These can usually be seen by close inspection of the entry or exit site under strong light and low power magnification. These important features can be "recovered" by casting the bullet's channel with a suitable silicon rubber product such as Mikrosil®. Although the dimensions of the rifling impressions on this casting are undersize, the general rifling characteristics of the responsible bullet, such as number of lands and grooves, direction of twist, and the *ratio* of the land widths to the groove widths, can be determined from such a casting.

Ricochet marks on this thick plastic material will also record the rifling characteristics of the responsible bullet. The only caveat that probably deserves mentioning is that one is looking at an impression of the *underside* of the bullet rather than the top of the bullet where the direction-of-twist determination is normally made. This means that rifling impressions visible in a ricochet mark in a truck bedliner (or similar material) that cant to the left are the consequence of right-twist rifling.

SUMMARY

Much is to be learned from the appearance, shape, and dimensions of bullet holes in various materials. In some situations these factors may relate to impact velocity, bullet shape, bullet hardness, bullet stability, and, of course, the nature of the material struck and the angle of intercept with the particular material. Other events useful in reconstruction arise out of the nature of the fracturing process in brittle materials, the ejection of the broken and pulverized material by the perforating bullet, the damage and deformation suffered by the bullet, and the transference of trace evidence between the projectile and the struck surface.

The deformation process associated with malleable materials such as sheet metal has its special attributes that are quite different from frangible materials such as sheetrock. The many examples presented in this chapter are merely a beginning point for any individual who is directly involved in the reconstruction of shooting incidents. These examples provide some insight, some concepts, and some expectations along with a basis for further study involving case-specific materials, firearms, and ammunition.

REFERENCES AND FURTHER READING

Haag, L.C., "The Measurement of Bullet Deflection by Intervening Objects and in the Study of Bullet Behavior after Impact," *CAC Newsletter* (January 1988).

Haag, L.C., "An Inexpensive Method to Assess Bullet Stability in Flight," *AFTE Journal* 23:3 (July 1991).

Haag, L.C., "Exterior and Terminal Ballistic Events of Forensic Interest," *AFTE Journal* 28:1 (January 1996).

Haag, L.C., "Bullet Penetration and Perforation of Sheet Metal," *AFTE Journal* 29:4 (Fall 1997).

Haag, L.C., "Sequence of Shots through Tempered Glass," *AFTE Journal* 36:1 (Winter 2004).

Laible, R.C., ed., *Ballistic Materials and Penetration Mechanics*, Elsevier Scientific Publishing Co., Amsterdam, Oxford, NY, 1980.

MacPherson, D., *Bullet Penetration—Modeling the Dynamics and Incapacitation Resulting from Wound Trauma*, Ballistic Publications, El Segundo, CA, 1994.

McJunkins, S.P. and J.I. Thornton, "Glass Fracture Analysis—A Review," *For. Sci.* 2:1 (1973).

Nennstiel, R., "Forensic Aspects of Bullet Penetration of Thin Metal Sheets," *AFTE Journal* 18:2 (April 1986).

Nennstiel, R., "Prediction of the Remaining Velocity of some Handgun Bullets Perforating Thin Metal Sheets," *For. Sci. International* 102 (1999).

Thornton, J.I., "The Effect of Tempered Glass on Bullet Trajectory," *AFTE Journal* 15:3 (July 1983).

CHAPTER 7

PROJECTILE RICOCHET
AND DEFLECTION

INTRODUCTION

Numerous articles have appeared over the last 30 years on the subject of projectile ricochet. These are listed in the "References and Further Reading" at the end of this chapter. These articles have largely contained the results for specific target materials and a few specific bullets or projectile types. Most of these articles have made little effort at setting forth any general properties or behavior for projectiles undergoing ricochet. This has been largely due to the substantial variety in bullet types, surfaces struck, the varying responses of struck surfaces, and variations in the post-impact behavior of such projectiles. Moreover, many of these papers arose out of specific case situations and therefore do not have general applicability to the overall subject of projectile ricochet. This chapter will provide the reader with some useful definitions as well as some general expectations for the behavior of projectiles upon impact with a variety of surface types and the sort of damage ricocheted bullets acquire during different types of ricochet events. The importance and reconstructive value of the appearance of ricochet damage on recovered bullets will be presented as well as that of the impact sites associated with these events.

Most of us have seen depictions of bullet ricochet in film and on television. A few of these have been good representations of the real world and certain laws of physics. Most, however, range from fanciful to farcical. What is most surprising is that some firearm examiners have seriously flawed notions about the behavior of projectiles when they strike various surfaces at low incident angle. In one recent report involving an injury sustained by a ricocheted bullet, an examiner of long experience included the following statement in his report on the matter: "After impact with the ground this bullet could have gone anywhere." This is, of course, simply incorrect and falls into the same category as the "anything is possible" statement. The range of possibilities may be large, but *anything* is *not* possible.

For example, it is not possible to be 40 years old without first being 20 years old. Through empirical testing and observation, we can learn to predict the general behavior of projectiles after shallow angle impacts and ricochets from specific surfaces. This chapter describes the ricochet process and some insight into what one can expect to see at the impact site, on the bullet and how the bullet will behave during its post-impact flight.

DEFINITIONS

The following definitions are used in this chapter and text:

Ricochet   The continued flight of a rebounded projectile and/or major projectile fragments after a low angle impact with a surface or object. (Figure 7.1a,b provides simplified profile views of projectile ricochet from two general types of surfaces.)

Deflection (as differentiated from ricochet)   A deviation in the projectile's normal path through the atmosphere as a consequence of an impact with some object. While it may be said that deflection always occurs with a ricochet, the term *deflection* is further refined for two types of impactive events during a projectile's normal flight path.

1. *Deflection* as a consequence of a *ricochet* is used to describe any lateral component of the ricocheted projectile's departure path relative to the plane of the impacted surface as viewed from the shooter's position and with the plane of the surface normalized to a horizontal attitude.

    It is the angle formed between the path of the departing projectile, subsequent to impact and the pre-impact plane of the projectile's path (see Figure 7.2, "plane of departure" vs. "plane of impact").

Figure 7.1

Simplified Profile View of Projectile Ricochet from (a) a Hard, Unyielding Surface; (b) a Yielding Surface

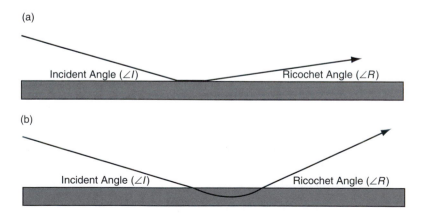

(a)

Incident Angle (∠I)   Ricochet Angle (∠R)

(b)

Incident Angle (∠I)   Ricochet Angle (∠R)

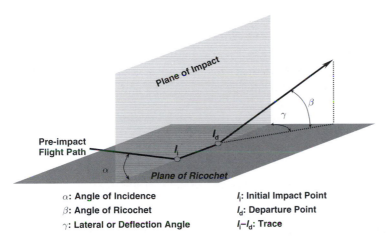

Figure 7.2

Detailed Diagram of Projectile Approach, Impact, Ricochet, and Deflection

Plane of Impact

Pre-impact Flight Path

β

γ

I_d

I_i

α

Plane of Ricochet

α: **Angle of Incidence**
β: **Angle of Ricochet**
γ: **Lateral or Deflection Angle**

I_i: **Initial Impact Point**
I_d: **Departure Point**
I_i–I_d: **Trace**

Source: Modified diagram from the work of Dr Beat Kneubuehl.

2. *Deflection* as a consequence of *perforating or striking an object* is used to describe deviations in *any* direction from the projectile's normal flight path as a consequence of perforating or striking an object rather than rebounding off surfaces. For example, a bullet may be *deflected* by passage *through* a tree branch, a windshield, or a panel of sheet metal. These would not represent instances of ricochet. Since such deflection can occur in any direction in the examples cited (up, down, right, or left), the clock position of such deflection is used to describe this form of deflection. As viewed from the shooter's position (or position directly behind the projectile at impact), 12 o'clock is taken as straight up relative to the horizontal plane at the location of the event, 3 o'clock directly to the right, 9 o'clock to the left, and so forth.

Incident Angle ($\angle I$)   The intercept angle described by the pre-impact path of the projectile and the plane of the impact surface at the impact site when viewed in profile (Figure 7.1a,b).

It is the angle formed between the path, of the projectile prior to impact and the plane of the impacted surface (see Figure 7.2, "plane of impact" vs. "plane of ricochet").

Note: This definition differs from the NATO method. For those who wish to convert from the forensic definition used here to the corresponding NATO angle, use the equation [90° − F.A.] = NATO $\angle$, where F.A. denotes the forensic angle.

Ricochet Angle ($\angle R$)   Using the same coordinate system as for incident angle, this angle is defined by the path taken by the ricocheted projectile (or major projectile fragments) as it departs the impacted surface with one additional qualification—the reference plane of the impact site is that *prior* to bullet impact even though in some situations the bullet is departing a much modified surface (e.g., water, sheet metal, soil) (Figure 7.1a,b).

It is the angle formed between the path of the departing projectile subsequent to impact and the plane of the impacted surface (see Figure 7.2).

Critical Angle   The incident (intercept) angle above which the particular projectile at a given impact velocity no longer ricochets from the impacted surface.

Pinch Point   The small area of surviving paint that was pinched between the initial contact point of a low incident angle projectile and a painted sheet metal surface. When present, the pinch point establishes the entry side of the ricochet mark. This phenomenon may, on occasion, also be seen on painted wood.

Bow Effect   The flow pattern of abrasive materials in soil, sod, and/or sand around the nose, ogive, and/or bearing surface of a bullet generated during penetration into and ricochet from such materials. This characteristic pattern is uniquely associated with ricochets from soil, sand, or sod where the bullet has entered into the substrate to some depth before departing the substrate. It is most noticeable on the ogive of the bullet, but may extend back along the bearing surface as well. This type of marking takes its name from the similarity of the flow pattern of water off the bow of a boat.

Lead-in Mark   The dark, elliptical transfer of material from a bullet as it makes its initial contact with a surface at low incident angle. The lead-in mark establishes the entry side of the ricochet mark. This phenomenon is a form of bullet wipe, but is transferred from only a portion of the bearing surface due to the shallow angle of intercept.

"Chisum" Trail   So named after Criminalist Jerry Chisum who first described it in the United States, this unique ricochet mark occurs on flat unyielding surfaces as the bullet departs the surface. It is caused by the right or left edge of a flattened bullet remaining in contact with the surface after the main body of the bullet has lifted off the surface. This asymmetrical extension of the ricochet mark will be on the left side if the bullet was fired from a firearm having left twist rifling and on the right side in the case of a right twist firearm.

Lead Splash   The impactive spatter and vaporization of lead with its subsequent downrange deposition with non-orthogonal impact angles. This is typically associated with lead bullets and semi-jacketed bullets possessing exposed lead points. The geometry of the deposition of lead splash can provide information on the direction of fire. The amount of lead splash is a function of impact velocity.

There is much to be learned from a careful examination of a ricocheted bullet as well as any suspected impact sites. Certain laws of physics must be obeyed during the ricochet process and the subsequent flight of the projectile. For example, the steeper the incident angle, the greater the forces acting on the bullet during impact and the greater the "work" performed on the bullet throughout the impact and ricochet process. This, in turn, results in greater velocity loss and greater deformation to the bullet than would occur if the same bullet struck the same surface at the same impact velocity but at a shallower angle. With increasing incident angle there will be a corresponding increase in evidence transfer and effects at the impact site. If the impacted surface is one that yields to the bullet's advance, the greater will be the deformation or disruption to this material as incident angle increases. Ultimately with increasing incident angle the projectile will no longer ricochet. Rather, it will either fragment into numerous pieces or it will penetrate/perforate the substrate. This important angle is known as the *critical angle*. For any ricochet to take place the incident angle *must* be less than the critical angle for the particular bullet and substrate.

If one isolates the parameter of impact velocity and continually increases this parameter while holding all others constant (i.e., incident angle, bullet, and substrate), the same previously described general effects will take place. These facts are quite logical and have a strong common sense quality to them. Our every day observations of the damage sustained by vehicles that graze a concrete divider at highway speeds vs. ones that strike such barriers at comparable speeds but at much more acute angles (e.g., −45°) provide simple examples of ricochet to which juries can relate. The vehicle represents a projectile and the concrete divider an impacted surface. At a shallow grazing angle a vehicle loses very little speed, sustains minimal damage and undergoes only a slight directional change as it "ricochets" off the divider. Likewise the concrete suffers very little damage. Consider now the consequences of a 45° impact at typical highway speeds. Substantial damage will occur to both the vehicle and the concrete divider. Velocity losses will likewise be much greater than the grazing example.

But bullets are small and are traveling much faster than automobiles. For example, the relatively low velocity of 900 fps for a common pistol bullet translates to over 600 mph. They can also impact a much greater variety of materials than the concrete dividers and barricades used in the vehicle-as-a-projectile analogy. As a consequence, some examiners confronted with ricochet questions or issues tend to give up with claims of "It's too complicated" or "There are too many variables." The illustrations and data presented in this chapter will refute such statements and provide the reader with a sound basis for understanding and evaluating possible ricochet issues.

SOME GENERAL PRINCIPLES, OBSERVATIONS, AND COMMENTS

The reflection of light from a first-surface mirror can be viewed as the ideal model for ricochet. In this situation, the incident angle equals the ricochet angle; there is no lateral deflection and no loss of velocity.

The behavior of a billiard ball rebounding from the edges of a pool table comes close to this ideal ricochet model. As before, the incident angle is, for all practical purposes, equal to the ricochet angle, very little velocity is lost during the ricochet process and neither the impacted surface (edge of the pool table) nor the projectile (billiard ball) sustains any damage during the impact. Unfortunately, bullets seldom behave this way when they ricochet yet lay people (form whom juries are chosen), and possibly others including lawyers and judges may have the billiard ball model in mind when the subject of ricochet arises.

The closest real-world example of such a ricochet would be represented by a steel BB or steel shot impacting a smooth, marble, or steel surface at low incident angle. These projectiles are much harder than most bullets and have a much higher coefficient of restitution than spheres composed of lead and of the same weight or diameter. By way of example, standard 0.173 in., 5.25 gr. steel BBs ricocheted from smooth, *un*yielding stone at a fixed incident angle of 12.5° and average impact velocity of 276 fps departed with an average velocity of 256 fps and a ricochet angle of 8.9° ± 0.5° for 10 shots. No visible damage was sustained by the smooth stone and only a faint blemish was noted on the recovered BBs.

In actual experience one or more of the following events will take place with the ricochet of common projectiles from a surface:

* The projectile will *lose* some of its *velocity* as a consequence of the impact and ricochet.
* The projectile will usually depart at some angle other than the incident angle. This ricochet angle is usually *less* than the incident angle for most, but not all, hard surfaces.
* The projectile will usually, but not always, undergo *deformation* and/or *damage* and possibly *breakup/fragmentation* during impact. An exception to this is low angle impact and ricochet from water. This often results in destabilization of the bullet but not deformation.
* The impacted surface may undergo deformation (malleable surfaces such as sheet metal) or breakup at the impact site (frangible surfaces such as cinder blocks).
* Mutual trace evidence transfers will often take place between the projectile and the impacted surface with matter from the impacted surface becoming embedded in the ricocheted bullet and traces of the bullet left at the impact site.

Impact surfaces can be categorized into several types that are useful in making some very general predictions regarding projectile ricochet. These substrate categories are:

un*yielding* surfaces
Examples: concrete, stone tile, steel plate
yielding surfaces with subdivision into homogeneous and non-homogeneous
Examples: sand, sod, wood, sheet metal, sheet rock, asphalt
frangible, yielding surfaces with the same subdivision as before
Examples: cinder block, bricks, concrete
liquid surfaces (water—a special case of a homogeneous, yielding surface).

The nature of the damage to the ricocheted projectile and the impacted surface are diagnostic in a variety of ways some of which may require empirical testing to properly evaluate and interpret.

FACTORS CONTROLLING OR AFFECTING PROJECTILE RICOCHET

For incident angles sufficiently low to permit a projectile to ricochet from a particular surface, the ricochet angle may be affected by:

incident angle
impact velocity
bullet shape (round nose vs. wadcutter vs. hollow point vs. truncated cone, etc.)
bullet weight
bullet hardness
bullet center of gravity location
impact surface hardness
response of the surface to bullet impact (the substrate either yields or does
    not yield).

This last parameter (surface response to impact) has a very important effect on ricochet angle. For example, a bullet impacting smooth, un*yielding* concrete vs. concrete that yields (fractures and leaves a crater) will typically result in signifi-cantly different ricochet angles keeping all other variables the same. In this example, the bullet departing from the cratered concrete will do so at a higher angle than a bullet that ricochets with*out* making a crater in the concrete. This is also true with other materials that behave in a similar manner, i.e., low ricochet angles result for unyielding surfaces and higher ricochet angles occur when the surface yields to the bullet's advance.

FACTORS CONTROLLING OR AFFECTING LATERAL PROJECTILE DEFLECTION DURING RICOCHET

Lateral or side-deflection (as a consequence of ricochet from a homogeneous material) depends largely on the direction of twist of the gun that fired the bullet with the magnitude of any deflection controlled by the length of contact with the substrate and the twist rate of the responsible firearm. A bullet striking a smooth marble floor in a bank will experience very little deflection because its contact with the marble is very brief in time and short in distance (typically about 1 in. in length). The same bullet fired at the same shallow incident angle into still water or smooth sand will remain in contact with the water or sand for a greater distance and be deflected significantly (e.g., 5–10° and possibly more) and in accordance to the direction of twist of the rifling of the responsible gun. Exceptions to this arise from inhomogeneities or non-uniformity in the substrate at the impact site. A representative example of such a substrate is asphalt used in roadways and parking lots. Asphalt is composed of relatively soft bituminous material containing stones or aggregate of various sizes and shapes. Bullets striking asphalt show much more variation in ricochet angle than more homogeneous materials. For example, encountering a substantial stone on the right side of a bullet's nose as it enters otherwise soft asphalt is likely to result in a left deflection regardless of rifling twist direction. Any significant bullet yaw at the moment of impact also stands to play a role in deflection particularly with yielding substrates.

In summary, lateral deflection during ricochet will be quite small or inconsequential for hard, unyielding surfaces and can be large for yielding surfaces.

PROJECTILE IMPACTS TO HARD, UNYIELDING SURFACES

Figure 7.1a provides the operative model for this event. During contact and interaction with a surface such as smooth marble, granite, concrete, or thick steel, the softer bullet sustains a conspicuous flattened area on its bearing surface that will extend out onto the bullet's ogive as incident angle increases. This flattened area will be quite smooth when the struck surface is smooth (e.g., polished marble) and heavily striated where the surface is abrasive (e.g., concrete). Characteristic mineral grain inclusions, particularly with concrete and comparable substrates, can easily be seen and characterized by SEM/EDX analysis. This examination also allows the direction of deposition of such mineral inclusions to be seen in excellent detail. With a normally spin-stabilized bullet, this pattern of deposition will be from front to back and aligned with the long axis of the bullet. A bullet that has struck some intervening object prior to impact and ricochet will usually do so in a yawed or destabilized orientation. In this situation the pattern of impact striae and mineral inclusions will be at some angle other than the longitudinal axis of the bullet. Figure 7.3 shows the impact sides of two recovered bullets that struck polished granite and common concrete at an incident angle of $-10°$ and while in normal, nose-first flight. If no significant, subsequent deformation occurs from post-ricochet impacts, the plane of the flattened area relative to the long axis of the bullet can provide a rough index of the incident angle (Figure 7.4).

The corresponding impact marks can be difficult to locate on smooth, shiny surfaces such as polished marble or granite. Examination of such surfaces with oblique light in a near-dark condition may be necessary. Ricochet marks on concrete, flagstone walkways and thick steel structures are more readily discernible. Examples of such marks showing the "Chisum trail" are illustrated in Figure 7.5.

If the struck object or surface cannot be collected, confirmation of the mark as bullet-caused can be accomplished through the copper and/or lead test carried out by the transfer method described in Chapter 4.

The ricochet angles from hard, unyielding surfaces remain low (or the order of 1–2°) and close to the plane of the struck surface for all incident angles up to the critical angle. This fact is of considerable importance in searching for a bullet that has ricocheted from such a surface and for any effort to reconstruct the pre-impact flight path of a recovered bullet that has ricocheted from such a surface. For example, the fact that one might locate an embedded bullet in a vertical wall 1 foot above the surface of a floor at a distance of 50 feet from the ricochet site

Figure 7.3

Bullets Ricocheted from Smooth Concrete and Polished Granite

Three, 124 gr., 9 mm bullets are shown in this figure. The bullet on the left was taken from a water recovery tank and is in pristine condition. The other two bullets struck smooth, flat slabs of concrete and polished granite with an incident angle of $-10°$ and with nominal impact velocity of 1000 fps.

Impact with the polished granite [middle bullet] has flattened one side of the bullet and essentially 'ironed' out the rifling impressions.

Impact with the more abrasive concrete has obliterated the rifling impressions and heavily striated the area of flattening.

Figure 7.4

Relationship between Incident Angle and Deformation to Ricocheted Bullets: Profile View of a Series of Bullets Ricocheted from a Hard, Unyielding Surface

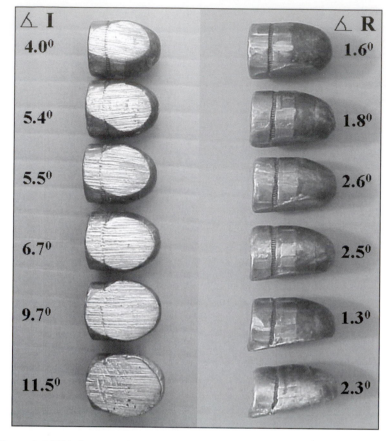

These six LRN bullets were fired into smooth flagstone at the incident angles listed on the left side of this figure. The flattened, impact sides of the bullets are shown on the left. The profile views of each bullet appear to the right along with the ricochet angles associated with each bullet.

It should be noted that the ricochet angles are all around 2° even though there has been a nearly 3-fold change in incident angle (4° to nearly 12°). This is typical behavior for bullets ricocheting from hard, unyielding surfaces.

An estimate of the incident angle can be made by drawing lines through the longitudinal axis of the bullet and along the plane of the impact damage with the bullet oriented in the profile view.

allows the ricochet angle to be calculated from the tangent relationship (1 foot of height after 50 feet of post-impact flight) yielding a result of 1.1°. But this angle tells us very little about the incident angle (other than it was less than the critical angle) and shots with this particular ammunition at incident angles of 2°, 4°, 6°, 8°, and 10° were later found to produce similar ricochet angles (Tables 7.1–7.3). The important and useful parameters in this situation are—

1. the appearance and dimensions of the impact mark and
2. the deformation suffered by the bullet during impact and ricochet.

The five impact sites in Figure 7.5 were all created by .45 Automatic bullets that struck this smooth stone at about −15°. Their direction of travel was from below and they departed at the top of this figure.

Four of the bullets were FMJ-RN bullets. The fifth bullet (lower right impact site) was a JHP bullet. All weighed 230 gr. and had impact velocities on the order of 850 fps. The two impact sites at the top of this figure were fired from a left twist barrel as evidenced by the asymmetrical extension of the bullet transfers on the left side of the mark. The remaining three were fired from a right twist barrel and display the same effect at the upper right of each impact mark.

Both of these parameters will vary in a noticeable and reproducible way with incident angle but it will once again require empirical testing to establish the relationships between incident angle and these parameters as well as the uncertainty limits associated with them. This was previously illustrated in Figure 7.4.

Table 7.1

Ricochet of 9 mm Luger Bullets from Smooth Concrete (Incident Angles = −5° and −10°)

Ammunition	Incident Angle = −5°	Incident Angle = −10°
Winchester 115 gr. FMJ-RN	∠R = 1.5°/1.6°/1.5° = Ave. 1.5°	∠R = 1.6°/1.2°/1.4° = Ave. 1.4°
Federal 124 gr. FMJ-RN	∠R = 1.1°/1.8°/1.6° = Ave. 1.5°	∠R = 2.4°/1.6°/1.2° = Ave. 1.7°
Winchester 147 gr. JHP	∠R = 1.5°/1.6°/1.5° = Ave. 1.5°	∠R = 1.6°/1.2°/1.4° = Ave. 1.4°

A Ruger P-85 9 mmP pistol was mounted in a Ransom Rest and positioned so as to create incident angles of −5° and −10° into smooth concrete for multiple shots for each of these three weights of 9 mm bullets. Downrange witness panels were used to measure the ricochet angles.

Table 7.2

Ricochet of 9 mm Luger Bullets from Smooth Steel (Incident Angle = −10°)

Winchester 115 gr. FMJ-RN (pdt. Q4172)
Average ∠R = 1.4° ± 0.1°   n = 12
Ave. velocity loss = 47 fps (4.4% of average impact velocity of 1061 fps)

Russian 115 gr. FMJ-RN (Steel Jacket-Lead Core)
Average ∠R = 1.4° ± 0.1°   n = 7
Ave. velocity loss = 71 fps (5.9% of average impact velocity of 1207 fps)

Federal 124 gr. FMJ-RN (M882)
Average ∠R = 1.4° ± 0.1°   n = 10
Ave. velocity loss = 73 fps (6.3% of average impact velocity of 1156 fps)

Winchester 147 gr. JHP (pdt. XSUB9MM)
Average ∠R = 0.9° ± 0.1°   n = 5
Ave. velocity loss = 52 fps (5.1% of average impact velocity of 1020 fps)

A Ruger P-85 9 mmP pistol was mounted in a Ransom Rest and a heavy steel plate positioned downrange to create an incident angle of −10° for multiple shots with each of these four types and three weights of 9 mm bullets. Downrange cardstock witness panels at measured distances from the impact sites were used to calculate the ricochet angles of each shot. The Oehler M43 PBL system and Doppler radar was used to obtain the various impact and post-impact velocity values.

Tables 7.1 and 7.2 provide the results of ricochet tests with a number of common 9 mm bullets upon their impact with smooth steel and concrete at selected incident angles. A brief study of these tables reveals several very interesting things. The ricochet angles are all very low and very reproducible. The velocity lost by these bullets during the ricochet process is also low. There is little difference in the ricochet angles for these bullets fired into smooth concrete at substantially different incident angles (5° vs. 10°).

The previous example of a bullet ricochet from a smooth marble or concrete floor illustrates another point and that is the geometric constraint imposed by projectile ricochet and the scene. With the consistently low ricochet angles, a victim struck in the head or upper body would have to be at some considerable distance from the impact site *if* he were standing at the time he was struck and *if* he

Table 7.3

Ricochet of 0.173 in. Diameter, 5.25 gr. Steel BBs from Smooth Stone Incident Angle Fixed at −12.5° Impact Velocity Varied as Shown

Group 1
(Low velocity)
Ave. V_{impact} = 276 ± 4 fps (N = 10)
Ave. $V_{ricochet}$ = 256 ± 4 fps (N = 5)
% Velocity loss = 7.2%
Ave. $\angle R$ = 8.9° ± 0.5° (5.6% C.V.)

Group 2
(Medium velocity)
Ave. V_{impact} = 400 ± 3 fps (N = 10)
Ave. $V_{ricochet}$ = 374 ± 6 fps (N = 5)
% Velocity loss = 6.5%
Ave. $\angle R$ = 9.1° ± 0.7° (7.7% C.V.)

Group 3
(High velocity)
Ave. V_{impact} = 528 ± 3 fps (N = 12)
Ave. $V_{ricochet}$ = 487 ± 2 fps (N = 5)
% Velocity loss* = 22%
Ave. $\angle R$* = 6.9° ± 0.2° (2.9% C.V.)

* *Note*: Some cratering of the stone was evident with these impacts.
A Daisy Model 880 Powerline air rifle was mounted in a machine rest and positioned to create a fixed incident angle of −12.5° with a section of smooth stone. Selected impact velocity levels were achieved with a specific number of pumps. An Oehler M43 PBL chronograph system was used to measure the pre- and post-impact velocities of these standard steel BBs. A downrange cardstock witness panel was used to calculate the ricochet angle based on the tangent function from the height of the BB hole and its distance from the impact site.

were on the same plane as the impact site. Likewise the approximate incident angle derived from the appearance of the impact site and ricochet damage to the bullet together with the constraints of the scene will often allow certain scenarios to be excluded and others included. Consider the interior of an airport hangar that is 100 feet in length with a smooth concrete floor. The ricochet mark is found at the 50-feet midpoint of the floor and a gunshot victim with an irregular entry wound to the forehead produced by the ricocheted bullet is found next to the wall downrange of the ricochet site. The shooter claims that he thought that the decedent was a burglar and that the wrench later observed in the decedent's hand was a gun. The shooter goes on to claim that he fired a warning shot into the floor immediately in front of his position while his presumed adversary was standing near the distant wall. The downward angle demonstrated by the shooter for this warning shot is on the order of 30–45°. This account is easily refuted with a little thought and a few test shots. Given the previously described ricochet behavior of bullets striking hard, unyielding surfaces, the victim could not be standing when struck. The height of his entry wound when standing would require ricochet angles of 6° or more. Rather the victim had to be down and nearly prone on the floor at the time of his fatal injury. The critical angle for this gun–ammunition–substrate combination is found to be 15° whereupon the bullet fragments into many pieces and produces a small crater in the concrete. The recovered bullet and shape of the impact site show an incident angle on the order of $8° \pm 3°$, a value that is also at odds with the shooter's account and that can be used to estimate his shooting position for selected gun heights.

PROJECTILE IMPACTS TO FRANGIBLE MATERIALS

The most common examples of this type of material are cinder blocks used in home and building construction, stepping stones cast from mortar, and some poorly fired bricks. These materials are more homogeneous than properly formulated and cured concrete but not as resistant to bullet impact. This type of surface is relatively hard and unyielding to bullet impacts but only up to a point. Prior to this point they will respond to bullet impacts in the same manner as hard, unyielding surfaces by recording an impact mark composed of ablated bullet metal and producing low ricochet angles. When that failure point for a particular bullet/incident angle/impact velocity is reached, the material will shatter immediately below the initial impact site where the force is concentrated and at its greatest. If the incident angle is essentially orthogonal, the resultant crater will be symmetrical with its deepest point in the center just as one would expect. Little, if any, identifiable bullet metal will be found in this crater because the shattered substrate material containing any such transfers is dislodged and expelled from the final crater. In non-orthogonal impacts the projectile continues on into the crumbled material knocking and flaking additional material out of its departure path. So long as the material is essentially homogeneous in the struck area, the effect is often a crater with its deepest point displaced toward the *approach* or *entry* side as shown in Figure 7.6. This is

Figure 7.6

Diagram of a Projectile Impact Crater in a Frangible Material

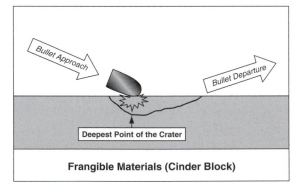

Frangible Materials (Cinder Block)

This drawing depicts the sudden impact of a bullet striking a relatively homogeneous frangible material as it approaches at a non-orthogonal angle. The shattering of the substrate extends the deepest immediately below the point of initial contact. The continued advance and ricochet of the bullet flakes away more of the material toward the departure side of the crater, but not to the depth of the initial disruption.

at first counter-intuitive until one considers the behavior of frangible materials and the forces involved. Ricochet angles are typically less than the incident angle but higher than they would have been had the substrate been not shattered under the bullet's initial impact. Since these are otherwise hard and abrasive materials in which the bullet has suffered rapid deceleration, the bullets usually experience considerable deformation and even fragmentation. If located, the bullet or fragments thereof will show heavy damage and numerous mineral inclusions associated with the struck object.

PROJECTILE IMPACTS TO SEMI-HARD OR SEMI-YIELDING MATERIALS

Still other materials fall somewhere in between hard, unyielding materials and yielding materials. Asphalt is the common example. It is also the most perplexing example of a shooting scene surface from which a bullet is believed to have ricocheted. There are multiple reasons for this. Bullet impact sites to asphalt are seldom conspicuous and an area of asphalt often has an assortment of small craters and gouges in it from other sources. Asphalt roadways and parking lots may continue to be used after a shooting incident and before a detailed search for a bullet impact site can be made. This can quickly modify the appearance of the impact site as can rain and street sweeping equipment. Chemical tests for lead or copper may fail on bullet impact sites in asphalt due to the dislodgement of the lead or copper-containing areas. Lead-in marks, lead splash, pinch points, and Chisum trails are not to be found in ricochet sites in asphalt. All of these missing phenomena combine to produce a rather gloomy forecast for success in locating a ricochet site in asphalt but there are several features that can be pointed out. The edges of a crater produced by a recent ricochet event will be sharp and fragile. These edges are quickly worn away with traffic and weathering effects. The newly exposed asphalt will look "fresh" compared to the surrounding, undisturbed asphalt. Most noteworthy will often be powdery deposits from one or more struck and fractured stones in the aggregate. This powdery material will usually be downrange of the struck or cracked stone and therefore provides directional information. This powder is a very good indicator of recent damage because it is quickly removed and lost with traffic passing over the site, street sweepers, rain, etc. Should one be so lucky as to see this effect, it should be photographed along with the inclusion of some form of orientation (direction). The author has, on occasion, observed small particles of bullet metal adhering to one or more struck stones in a ricochet mark in asphalt. These too should be photographed in place. Then efforts are made to collect them. Particles of dislodged asphalt may be found predominantly downrange of the impact site. The ejected asphalt possesses another useful reconstructive property and that is its ability to produce pseudo-stippling of the skin. This occurs when the entry wound is close to the impact site. For example, a gunshot victim found in a parking lot with a head wound surrounded by intense pseudo-stippling composed of asphalt particles had to be down at the moment he was shot given the very short range ballistic properties of such ejected material. This can be easily demonstrated with witness panels of suitable materials such as fresh pigskin mounted at selected distances downrange of test shots fired into asphalt at the incident angle needed to produce a comparable crater and damage to the ricocheted bullet.

Bullets associated with ricochets from asphalt have characteristic features. The damage will consist of coarse front-to-back gouges and striae with both mineral inclusions and tar-like smears from the organic constituents in asphalt. This material is soluble in a variety of organic solvents. This can be demonstrated with a cotton-tipped applicator moistened with toluene. With jacketed bullets, impacts with the small stone aggregate in asphalt often produce tears in the jacket. The overall coarseness and irregularity of the ricochet damage produced by asphalt distinguishes it from impacts with hard, unyielding surfaces such as concrete.

Because of its dual/non-homogeneous composition of soft yielding organic material and hard mineral material, ricochet angles vary widely and can even exceed the incident angle. By way of an example, five rounds of 124 gr. 9 mm FMJ ammunition were fired into moderately weathered asphalt at an incident angle of $-5.0°$ and resulted in ricochet angles of 2.1°, 4.5°, 5.3°, 5.7°, and 7.1°. As can be seen these results show a wide dispersion with values above and below the value of the incident angle. Table 7.4 provides some additional results for shots into asphalt at three selected incident angles.

Table 7.4

Ricochet of 9 mmP Bullets from Asphalt at Three Incident Angles

Incident Angle = −5°	Incident Angle = −10°	Incident Angle = −15°
2.1°/4.5°/5.3°/5.7°/7.1° = Ave. 4.9° ± 1.8°	3.8°/4.5°/5.6°/7.3°/7.7° = Ave. 5.8° ± 1.7°	5.4°/7.2°/9.5°/14°/14° = Ave. 10.0° ± 3.9°

A Ruger P-85 9 mmP pistol was mounted in a Ransom Rest and incident angles of −5°, −10°, and −15° were used to fire Federal brand 124 gr. FMJ-RN bullets into moderately weathered asphalt in a parking lot. Five rounds at each of the three incident angles were fired into the asphalt. The temperature of the asphalt was approximately 75 °F. The average impact velocity for these bullets was 1041 fps ± 10 fps. All fifteen strikes created visible craters in the asphalt although several of the −5° impacts were quite subtle.
The wide range of ricochet angles shown above is a common phenomenon with asphalt.
It should be noted that one of the −5° shots departed with an angle of 7.1°.

PROJECTILE IMPACTS TO YIELDING SURFACES

Figure 7.1b provides the operative model for this event. With these surfaces the ricochet angle is often *greater* than the incident angle. This is the complete opposite effect of ricochets from hard, unyielding surfaces. The results for some ricochet tests with the 147 gr. 7.62NATO bullet fired from an FAL rifle at fixed incident angle of 4° provide a useful illustration. (*Note*: This bullet is shown in Figure 6.19 in the previous chapter.) These FMJ-BT bullets impacted smooth dry sand and smooth concrete at approximately 2600 fps. They ricocheted intact but destabilized as evidenced by the yawed bullet holes in multiple downrange witness panels. The average ricochet angle from sand was 12.3° ± 1.3° for six shots and for concrete, the average ricochet angle was 1.9° ± 0.5°. The most dramatic examples of ricochet angles substantially exceeding the incident angle are achieved with water and fine sand but such elevated ricochet angles also can occur with soil, sod, sheet rock, and sheet metal. This, at first, defies logic but it occurs because the bullet is pushing or forming a departure ramp at its front underside due to its ogival shape and the even, yielding nature of these substrates. This results in the bullet departing from a very different surface that the one it first encountered. Because bullets spend considerably more time and distance interacting with yielding surfaces as compared to hard, unyielding surfaces the velocity losses are much greater. On occasion the author has found bullets ricocheted from soil and sand a few feet beyond the impact site. As with hard surfaces, the ricocheted bullets are destabilized and lose their remaining velocity rapidly because of their yawing or tumbling flight. They do *not* acquire the flattened area characteristics of low angle impacts with hard, unyielding surfaces for the simple reason that the struck surface *yields* to the bullet's advance. With materials like wood, sheet metal, sheetrock, or asphaltic tile flooring, the bullet may retain its basic shape and only show a slight scuffing or burnishing in the contacting area with the possible transference of some of the struck substrate. The deepest point of the indented ricochet mark is usually displaced towards the departure end of the mark. However, in some materials, and with certain incident angles, the indentation may be so symmetrical (with the deepest point at the midpoint of the indentation) that direction of fire is not obvious from the geometry of the ricochet mark itself. There are several phenomena that can establish the direction of fire in such ricochet marks. At low incident angles the initial contact between the bullet and the surface results in the production of a dark, elongated, and elliptical transfer of material. When present, this *lead-in* mark establishes the direction of travel for the causative bullet. Three examples of lead-in marks are shown in Figure 7.7.

With painted sheet metal another mark is often produced called the *pinch point*. This takes the form of a small area of surviving paint that was pinched

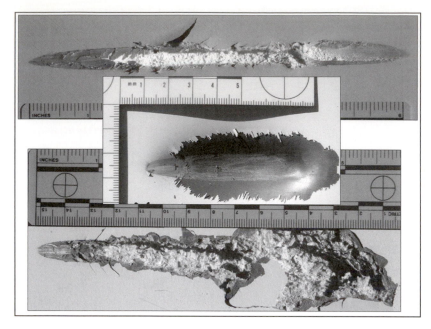

This assembly of three photographs shows low incident angle strikes and ricochet damage to painted drywall (top), automotive sheet metal (middle), and painted particle board (bottom). The dark gray lead-in marks produced by the .38-caliber LRN bullets are at the left edge of each mark showing the direction of travel for each as left to right. All three photographs are printed to the same scale.

between the bullet's initial contact point and the painted sheet metal surface. As with the lead-in mark, the pinch point is at the beginning of the ricochet mark. Figure 7.8 provides an illustration of a pinch point in sheet metal. The pinch point tends to occur more often with round nosed bullets and less often with truncated cone bullets.

There is yet another phenomenon that sometimes occurs in painted sheet metal. Some paints for reasons that are unclear fracture in a characteristic and reproducible way as the metal yields to the bullet's advance. These fracture lines may not be readily visible upon simple inspection but can be raised or enhanced by dusting with a fingerprint powder chosen to contrast with the particular color of paint. Such dusting with a fingerprint brush should only be undertaken *after* photographs and any chemical tests for bullet metal transfers are completed and other means of direction of travel determination have been exhausted. If present, these fine fracture lines simply trap some of the fingerprint powder as the fingerprint brush is gently worked back and forth parallel to the long axis of the ricochet mark. These fracture lines take the form of a bow wave or shock wave and, as with the lead-in mark and the pinch point, show the direction of travel of the responsible bullet. These enhanced fracture lines can, of course, be not only photographed but also lifted much like a fingerprint

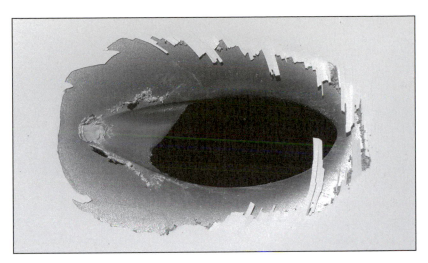

Figure 7.8

*The Pinch Point in a
Painted Metal Surface*

The pinch point is the small, circular area of surviving paint to the left of this elongated bullet hole in automotive sheet metal. This represents the site where the approaching bullet first contacted the metal.

using wide fingerprint tape. Once lifted the tape is transferred to an appropriate card, dated, marked for identification, and impounded. Figure 7.9 shows an example of these fracture lines at the margins of a ricochet mark in painted sheet metal after dusting with fingerprint powder. (*Note*: Do *not* attempt to use the MagnaBrush® on sheet metal made of steel.)

The fracture lines in the paint visible around the edges of the left half of this ricochet site have been rendered visible by dusting the area with black finger print powder. These fracture lines can be thought of like shock waves radiating out from a bullet in flight. They establish the direction of travel for this bullet as left to right. Note that in this example there is no lead-in mark or pinch point to aid in establishing the direction of travel.

Lead bullets and bullets with exposed lead at their noses upon low incident angle impact with many surfaces produce a spattering and vaporization of lead with its subsequent downrange deposition. These lead deposits are referred to as *lead splash* and, if necessary, can be rendered visible with the sodium rhodizonate test either by direct application or by the transfer technique. The geometry of the deposition of lead splash can provide information on the direction of fire. The amount of lead splash is also a function of impact velocity and the nature of the surface struck. Just as with asphalt, a victim located close to the impact site may sustain pseudo-stippling from lead splash and bullet fragments generated during the impact and ricochet process. With the sort of orthogonal strikes described in the previous chapter, lead splash can even extend back toward the source of the shot. This can be particularly confusing when such a bullet strikes a window pane after first passing through an adjacent curtain or pull-down shade. The deposits of lead splash on the *exit side* of the curtain or shade can look like close range gunshot residue.

Bullets ricocheted from soil, sod, and sand also acquire some interesting characteristics. Unless small stones are struck, these bullets often emerge intact and with minimal deformation. Here again the ricochet angles can often exceed the incident angles. The sites of such impacts can be exceeding difficult to locate. Bullets striking a grassy lawn seldom show an entry hole and even the exit can be difficult to find. The channel produced by bullets striking and ricocheted from such a surface is usually on the order of 5–10 in. in length and may only be located by carefully "combing back" the grass by hand or by feeling for the channel produced by the bullet. As these bullets plow their way through the abrasive mineral content of these substrates, the abrasive particles flow around the nose, ogive, and even bearing surface of the bullet leaving a pattern of abrasion that is reminiscent of the flow pattern of water off the bow of a boat hence the name—*bow effect*. The bullet in Figure 7.10 is a typical example. The presence of this effect is a dead give-away that the bullet has entered into soil or sand at a shallow angle, plowed through it for some distance, then emerged as a ricochet event. If the bullet completely entered the substrate, the abrasive effects will envelop the bullet. If the bullet has not been totally surrounded by soil, sand, or sod, then only one side of the bullet will display the bow effect. Evidence of late-term yawing as the bullet neared the end of its journey through the material can often be seen as the flow pattern changes direction on the bullet's surface and this new abrasive damage starts to overwrite the earlier damage. Careful examination of the embedded mineral grains and the flow pattern they possess may provide further useful information. This is most effectively accomplished with a SEM/EDX instrument. Just as with an earlier example for hard, unyielding surfaces, a bullet that passes through some intervening object then strikes and ricochets from soil or other surface will show a very different pattern of abrasion from its yawed impact with the ground. This is illustrated in Figure 7.11.

Figure 7.10

Two FMJ 9 mm Bullets Ricocheted from Soil

These two bullets were recovered after they entered relatively soft soil, plowed through the soil for about 6–8 in. then emerged. The abrasive action of the soil grains has effectively obliterated the rifling engravings on these bullets. It is the pattern of this abrasive action that gives rise to the term "bow effect." Examination of such bullets under the SEM will highlight this pattern and show the presence of numerous embedded soil grains.

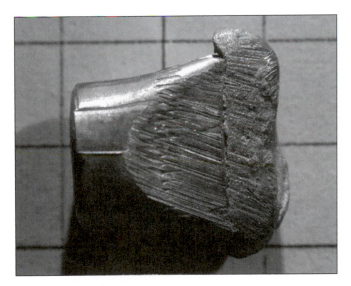

This hollow point bullet passed through a gunshot victim prior to striking a hard, unyielding surface (concrete) while in yaw. The bullet mushroomed as it passed through soft tissue, then impacted the hard, abrasive surface of the concrete. This is evidenced by the diagonal pattern of striae caused during the impact and ricochet process. (Photo courtesy of Michael Haag.)

THE POST-IMPACT FLIGHT OF RICOCHETED AND DEFLECTED BULLETS

The flight of bullets through the atmosphere is primarily governed by the forces of gravity and air resistance. This will be covered in much more detail in the Chapter 14 on True Ballistics but for the purposes here it is important to realize that a ricocheted or deflected bullet is typically destabilized and often deformed. These two conditions cause the bullet to lose velocity much faster than it otherwise would in normal, nose-forward, spin-stabilized flight. This substantial increase in velocity loss over time or distance has important implications insofar as the sort of distances serious or fatal wounds might be inflicted by a ricocheted or destabilized bullet. The author has carried out a number of measurements of the post-impact behavior and flight characteristics of some common bullets using Doppler radar and the Oehler M43 chronograph system. A couple of examples from these efforts should be helpful. Speer's 115 gr. 9 mm TMJ-RN bullet and Winchester's 147 gr. 9 mm JHP SilverTip bullet were fired into hard desert ground with scattered loose dirt and gravel on its surface at a nominal incident angle of $-1.5°$. The impact area was approximately 50 yards from the carbine used to launch these bullets. Pre- and post-impact velocities, drag coefficients, and other data were recorded for each bullet's flight for several hundred yards or until the bullet re-impacted the ground. A downrange witness panel consisting of thin cardstock showed that the bullets had not fragmented but were clearly destabilized. The very high and ever-changing drag coefficient values from the radar data also showed that they were tumbling in flight after ricocheting from the ground. Velocity loss values during ground impact for five shots with the 115 gr. TMJ-RN bullets were 82, 98, 230, 230, and 605 fps after an impact velocity of approximately 1150 fps for this bullet. The 147 gr. JHP bullets experienced velocity losses of 127, 216, 300, and 395 fps for four radar tracks and had an average impact velocity of 1140 fps at the 50 yd mark.

The very high aerodynamic drag on these tumbling and possibly deformed bullets caused them to lose velocity very quickly during their post-impact flight. From an evaluation of the data generated by the Doppler radar system, it was determined that the post-impact velocity of these bullets would fall to 200 fps after traveling approximately 130–260 yd. Had they not been fired purposefully into the ground, these bullets would have had velocities in excess of 1000 fps at 130 yd and 900 fps at 260 yd beyond the muzzle. This downrange velocity of 200 fps was chosen because it approximates the velocity necessary to produce a perforating injury to skin with subsequent penetration into the underlying tissues. These downrange distances for the 200 fps velocity are optimistic since the velocity loss during ground impact was not considered in carrying out the calculations. Similar results for velocity loss and post-impact behavior were obtained for .40-caliber and .45-caliber pistol bullets.

GUNSHOT WOUNDS FROM RICOCHETED AND DEFLECTED BULLETS

The other important consideration relates to the yawing or tumbling motion of most ricocheted and deflected bullets. Although such bullets can, on occasion, strike nose-first and produce a normal appearing entry wound, the more common situation encountered is a yawed or keyholed entry. This will usually be described by the pathologist simply as an irregular entry wound or an atypical entry wound. If the injury is in a clothed area, a detailed examination and testing of the bullet hole can provide useful information through the shape of the hole and the nature of any bullet wipe around the hole. Direct strikes by stable, nose-first bullets usually produce oval or round holes with bullet wipe around the margin. Ricocheted or deflected bullets typically produce atypical holes with incomplete or non-existent bullet wipe because of their prior contact with the intervening material. The depth of penetration in the victim will often be reduced compared to what one would otherwise expect for the particular bullet where the distance between impact site and victim is substantial (e.g., 50+ yd or more) due to the increased velocity loss. Reduced penetration can also occur because the bullet enters the victim with a yawing motion. A bullet fired directly into a body will maintain its nose-forward orientation for some distance before it begins to yaw whereas a ricocheted or deflected bullet enters the victim in an unstable and yawing condition. The latter situation means that such a bullet will be decelerated more quickly in tissue than a direct shot. The only complicating factor arises with hollow point bullets. A properly designed hollow point bullet fired directly into tissue will expand or mushroom. This, of course, greatly slows the bullet and limits penetration. The same bullet ricocheted, and that does not strike nose-first, will not expand and will likely penetrate more deeply than a properly expanded bullet. The resolution to this is the careful laboratory examination of the recovered bullet for ricochet damage and/or trace evidence acquired from an intervening object.

Most of these phenomena can be demonstrated with a few simple tools and materials. Figure 7.12 shows a section of flagstone, a series of cardstock witness panels, and a block of tissue simulant used to record the behavior and ricochet angles of three shots previously ricocheted off the flagstone. The ricocheted bullets passed through the multiple witness panels and entered the block of tissue simulant. The orientation of the ricocheted bullets at each downrange position as well as the ricochet angle can be derived from an inspection and height measurements of the bullet holes in the multiple witness panels. Nearly all of the bullet holes in this type of demonstration will show the particular bullet in a state of yaw but occasionally a perfectly normal-appearing round hole will be present showing that a ricocheted bullet can pass through a nose-forward orientation during its post-impact flight. This is an important phenomenon to witness because more than one pathologist and at least one ballistician of considerable experience have testified that a

Figure 7.12

Bullet Ricochet through Witness Panels

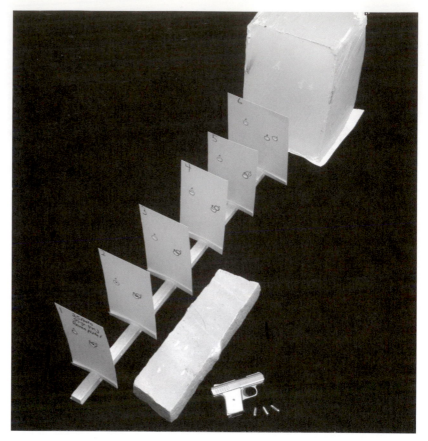

This figure shows an inexpensive apparatus for recording the post-ricochet behavior of fired bullets. The six numbered cardstock witness panels are mounted in thin slots in a strip of balsawood that has been glued to a metal strap to provide extra strength. The spacing shown here is 6 in. between each card. This assembly of witness panels is mounted at an appropriate position downrange of the "target" material—in this example, the rectangular section of stone.

A medium such as ballistic soap (shown above), ordnance gelatin, or a used Kevlar vest is positioned just beyond the last witness panel for recovery of the ricocheted bullet and/or wound ballistic evaluation. Multiple shots can usually be ricocheted through such an arrangement and the orientation of each ricocheted bullet evaluated through a careful inspection of the witness panels. The ricochet angles can also be determined from these panels by measuring the change in height between the first panel and the last panel for each bullet.

round (normal) entry wound precludes the possibility of a ricocheted bullet. The simple experimental design and equipment is illustrated in Figure 7.12 and the round bullets holes in several post-ricochet witness panels quickly refute this claim.

The ricochet damage to bullets and their penetration depths in the tissue simulant can also be ascertained through this simple demonstration. Several direct shots into the same tissue simulant with the same gun and ammunition and from the same distance can be used to compare penetration depths.

SHOTGUN DISCHARGES AND PELLET RICOCHET

The ricochet of shotgun pellets represents a special situation in the following respects:

- The projectiles (customarily lead pellets) are relatively soft and are usually damaged significantly upon impact, particularly with unyielding or abrasive surfaces.
- With close range impacts (e.g., −10 yd or less), the pellets often collide with each other during and immediately after impact and ricochet. This may complicate their ricochet behavior and result in the likelihood of some pellets fusing together during the ricochet process.
- Lateral deflection due to projectile rotation is a moot point as is the destabilization of the projectiles.
- The approximate incident angle can often be determined from the arcsine function of the ratio of the maximum and minimum diameters of oval pellet patterns on relatively flat surfaces.

APPROACH TO CASEWORK INVOLVING THE ISSUE OF RICOCHET

If a ricochet occurred in *this* case, ask yourself *what would you expect to see.*

- at the scene (impact site)?
- on the bullet?
- with or in the victim (bullet hole in clothing/wound appearance/penetration)?

Start with what is known (not in dispute) regarding:

- the recovered projectile (impact damage, weight loss, embedded trace evidence)
- the wound (entry wound appearance, satellite injuries, pseudo-stippling, wound track, penetration depth)
- the bullet hole (size, shape, bullet wipe, satellite defects)
- the scene geometry and behavior of ricocheted projectiles.

Design and carry out any empirical tests necessary to evaluate or illustrate the post-impact behavior of ricocheted projectiles of the type involved in the incident under investigation.

Any account of a ricocheted shot resulting in an injury must consider:

- the location and the nature of the injury (and any bullet hole in clothing)
- the nature of the surface(s) between the gun and the victim

- the appearance of the bullet (if recovered)
- the distance between the gun and the victim
- the geometric constraints imposed by the scene, and
- the ricochet behavior of the bullet/surface combination.

SUMMARY AND CONCLUDING REMARKS REGARDING RICOCHET

Efforts to predict the specific ricochet behavior of projectiles during and after impact with a surface must consider a number of variables, some of which play major roles in bullet behavior while others have minor roles.

Measuring ricochet angle for a particular bullet/incident angle/impact velocity/surface does *not* always allow one to determine, by calculation, the ricochet angle for some new incident angle nor is there necessarily a linear relationship between a series of values for $\angle I$ vs. $\angle R$.

Empirical testing will often be necessary to gain some insight into the ricochet behavior for a particular bullet–surface combination before any meaningful back-extrapolations are carried out.

The general behavior of bullets ricocheted from hard, unyielding surfaces and those ricocheted from yielding surfaces can be useful in searching for the ricocheted bullet or subsequent downrange impact sites.

The appearance and nature of impact damage to recovered bullets can provide considerable insight into the pre-impact orientation of the bullet and the type of surface struck as well as the composition of the struck surface.

The approximate angle of incidence can often be estimated from the impact damage suffered by the bullet.

An examination of the impact site can likewise provide information about the design and composition of the projectile, the approximate angle of incidence and possibly information about the responsible firearm.

PERFORATING PROJECTILES AND PERFORATED OBJECTS

The behavior of bullets that have struck objects (but not ricocheted from them) is a necessary subject to be included in any discussion of ricochet but there are some distinct differences between the two events (e.g., *deflection* as previously defined and discussed). Some of the phenomena described for ricocheted bullets are also present with bullets that strike or perforate objects. These include sudden velocity loss, bullet deformation and destabilization, trace evidence transference, and changes (damage) to the impacted surface. In the case of *perforating* bullets, ricochet angle is replaced by definition 2 of *deflection angle*—a departure angle, commonly stated in degrees and position on the clock face, from the normal, pre-impact flight path of the bullet. For most objects struck in a *grazing* manner (e.g., branches, small limbs), the direction of deflection generally correlates with the side struck, e.g., a left graze to an erect-standing sapling will usually deflect toward the 9 o'clock direction and result in a destabilized bullet. The direction of twist seems to play no significant role in this situation.

Perforated *objects* can be subdivided into "thin" and "thick." Although no numerical values can be assigned for these distinctions, a "thin" object would be represented by sheet metal or panes of common window glass. In such objects the track of the bullet in the object is significantly *less* than the overall length of the bullet.

Examples of "thick" objects would be perforating strikes to human bodies, the trunks of small trees and containers of liquid. In these situations the track of the bullet in the object substantially *exceeds* the overall length of the bullet.

These distinctions are made because it appears that twist rate, pre-impact yaw, and/or bullet stability as well as entry angle may have some influence in deflection direction with strikes to "thick" objects.

Another group of objects falls somewhere between "thick" and "thin" forms. Common examples are fence boards and sheet rock. The tracks through these objects provide a marginal inference as to the bullet's pre-impact flight path. Examples of this will be found in the next chapter. Bullet deflection in these materials, on the other hand, is minimal to non-existent as a practical matter.

Bullet yaw at impact also appears to play a role in the direction of deflection as well as the magnitude of any deflection in "thick" objects. This means that bullets striking "thick" objects at close range, where bullets are seldom fully stabilized might do so with substantial yaw and thereby experience more deflection than when striking the same objects at greater distances. The classic example of this is illustrated on pages 406 and 407 of the 3rd edition (1966) of *Hatcher's Notebook* where a .30-'06 bullet, after traveling 200 yd, penetrated 32.5 in. of oak. This bullet remained point-forward throughout its track in the wood. Another shot with the same ammunition, fired from only 50 feet, resulted in only 11.25 in. of penetration with considerable deflection. This same effect can be observed with close range shots (a few feet or yards) into ballistic gelatin as compared to long range shots (50–100+ yd) using the same bullet and firearm. Delayed yaw to little measurable yaw at all will typically take place with the long distance shots whereas rapid yaw will usually take place with shots fired at close range.

Several figures have been included at the end of this chapter depicting simple setups for studying the behavior of ricocheted and deflected bullets and for measuring ricochet and deflection angles (Figures 7.13 and 7.14). The materials shown in these figures require a minimum of expense, a relatively small investment in time and the use of a pocket calculator with trigonometry functions.

Figure 7.13

Materials for Basic Ricochet Tests and Measurements

From left to right, this figure shows a pistol secured in a *Ransom Rest*® with an inclinometer positioned on the barrel for the selection of the desired incident angle. A smooth stone slab has been leveled with a bubble level. This level also has a built-in and co-aligned laser. The tape measure will be used to measure various distances depending on whether one or two witness panels are used. The frameworks for the cardstock witness panels at the right are constructed from plastic pipe.

There is no single mandated procedure for the measurement of ricochet angles. If only one (1) downrange witness panel is used, the distance from the bullet impact site to the base of the witness panel must be measured as well as the height of the ricocheted bullet hole above the plane of the target material. In this method, the tangent function is used to calculate the ricochet angle by dividing the true height of the bullet hole by the distance from the bullet impact site.

The use of two (2) witness panels obviates the need for identifying the specific bullet impact site and measuring the distance to the first witness panel. In this method the height *change* between WP1 and WP2 positioned a known distance apart is used to calculate the ricochet angle through the arctangent function. The two witness panel method also provides more information on post-impact bullet behavior.

All of the items depicted in this figure have been grouped much closer together for photographic reasons and would actually be separated much more than shown here. Other items that can be integrated into this technique are one or more chronographs and a bullet recovery medium (loose Kevlar in a box, ballistic soap, or ordnance gelatin) behind the final witness panel.

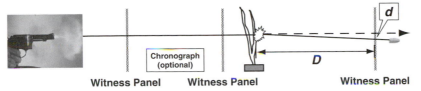

This diagram illustrates a simple and inexpensive setup for measuring bullet deflection and post-impact bullet behavior. The target material (in this case a tree branch) is mounted in a secure fixture. One or more cardstock witness panels are mounted downrange of the target. (One is shown in this figure.) Two cardstock witness panels are positioned approximately 3–4 feet apart and in front of the target material. A Ransom Rest is preferable for the firing of the gun but not essential. A shot is fired at the target and through the two pre-impact witness panels. If the bullet misses the target, the bullet holes are marked in some way so they will be disregarded. Such a missed shot can be useful, however, in making a subsequent sighting correction. When a shot hits the target material, a small laser pointer is directed through the two bullet holes in the pre-impact witness panels and on to the downrange witness panel where the intercept point of the laser beam is marked. This is represented by the dashed line in this figure. The distance (d) to the bullet hole associated with the strike to the target is measured and its clock direction noted. The deflection distance d is divided by the distance between the strike of the target and the downrange witness panel (D) followed by the use of the $\tan^{-1}$ function on a pocket calculator. This calculation gives the degrees of deflection experienced by the bullet.

Multiple downrange witness panels at selected intervals will allow one to study the yawing behavior of the destabilized bullets.

CONCLUDING COMMENTS

The information presented in this chapter and gained from some of these simple shooting experiments should provide forensic examiners with some approaches and methodologies for measuring bullet performance and behavior as well as some ability to predict general bullet behavior after ricochet or deflection. The interpretation of bullet impact marks, damage to bullets due to such impacts and trace evidence considerations have also been illustrated for several common examples of bullet ricochet. Nonetheless it must be reaffirmed that a critical assessment of either incident angle or ricochet angle, the limits of deflection, velocity loss, or other parameter of interest in casework *will* require some empirical testing. The experimental design should evaluate the role of any variables that stand to influence the outcome of the impactive event(s) under consideration.

REFERENCES

"Bouncing Bullets," *FBI Law Enforcement Bulletin*, Vol. 38 (October 1963) pp. 1–9.

Burke, T.W. and W.F. Rowe, "Bullet Ricochet: A Comprehensive Review," *J. For. Sci.* 37:5 (September 1992) pp. 1254–1260.

Gold, R.E. and B. Schecter, "Ricochet Dynamics for the Nine-Millimetre Parabellum Bullet," *J. For. Sci.* 37:1 (January 1992) pp. 90–98.

Haag, L.C., "Bullet Ricochet: An Empirical Study and Device for Measuring Ricochet Angle," *AFTE Journal* 7:3 (December 1975) pp. 44–51.

Haag, L.C., "Bullet Ricochet from Water," *AFTE Journal* 11:3 (1979) pp. 26–34.

Haag, L.C., "Bullet Impact Spalls in Frangible Surfaces," *AFTE Journal* 12:4 (1980).

Haag, L.C., "The Measurement of Bullet Deflection by Intervening Objects and the Study of Bullet Behavior After Impact," *AFTE Journal* 19:4 (1987) CAC Newsletter (January 1988).

Haag, L.C., "An Inexpensive Method to Assess Bullet Stability in Flight," *AFTE Journal* 23:3 (July 1991).

Haag, L.C., *Firearms Trajectory Analysis Manual*, California Department of Justice-California Criminalistics Institute, Sacramento, CA (1996, 1997, 1998).

Haag, L.C., "Bullet Penetration and Perforation of Sheet Metal," *AFTE Journal* 29:4 (Fall 1997) pp. 431–459.

Haag, L.C., Ricochet Workshops: AFTE 2002 SWAFS 2002 CAC/NWAFS 2003.

Haag, L.C. and M.G. Haag, "Projectile Ricochet," Chapter 5, *Forensic Shooting Scene Reconstruction Course Manual* (October 2002/November 2004).

Hartline, P.C., G. Abraham and W.F. Rowe, "A Study of Shotgun Pellet Ricochet from Steel Surface," *J. For. Sci.* 27:3 (July 1982) pp. 506–512.

Hatcher, J.S., Hathcer's Notebook, the Stackpole Co., 3rd edn, Second printing, Harrisburg, PA (1966).

Houlden, M.A., "The Distribution of Energy Among Fragments of Ricocheted Pistol Bullets," *J. For. Sci. Soc.* 34(1) (1994) pp. 29–35.

Janssen, D.W. and R.T. Levine, "Bullet Ricochet in Automobile Ceilings," *J. For. Sci.* 27:1 (January 1982) pp. 209–212.

Jauhari, M., "Bullet Ricochet from Metal Plates," *J. of Crim. Law, Criminology and Police Sci.* 60:3 (September 1969) pp. 387–394.

Jauhari, M., "Bullet Ricochet," *The Indian Police J.* 16:3 (January 1970) pp. 43–47.

Jauhari, M., "Mathematical Model for Bullet Ricochet," *J. of Crim. Law, Criminology and Police Sci.* 61:3 (1970).

Jauhari, M., "Approximate Relationship Between the Angles of Incidence and Ricochet for Practical Application in the Field of Criminal Investigation," *J. of Crim. Law, Criminology and Police Sci.* 62:1 (1971) pp. 122–125.

Jordan, G.E., D.D. Bratton, H.C.H. Donahue and W.F. Rowe, "Bullet Ricochet from Gypsum Wallboard," *J. For. Sci.* 33:6 (November 1988) pp. 1477–1482.

Kneubuehl, B.P., "Das Abprallen von Geschossen aus forensischer Sicht," doctoral thesis, University of Lausanne-Institute of Police Science and Criminology, Thun, Switzerland (1999).

McConnell, M.P., G.M. Triplett and W.F. Rowe, "A Study of Shotgun Pellet Ricochet," *J. For. Sci.* 26 (October 1981) pp. 699–709.

Mitosinka, G.T., "A Technique for Determining and Illustrating the Trajectory of Bullets," *J. For. Sci. Soc.* 11:1 (1971) pp. 55–61.

Nennstiel, R., "Study of Bullet Ricochet on a Water Surface," *AFTE Journal* 16:3 (July 1984).

Nennstiel, R., "A Fatal Bullet Ricochet," *AFTE Training Seminar*, St. Louis, MO (July 2000).

Rowe, W.F., T.W. Burke and R. Griffin, "Bullet Ricochet from Concrete Surfaces: Implications for Officer Survival," *Jour. of Pol. Sci. and Admin.* Vol. 16 (1988) pp. 264–267.

Salziger, B., "Beschuss von PKW-Frontscheiben," *Der Auswerfer 3. Ausgabe BKA Wiesbaden* (March 1997) pp. 35–38.

Salziger, B., "Geschossreflektion an harten Oberflächen," *Der Auswerfer 4. Ausgabe BKA Wiesbaden* (September 1997) pp. 69–71.

Salziger, B., "Reflektion von .22 l.r. Geschossen an harten Oberflächen," *Der Auswerfer 4. Ausgabe BKA Wiesbaden* (September 1997) pp. 45–55.

Wohlwend, C. and S. Weber, "Das Abprallen von Geschossen auf Asphalt," *Der Auswerfer 13. Ausgabe BKA Wiesbaden* (May 2003) pp. 13–20.

CHAPTER 8

THE PRINCIPLES OF "TRAJECTORY" RECONSTRUCTION

BULLET PATHS IN FIXED OBJECTS

It is unfortunate that the tracking and back-extrapolation of bullet paths through various materials has come to be called "trajectory" reconstruction but there seems little likelihood of correcting this misnomer. Even the various straight probes used for the procedures described in this chapter are often called "trajectory" rods. A true trajectory is a curved path. Such flight paths and their reconstruction are illustrated and discussed in Chapter 14 dealing with long distance shootings. The distances involved in most shooting cases are on the order of a few feet to 10–20 yd, e.g., shots fired inside a room to shots fired across a back yard or from the street to the front of a house. As a practical matter, the flight paths of projectiles over these distances are effectively straight lines as the amount of curvature in the flight path of even low velocity bullets amounts to less than an inch at these distances. The uncertainty in our ultimate measurements is usually greater than this so considering short range trajectories as straight lines is not particularly troublesome.

THE LOCATION OF THE BULLET HOLE AND ANGULAR COMPONENTS OF THE PROJECTILE'S PATH

Whether one uses probes, string lines, a laser, surveying equipment, or some other technique there are three elements that remain constant and critical to the reconstruction of a bullet's pre-impact flight path. These are enumerated below.

1. The location of the entry bullet hole or impact site.
2. The vertical angle or component of the reconstructed bullet path.
3. The azimuth angle (or "compass" angle) of the reconstructed bullet path.

The location of the bullet's entry point must be associated with one or more locatable reference points. The location of a bullet hole in the wall of a residence, for example, would be satisfied by a description of the wall (e.g., the east wall of the living room), the height of the bullet hole above the floor and the distance of the bullet hole from a referenced corner (e.g., 9.0 feet north of the southeast corner and 20 in. above the floor).

Every bullet path can be resolved into two angular components. The author has chosen the term vertical angle to describe any ascending or descending angle possessed by the projectile as it penetrates or perforates one or more objects. A downward angle of travel is given a negative (−) sign and an upward angle of travel, a plus (+) sign. A projectile with no discernable upward or downward component relative to the normal horizontal plane would be described as having a 0.0° vertical angle and a bullet fired straight down into the floor or other substrate would have a −90° vertical component.

Once the best estimate of the bullet's path has been established with an appropriate tool such as a trajectory rod, the vertical component can be measured in a number of ways. Digital inclinometers, angle finders, special protractors, mitering gauges, surveying equipment, laser devices, and/or simple measurements of height changes along a selected distance out from the bullet hole can be used to ascertain this angular component of the projectile's path (review Figure 7.6 in the previous chapter).

The azimuth component can be thought of like a compass direction as one views the bullet's flight path from above. Others have referred to this component as the side angle or the lateral angle. In measuring this angle it is often easiest to use the struck surface as the plane of reference with that plane being defined as 0.0°. A shot fired straight (orthogonally) into the wall would have an azimuth component of 90° and a vertical component of 0.0°. *Note*: It should be pointed out that this is contrary to the NATO method of defining incident angles but is more useful in shooting scene reconstruction (see Figure 8.5 for the distinction).

THE USE OF STRING LINES

String lines are inexpensive, relatively easy to see, and easy to photograph. Kits are available from selected forensic equipment suppliers that contain various bright colors of string for this purpose. One supplier even offers a string that contains reflective particles for use in nighttime scenes. Their primary shortcomings are that they are non-rigid and therefore sag as soon as the horizontal distances exceed a few yards with one exception. String may also be the best demonstrative tool for representation of a trajectory over relatively long distances if the investigator can take a photograph looking straight down over the area. From this vantage point, any sag in the string is not seen. A helicopter or fire truck ladder may be used to gain the height needed over an area.

A string line also requires an anchor point that is sufficiently firm that the string can be drawn taught. String can be difficult to thread though multiple bullet holes and great care must be taken that it passes through the center of the bullet holes without experiencing a change in direction as a consequence of pressing against the edge of an entrance or exit hole. On the other hand colored string held an advantage over a probe inserted in a bullet's path through the center console of a vehicle for illustrating an officer's approximate shooting position. In this particular case the string was anchored at the final resting point of the bullet, drawn back through the center of the bullet hole in the plastic console, out through the open side window, through the bore of the officer's 9 mm semi-automatic pistol and finally out through the ejection port where it was anchored by the pistol's slide as the officer held the pistol in his recollected shooting stance. Photographs were taken at a right angle to the string line that included the officer, the string line, and the decedent's vehicle and later used to illustrate the approximate path of the bullet and the associated separation distance between the shooter and the side of the vehicle.

THE USE OF PROBES AND "TRAJECTORY" RODS

Rigid probes ranging from cleaning rods and wooden dowels to professionally manufactured rods specifically for the purpose of shooting reconstruction have been used to probe and illustrate the paths of penetrating and perforating projectiles. Cleaning rods are not recommended particularly if gunshot residues such as bullet wipe around the margin of the entry bullet hole are of importance. Photographs and any chemical tests for bullet metal that are to be conducted should be done prior to any probing of a bullet hole. Various diameters, colors, and compositions of trajectory rods have been offered through law enforcement suppliers. The most elaborate of these is a set of threaded and plastic-coated metal rods available in several diameters and colors that can be assembled to almost any reasonable length. Each set comes with centering cones to assist in aligning the probe in the center of each bullet hole. One section possesses a bullet-like tip to aid its insertion in tight-fitting bullet tracks in materials such as wood and rubber. This same set of rods will also accept a small, cylindrical laser that can be used to forward-project or back-extrapolate the path of the bullet once the rod is properly positioned in the bullet track and co-alignment of the laser with the rod has been assured. Even something as simple and inexpensive as hollow brass rods of various diameters available in hobby shops can be used so long as any necessary tests for copper around the margin of the bullet hole have previously been carried out. These inexpensive hollow rods allow the investigator to peer through the rod and locate possible shooting positions and downrange bullet impact sites. Portable lasers have also been passed through such rods. This can be particularly helpful when dealing with bullet paths through wood where the wood fibers otherwise get in the way of the laser beam.

PROCEDURES—VERTICAL ANGLE DETERMINATION

Properly executed, a photographic method provides the simplest technique for documenting and later measuring both the vertical and the azimuth components of the projectile's path. This method can also be employed as (and is strongly recommended as) an adjunct to any other techniques employed in determining these two angular components.

There are three critically important steps for the photographic method. The camera lens must be at the same height as that portion of the trajectory rod or string line, and it must be perpendicular to the vertical plane in which the line or probe lies. These two factors are of critical importance. Failure to do this is the most common mistake in this method, the effect of which will be illustrated later. A suitable vertical reference line must also be in the field of view. In some scenes there will be natural reference lines in the immediate background and so long as they are in adequate focus they will suffice. A preferable technique is to drop a plumb line through the center of the field of view with this line just touching the probe or string line. The known vertical reference line in the resultant photograph is subsequently used to measure the vertical angle of the probe or string line. Figure 8.1 illustrates the proper orientation of the camera for a trajectory rod passed through the entrance and exit holes

Figure 8.1

Proper Profile View of a Trajectory Rod through a Perforating Bullet Strike to an Interior Residential Wall

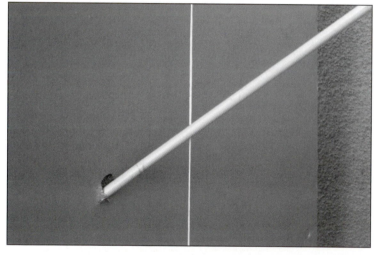

The camera is at the same level as the trajectory rod and at a right angle to the rod. A plumb bob and line have been brought in immediately adjacent to the rod. Although slightly out of focus, the corner of the room can also serve as a vertical reference line. The vertical component of this bullet's trajectory can be measured with a protractor from an enlarged view of this photograph. As defined in this chapter, the path of this perforating bullet has a vertical component of $-36°$ to $-37°$.

in a simulated residential wall composed of 5/8 in. sheetrock, a 4 in. airspace and 3/8 in. exterior wooden siding. The vertical component of the probe was determined to be −37°. This was accomplished by enlarging this digital photograph on a flat screen monitor and measuring the relationship of the probe to the vertical reference line with a common protractor. Figure 8.2 a,b, and c shows the effect of taking photographs from popular but improper positions if one wishes to ascertain the vertical component of the projectile's pre-impact

(a)

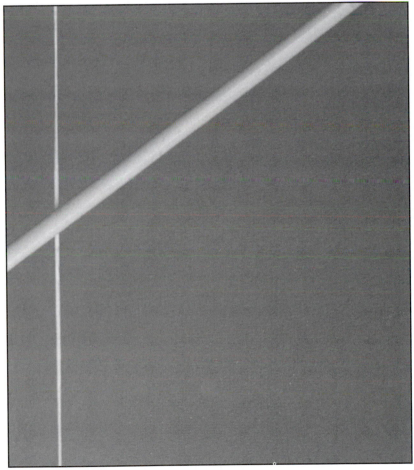

Figure 8.2(a)

Improper View of the Trajectory Rod through the Perforating Bullet Strike previously shown in Figure 8.1

In this view the camera is flush against the bullet-struck wall and at the same level as the trajectory rod. As in Figure 8.1, plumb bob and line have been brought in immediately adjacent to the rod but this camera position provides a slightly erroneous value for the vertical component of this strike. This camera position would be correct if this strike had no right-to-left azimuth component.

Figure 8.2(b)

Improper View of the Trajectory Rod through the Perforating Bullet Strike previously shown in Figure 8.1

(b)

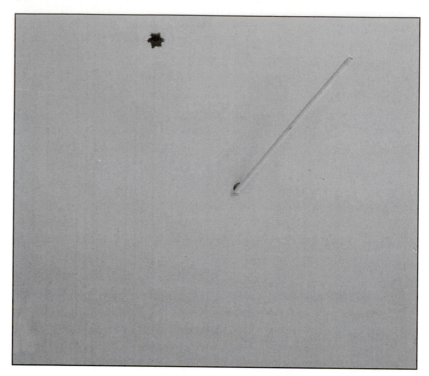

In this view the camera is directly in front of the bullet hole in the wall and at the same level as the trajectory rod. Although a popular view in many scene photographs, it provides an erroneous value (ca. −55°) for the vertical component of this right-to-left and downward strike.

flight path. The camera in Figure 8.2b is at the same height as the probe and directly out in front of the wall. In this view the apparent vertical angle is −53° for an error of 16° (or 43%) as a consequence of this improper camera position. The camera in Figure 8.2c has been positioned against the wall and at the same height as the probe. In this view the apparent vertical angle is −41° for an error of 4° (or 11%) as a consequence of this improper camera position.

Alternatively, an inclinometer can be carefully positioned on the trajectory rod (or carefully juxtaposed with a string line or laser line) and the vertical angle simply read from the device. This is depicted in Figure 8.3. The vertical angle can also be determined with a zero-edge protractor* positioned vertically

*A zero-edge protractor is one that lacks any protruding legs or tabs and will allow one to place its baseline edge directly against the plane of the struck surface. Such a protractor is shown in Figures 8.4a,b.

(c)

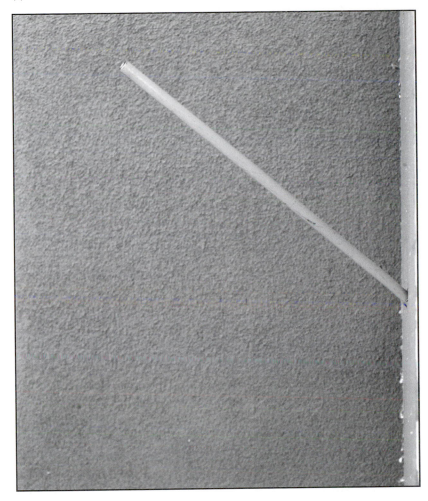

Figure 8.2(c)

*Improper View of the
Trajectory Rod through
the Perforating Bullet
Strike previously shown in
Figure 8.1*

In this view the camera is flush against the bullet-struck wall and at the same level as the trajectory rod. The surface of the struck wall and the intersecting corner of the wall beyond the trajectory rod provide a vertical reference line but this camera position provides an erroneous value (ca. −40°) for the vertical component of this strike because it is not perpendicular to the rod.

with the index mark at the junction of the probe and the bullet hole after which the protractor is carefully hinged over so that it is in full contact with the probe. It is critical that the zero edge of the protractor be oriented vertically, and the flat plane of the device be in the same vertical plane as the probe or string. The degrees above or below the protractor's horizontal line are then read at or under the probe.

Figure 8.3

*Vertical Angle
Determination with a
Digital Inclinometer*

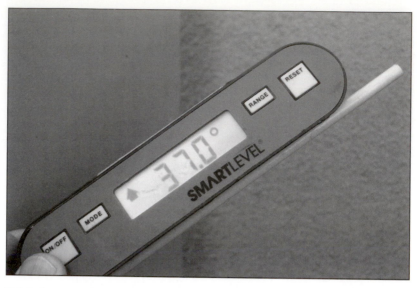

This is the same view as shown in Figure 8.1. A digital inclinometer has been carefully juxtaposed along the trajectory rod and photographed to memorialize the vertical component of the rod. An inexpensive angle finder available in most hardware stores can be used in the same manner.

PROCEDURES—AZIMUTH ANGLE DETERMINATION

A zero-edge protractor, a plumb bob and line, and a means of leveling the protractor are needed to measure the azimuth angle on site. The photographic method in this case is usually much simpler but, as before, the azimuth angle will have to be derived from the photograph or image at a later time. The azimuth angle can also be thought of as the shadow or track that the bullet would have cast on the floor or ground when viewed or illuminated from directly above. In fact this is what one is accomplishing with a photograph taken from directly above a probe extending out of the bullet's path in a wall or other structure. It is this line or path that will ultimately be drawn in a plan view, scale diagram of the scene. If the bullet hole is high on a wall then the azimuth component can be derived from an upward photograph taken with the camera once again against the wall and the lens directly under the probe or string line. The junction of the ceiling and the wall can serve as a subsequent reference line so long as it is in reasonable focus. Figure 8.5 depicts a downward view of the same probe in the previous figures. The junction of the carpeted floor and the wall becomes the reference line for the subsequent derivation of the azimuth angle. This angle derived from this photograph was determined to be

(a)

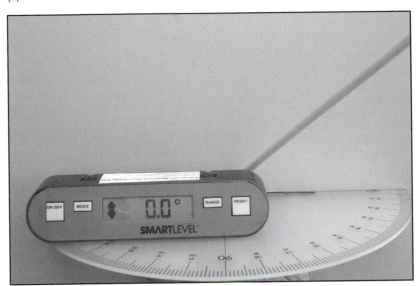

The zero-edge protractor is leveled and butted up against the trajectory rod with the centering mark of the protractor at the appropriate edge of the rod in preparation for the dropping of a plumb bob and line as shown in the next figure.

(b)

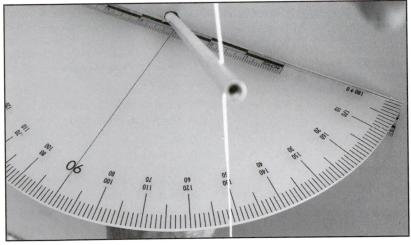

That position along the trajectory rod where the plumb bob just touches the curved edge of the protractor is located and azimuth angle read from the protractor (52° in this illustration).

Figure 8.4
(a) The First of Two Steps for Azimuth Angle Determination with a Zero-Edge Protractor.
(b) The Second of Two Steps for Azimuth Angle Determination with a Zero-Edge Protractor

Figure 8.5

The Photographic Method for Determining the Azimuth Component of a Trajectory

In this view the camera is flush against the wall and directly above the trajectory rod. The junction of the wall and the floor serve as a reference line for this photographic method of azimuth angle measurement.

54° right to left as one faces this wall. The NATO angle would be equal to 90° minus 54°, or 36°.

If, in a real case, the probe and the junction of the wall and floor cannot be kept in reasonable focus, something like a yardstick should be leveled and taped to the wall just below the probe. This will provide the necessary reference line and will be in good focus in the resultant photograph.

Warning: The technique of placing the protractor in contact with the probe, as one would do to measure a vertical angle, will not give the correct result for the azimuth angle. The use of the zero-edge protractor for azimuth angle requires several steps. First, position it flush against the wall and parallel to the floor (i.e., in the horizontal plane) with the central index mark at the same side of the probe along which the angle is to be determined. This arrangement is shown in Figure 8.4a. Now carefully bring in a plumb bob and line to the point that it just touches the probe while also just touching the outer edge of the protractor with the degree marks and read off the azimuth angle (see Figure 8.4b). This is much easier to accomplish with the help of a colleague or assistant. A mitering gauge can also be used in essentially the same way but these are usually more difficult to maintain in a flat orientation while one of the arms is flush with the wall.

The azimuth angle can also determined by dropping a plumb line to the floor and marking this spot at some distance out from the wall and at the end of a probe (e.g., 2 feet or 24 in. out). This can also be done if one is using a laser line directed through the entry and exit holes. The distance from this spot to the spot on the floor immediately below the entry hole represents the hypotenuse of a right triangle that is completed by measuring the direct perpendicular distance to the wall across the surface of the floor. A right angle square is also useful in insuring that this direct line to the wall is truly perpendicular to the surface of the wall. The legs of this right triangle and the use of the tangent function on a pocket calculator will provide the azimuth angle. Using the 24 in. out example for the simulated wall and probe in the previous figures, the direct distance to the wall is found to be a little less than 19 1/2 in. and the distance from this point back over to the point directly below the bullet is found to be 14 1/8 in. A review of the important properties and trigonometric relationships of right triangles is shown in Figure 8.6. The tangent relationship for the legs of a right triangle is the most useful trigonometric function for the reconstruction of shooting incidents. In this example the tangent of the angle we wish to determine is equal to 19.5 divided by 14.1. This division gives the number 1.38. This number is entered on a pocket calculator, the shift key depressed followed by the $\tan^{-1}$ key. An angle of 54° will now appear on the display.

In addition to its use in calculating the azimuth angle as just described, it can also be used for the vertical angle. This is particularly handy when lasers

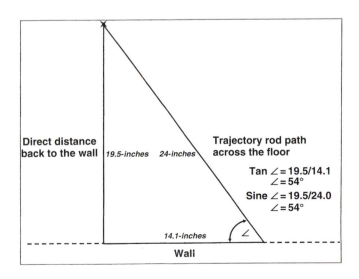

Figure 8.6

Useful Trigonometric Relationships for Right Triangles

are used and have been positioned somewhere in the room so that the laser beam passes through the entrance and exit holes. A plumb bob and line is dropped to the floor at any distance out from the bullet hole in the wall and this spot on the floor marked as in the previous example. The height of the beam above the floor at this point is measured and recorded. The distance from this point back across the floor and under the laser beam to the wall immediately below the bullet hole is measured. The difference between the height of the bullet hole above the floor and the height just measured divided by the distance value represents the tangent of the vertical angle. The arctangent ($\tan^{-1}$) on a pocket calculator will yield the value of the vertical angle. Lasers have another advantage in this application in that they can be set up at much greater and more useful distances out from the bullet hole than probes. Improved accuracy and reliability are gained by dropping a plumb bob to the floor 10 or 15 feet out than the mere 2 feet out used in the previous example for azimuth angle determination. Recall that the photographic method and inclinometer gave values of $-37°$ for the bullet track through the simulated wall. A co-aligned laser attached to the trajectory rod is beamed across the living room and a plumb line and bob passed through the beam at a distance of 5 feet (60 in.) out and across the floor from the wall below the bullet hole. The height above the floor of the laser beam striking the plumb line is found to be 65 in. Since the bullet hole is 20 in. above the floor, the height change over the 5-foot distance across the floor is 65–20 or 45 in. Dividing 45 in. by 60 in. gives 0.75, the arctangent of which is 36.9°. This example is diagramed in Figure 8.7.

Figure 8.7

The Use of the Tangent Function to Calculate the Vertical Component of a Perforating Projectile's Flight Path

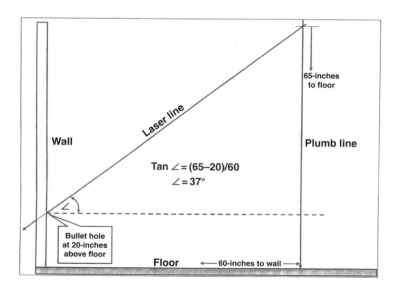

A final description of the shot through the simulated wall cited above might read: "The bullet struck the east wall 20 in. above the floor at a distance of 9.0 feet north of the southeast corner of the room. The vertical component of the bullet's path through the wall was approximately 37° downward (−37°) and the azimuth component was approximately 53° out of the plane of the struck wall with a right-to-left track (SW to NE) as one views this wall."

NON-PERFORATING BULLET PATHS

There are situations where a projectile perforates the first layer or panel of a wall, ceiling, cabinet, etc. but fails to produce an exit hole. Many interior walls are mounted on 2×4 studs (wooded boards) that are secured to a solid block or brick wall. In other situations the bullet may have come to rest in one of these 2×4 studs, the floor inside the wall or other substantial structure after perforating the outer panel or layer. Many of these walls are filled with loose fiberglass insulation. Even in the absence of such insulation it can be difficult to follow the path of the bullet with a small light or probe. For these situations some sort of view window is needed that does not disturb the entry bullet hole or the subsequent impact point(s) or final resting place of the bullet. This can be accomplished with a relatively simple technique and a little bit of effort. After the location of the entry hole is thoroughly measured and documented, an adjacent viewing window can be cut out with a small keyhole saw or a sharp utility knife. Such an opening should be large enough for subsequent good viewing and photography. This window should allow the embedded bullet or downrange impact site to be located. At this point an appropriate probe can be carefully passed through the entry bullet hole and either placed against the embedded bullet, the next bullet hole, or the bullet impact site. Once this is accomplished and documented, one or more of the previous methods can be applied to ascertain the angular components of the projectile's flight. A slight alternative to this approach involves cutting out the area containing the bullet hole itself. Either way, some sort of hole must be cut to recover the projectile and one might as well collect the actual bullet hole in the process. To do this, a pair of strings is pinned to the wall above and to the right and left of the bullet hole. These are brought across the bullet hole and taped at the bottom ends to form an "X" at the center of the hole. Locator marks for the 8 o'clock and 4 o'clock positions of the secured strings are made. This technique is shown in Figure 8.8a. The strings are temporarily removed after which an irregular area containing the bullet hole, on the order of 4–8 in. "square," is cut out of the wall. This excised section is marked for identification and impounded.

Figure 8.8

(a) Preparatory Step for
Bullet Path Tracking
and Bullet Recovery in a
Non-Perforating Strike to
a Wall; (b) Reconstruction
Step for Bullet Path
Tracking in a
Non-Perforating Strike
to a Wall

(a)

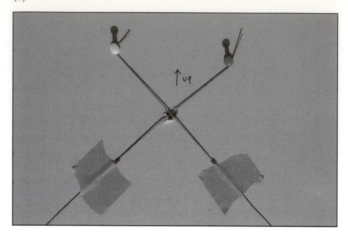

An "X" is temporarily formed across the entry bullet hole with a pair of strings. Black index marks (just above the pieces of masking tape) are added so that the proper position of the strings can be re-established after a section containing the entry bullet hole is cut out. The tape holding the strings is temporarily removed so that they do not interfere with the cutting out of a section of sheetrock containing the bullet hole.

(b)

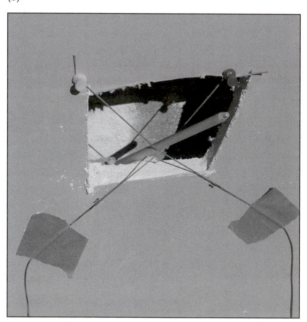

The section of sheetrock containing the entry bullet hole has been removed and the two strings re-positioned with the intersection point representing the former center of the bullet hole. Once the subsequent downrange impact point of the bullet has been located, a laser beam or trajectory rod can be aligned with these two points and the angular components of the bullet's pre-impact flight path determined by one or more of the methods previously described.

The intentional irregularity in its shape comes into play if there is a need to remount the cutout section or as an aid in later remembering its original orientation in the wall. Now the strings are re-positioned using the pair of locator marks and secured as before (see Figure 8.8b). Once the downrange impact point or embedded bullet inside the wall is located, a suitable probe or string line can be positioned in such a way that it passes through the saddle of the "X" and out into the room or area from which the bullet was fired. Once the probe or string line is properly positioned, any of the various techniques previously described can be used to determine the azimuth and vertical components of the projectile's pre-impact flight path. Lasers are especially useful in these situations involving non-perforating bullets as will be seen in the next section of this chapter.

THE USE, ADVANTAGES, AND LIMITATIONS OF LASERS

A variety of small, portable, battery-powered lasers have become available ranging from simple shirt pocket pointers to units with integral inclinometers, theodolites, and optical sighting devices. Laser beams follow straight lines, they have no weight, they do not sag, and they do not and cannot alter the bullet hole in any way. In all applications, lasers can be thought of as an alternative to probes and string lines with some additional advantages. For example, scene investigators can step into and out of the laser beam without disturbing the setup of the laser or the bullet holes through which it passes. If we know where a gunshot victim was struck by a bullet that first perforated a door, window, curtain, etc. and the general path of his gunshot wound, we can position a suitable stand-in for the victim in those orientations that will satisfy the reconstructed flight path of the bullet at the scene. Moveable objects that have been struck or perforated by a bullet (a hollow core door, a plastic patio chair, a box of cereal on a kitchen counter, etc.) can be moved in and out of the laser beam. Another major advantage of lasers is that they cannot alter the physical items that were struck by bullets. For example, a tempered glass side window of a car, a window screen, or a curtain may have been perforated by a bullet. The use of any physical object to attempt a trajectory reconstruction would disturb the item. The use of the laser becomes the method of choice for these delicate materials.

One of the special applications of these devices is their use and value in attempting to locate downrange impact sites and bullets from perforating shots through materials. In these situations an external laser is directed through the center points of the entrance and exit holes. This requires some trial-and-error effort and can be the source of some aggravation. It is best accomplished with the laser mounted on a fully adjustable tripod with a gimbaled head. An alternative is the use of the specially designed trajectory rods that will accept a screw-on laser. However, it is critical that this laser is co-aligned with the trajectory rod before one relies on the projected beam of such a laser. In either arrangement, a piece of white paper or cardstock is then used to track the laser beam as it exits the perforated wall or material. Even though destabilized, deformed, and even tumbling, exiting bullets will follow relatively straight paths over short distances out to as much as 10–20 yd. The reverse application is also very useful. Passing a laser beam through an exit hole and out of the entrance hole allows one to find possible to likely shooting positions in indoor scenes. Another advantage arises in this situation. The shooter or a suitable substitute for the shooter can take various shooting positions in the actual room so that the laser beam passes right up the barrel of the gun or skims across the sights of the gun. This is not easily accomplished with string lines or probes (with the previous exception of the specially designed probes that are threaded to accept a special laser). These positions can be photographed and the distance out from the struck surface as well as the height of the gun can be measured. In outdoor scenes this technique can be quite helpful in locating the source of the shot and aid in the search for additional evidence such as expended cartridge cases. The ultimate system employs two lasers aligned and pointing in opposite directions allowing for downrange projections and back-extrapolations from a single setup.

The situation described in the previous section for non-perforating shot is especially amenable to laser usage. After the location of the entry hole is thoroughly measured and documented, the same choices of viewing window are considered. The laser is subsequently positioned in such a way that the beam passes through the bullet hole or across the saddle of the "X" in the crossed-strings technique depicted in Figure 8.8b and on to the embedded bullet or bullet impact site within the wall. Once the laser is properly positioned, any of the various non-photographic techniques previously described can be used to determine the azimuth and vertical components of the projectile's pre-impact flight path.

Outdoor scenes and situations where there are substantial distances between two or more bullet holes or impact sites caused by the same shot are much more amenable to laser usage than probes or string lines. Consider a situation where there is a bullet hole in a thin wooden fence followed by the bullet embedding itself in the stucco wall of a house 100 feet downrange. Even the distance between a bullet-perforated front door and a subsequent bullet hole at the end of a hallway beyond the door would be very difficult to work with using string lines or probes. The alignment of a two-directional laser on a suitable tripod with the bullet hole in the fence and the impact in the stucco wall would be the method of choice in this situation.

There are a few drawbacks to lasers. The laser beam can be difficult to see in bright light and can only be photographed in very subdued to nearly dark conditions. The special technique required to do this is described in the Appendix of this book under Laser Photography.

SOME THOUGHTS ON THE RECONSTRUCTED ANGLES

The examples described and illustrated here involved perfectly perpendicular walls and flat, level floors. Real scenes are seldom so simple and changes in heights between rooms, the inside floor, and the outside ground must be dealt with. The trueness of walls and floors should be checked at some point with a level or inclinometer and right angle square and any deviations noted. These can be addressed later so long as they are recorded in an understandable way, e.g., the east wall leans away (toward the east) by 1° off true vertical and the floor of the living room slopes down toward the east $-1.0°$ It may be necessary to enlist the help of a surveyor for outdoor scenes with complicated and substantial changes in terrain.

An interior wall made of sheetrock was used as an example in this chapter. It is nearly ideal in that very little to no measurable deflection occurs when bullets perforate this material. Nonetheless, there will still be some uncertainty in our reconstructed trajectories. Some of this arises out of the very short distances between two reference points or out of our inability to identify the true centers of multiple bullet holes, or to perfectly center a probe, string line, or laser line through these holes. There is the human factor of one examiner to another measuring the same angle differently by a degree or two. Other intermediate materials may induce varying amounts of deflection in a totally random manner. Carrying out the reconstructive techniques described in this chapter involves a certain degree of acquired skill. Participants in the author's reconstruction course measuring multiple bullet paths through a simulated wall comparable to that illustrated here obtained values that varied by about $\pm5°$ around the average for the group. From this practical exercise, $\pm5°$ is a reasonable uncertainty level for any numerical value for most bullet paths derived from actual scene work. The end result is that while we may draw a single line across the floor of a plan view of a room representing the azimuth angle and a similar line for the vertical component in a profile view of the room, we should be prepared to discuss or illustrate these components as a cone rather than a line. This cone is formed from our best estimate of the bullet's flight path plus and minus our uncertainty limits. A more critical assessment of any particular perforated material and the uncertainty of our measuring technique can always be evaluated through empirical testing if the need arises.

Aside from the geometric constraints of any particular shooting scene, the more interesting and useful angular component of a bullet's flight path is the vertical component. In most scenes, steeply rising or descending angles quickly place the firearm very close to the struck object or surface. Consider a bullet hole in an interior wall that is only 24 in. above the floor. The track is found to be upward at +45°. The tangent of 45° is 1 which means the back-extrapolation of this bullet's path intercepts the floor 24 in. from the point directly under the bullet hole in the wall. The gun in this case had to be held near or essentially on the floor to fit within the geometry of the scene. Such a shot could not have been made by a standing subject. Steep downward angles accomplish the same thing. If the entry bullet hole is inside a confined location, back-extrapolation of such angles quickly results in very awkward or impossible gun positions for even a standing shooter in the absence of ladders, stairways, or other elevated sites. The example used in this chapter with the −37° vertical angle for a bullet hole 20 in. above the floor quickly leads back to gun heights on the order of 5–6 feet at similar distances from the wall. A simple calculation for the height of a shot from a doorway located 12 feet away from the bullet-struck wall produces a gun height of over 10 feet in a room that has an 8-foot ceiling. The same situation arises in an outdoor shooting scene. One must consider the following with steep vertical angles: a shot from very close range, a shot from an elevated location (such as a multistory building or a hill), or a bullet on the descending path of a very long trajectory. This latter situation will be discussed in Chapter 14. Relatively flat vertical angles on the other hand and the absence of gunshot residues around the bullet hole often leave open a considerable range of distances from which the shot could have occurred.

TECHNIQUES, TOOLS, AND SUPPLIES USED IN TRAJECTORY RECONSTRUCTION

There are presently very sophisticated pieces of equipment and instrumentation available to make many, perhaps most, of the measurements described in this chapter. No doubt more will appear in the years to come. But the purpose of this chapter was to provide the basic principles and concepts for short-range trajectory reconstruction. Although simple, these techniques work; they are relatively easy to understand and illustrate and they utilize inexpensive devices and equipment. Figure 8.9 provides an assortment of tools, devices, and materials that have been used by the author in the reconstruction of bullet trajectories. There is considerable redundancy displayed in this figure and it is often the case that only two or three of the items shown need actually be used. Yet it is difficult to resist an occasional trip through the tool section of a large hardware store or a yearly encounter with a representative of one of the major law

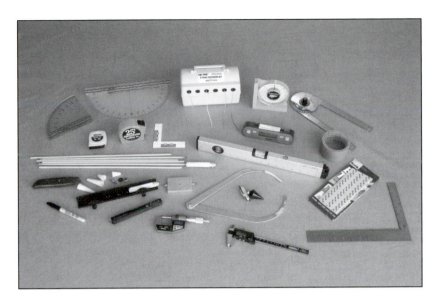

Figure 8.9

Basic Tools for Use in Trajectory Reconstruction

Starting in the upper left: zero-edge half protractor, zero-edge protractor, dispenser of colored strings, angle finder, mitering gage, tape measures, ABFO scale, digital inclinometer, bubble level with built-in laser, masking tape, utility knife, trajectory rods (one with laser attached), utility knife, centering cones, tripod mount for holding two opposed lasers, tripod mount for 1 to 2 threaded lasers, calipers, plumb bob and line, re-useable adhesive, marking pen, laser pointer, digital micrometer, digital calipers, right angle square.

enforcement supply catalogs. These invariably result in the addition of another useful gadget to the author's field kit.

SUMMARY AND CONCLUDING COMMENTS

The use of photography for documenting, illustrating, and subsequently measuring the vertical and azimuth components of a bullet's track through walls and other objects is highly desirable and strongly recommended as a cross-check of any other techniques employed for these purposes. Properly positioned and executed, photographs are relatively easy to accomplish if the simple steps described in this chapter are followed. There should be no rush to recover a projectile inside a wall or other similar space and there is no justification in routing out a bullet hole in an effort to do so. This bullet hole, its shape, its location, its chemistry, and the path of the bullet that caused it may be of greater importance than the bullet itself and they are of critical importance in reconstructing the shooting. The bullet will be waiting for you at the end of your efforts. Time is on your side. The actual incident probably took place in a few seconds. You have hours and perhaps days if it becomes necessary. A deliberate and thoughtful assessment of each shooting scene followed by a strategy tailored to the particular situation is required. The scene should only be released when all the data and measurements necessary to reconstruct the flight path(s) of the shot(s) have been gathered and all the appropriate photographs have been taken to the point that a mock-up of the room or scene could be constructed if necessary in the event some future issue or hypothetical scenario is raised.

REFERENCES AND FURTHER READING

Cashman, P.J., "Projectile Entry Angle Determination," *JFS* 31:1 (January 1986) pp. 86–91.

Garrison, D.H., Jr., "Reconstructing Drive-By Shootings from Ejected Cartridge Case Location," *AFTE Journal* 25:1 (January 1993) pp. 15–20.

Garrison, D.H., Jr., "Reconstructing Bullet Paths with Unfixed Intermediate Targets," *AFTE Journal* 27:1 (January 1995) pp. 45–48.

Haag, L.C., "The Measurement of Bullet Deflection by Intervening Objects and the Study of Bullet Behavior after Impact," *CAC Newsletter* (January 1988).

Haag, L.C., "Portable Laser-Theodolite System for Use in Shooting Scene Reconstruction," *AFTE Journal* 23:1 (January 1991).

Haag, L.C. and M.G. Haag, "Shooting Scene Reconstruction Course and Manual," Gunsite Training Facility (October 2002, November 2004).

Hueske, E.E., "Lateral Angle Determination for Bullet Holes in Windshields," *SWAFS Journal* 27:1 (February 2005) pp. 39–42.

Stone, R.S., "Calculation of Trajectory Angles using a Line Level," *AFTE Journal* 25:1 (January 1993) pp. 21–24.

Trahin, J.L., "Bullet Trajectory Analysis," *AFTE Journal* 19:2 (April 1987).

CHAPTER 9

THE SHOOTING OF MOTOR VEHICLES

Automobiles (and most other vehicles to include trucks, trailers, and mobile homes) present special reconstructive problems. Their construction and makeup present a wide variety of surfaces susceptible to bullet impact. These include frangible materials such as the laminated glass in windshields and tempered glass in side and rear windows, malleable surfaces (sheet metal), heavy, unyielding metal structures (support members, the frame, engine, axels, etc.), various composites, soft materials (plastic, fabric, rubber moldings, insulation), tires, painted and unpainted surfaces. Each of these materials responds in different ways to impacting projectiles. The varied shapes and compound curvatures common in automobiles add another complicating factor when bullets strike such areas.

Vehicles are often in motion and/or are not on level surfaces when struck. The surface upon which they come to rest may differ substantially from the surface where they were struck by one or more projectiles. In some cases it may not be possible to determine the exact location or orientation of a vehicle when it was struck by gunfire. The positions of doors and/or windows may change during or after a shooting incident. Other changes such as the loss of shattered glass from struck side windows and/or rear windows may occur after the shooting portion of the incident. Cracks in one or both layers of laminated windshield glass often continue to propagate during the post-incident interval due to stresses, movement of the vehicle, or other forces and events. Tires may deflate slowly after being shot and go flat at some location other than that where the shooting incident took place. Finally, vehicles are often moved from the shooting scene before they can be examined for the purpose of shooting reconstruction. In those cases where the vehicle was stationary when shot then moved to an examination bay, a thorough review of any and all photographs and videotapes of the vehicle at the scene of the shooting may allow questions to be answered regarding its orientation at the time of the shooting or the location of bullet holes in tempered glass that has since fallen out.

These matters may seem to make the whole question of vehicle shooting reconstruction a daunting to even futile endeavor. There is much that can be done however as long as the examiner keeps these matters in mind as the work progresses and he or she adheres to the fundamental principles of reconstruction previously articulated in Chapter 8.

VEHICLES AT A SCENE

It is unlikely that a shot vehicle is going to be left at a scene sufficiently long to carry out *all* of the measurements that one might desire. For this reason its location and orientation at the scene should be documented by measurements and photographs. These photographs should include straight-on front, rear, and side views and should include a known frame of reference such as a plumb line or surveyor's rod positioned in the vertical plane so any tilt or listing of the vehicle at the scene can later be determined. This is of particular importance if the vehicle was shot at this location. Alternatively, an inclinometer can be placed on selected sites such as the edge of the front and back windows, the center of the hood and truck, and the particular angles noted and each setup photographed. If there has been a failure to do this, it may be possible to locate one or more fixed references in the scene photographs. In this situation one will need to return to the scene, measure the actual vertical or horizontal angle formed by the surrogate reference feature (e.g., the edge of a building, a telephone pole) and use this feature to estimate any desired angles for the evidence vehicle in the scene photographs. During the original scene processing, the positions of the tires should be marked in some way so that they can later be repositioned in the examination bay or other suitable area. Stripes of bright fluorescent spray paint on the sidewalls of each tire that continue down onto the roadway are useful for this purpose. These will usually survive on paved surfaces for days to months should the need arise to return and re-position the vehicle at the scene. Data regarding the roadway, parking lot, or any other surface occupied by the vehicle when shot should be gathered at some point before any significant changes (such as re-surfacing of the street or parking lot) have taken place. This may require the assistance of a surveyor or someone from the streets or highway department or it may be as simple as taking a few inclinometer measurements or checking several representative areas with a carpenter's level. In situations where a vehicle was in motion, the assistance of a qualified expert in accident reconstruction may be appropriate to address questions about speed, acceleration, or deceleration. In a few instances the author has requested the videotaping of the actual vehicle driven through the shooting zone at differing speeds so the natural movements of the vehicle could be later studied in a frame-by-frame manner.

If an injured or deceased subject has been removed from any seat that is adjustable, its position should *not* be altered. Prompt inquiry of any paramedics involved should be made regarding any changes they may have made in the seat or seatback positions. These concerns also apply to any adjustable windows in the vehicle. Specific photographs showing the seat and window positions should be included in the scene photographs. It is suggested that strips of tamper-proof evidence tape or masking tape be placed over these adjustments and placed in such a way that any changes in the seat or window positions would disturb or break the tape.

Tempered glass rear and side windows that have been shattered but still largely in place must be thoroughly photographed with a suitable scale so the approximate center point of any bullets holes can be determined. Concerted efforts should be made to secure the glass in place as much as possible as soon as photo documentation has been completed. Both sides of the shattered glass should be reinforced with strips of wide, clear tape. Other scene investigators have used sheets of clear, adhesive film for this purpose. Neither technique will guarantee that the shattered window will survive transport from the scene to the impound area or laboratory, consequently scene documentation is of critical importance.

There are several determinations that need to be made in those instances where the tempered glass in side windows appears to be totally missing as a result of one or more projectile strikes. A search of the interior of the vehicle and the scene should be made in an effort to find any portions of the window that have surviving radial fractures and flaking near one end (from cone fracturing). These are associated with the first strike to the glass. The conical spalling or flaking effect will allow the direction of fire to be established after the laboratory determines which surface represents the exterior or interior side of the glass. The rubber molding at the top of the window frame should also be carefully examined for any trapped pieces of tempered glass. If present, these establish the window as being in the up (closed) position when the window was shattered. Any such glass fragments in this location should be photographed then carefully removed and marked in some way that shows which surface was the exterior or interior side of the fragment. In the absence of such fragments some further effort will be necessary. Before the vehicle is released and after any other reconstructive efforts have been completed on the door that contained the shattered window, the interior door panel should be removed or loosened sufficiently to see and photograph the position of the window carrier. There will invariably be some remaining glass in the carrier. The interior surface of several representative pieces of this glass should be marked in some way then removed from the carrier and impounded for submission to the laboratory. Without these reference fragments of glass, they may be unable to determine the direction of fire. At this point the reader might find the "Vehicle Data" checklist included in the Appendix useful as a guide and reminder for documenting matters that may not seem important at the onset of an investigation but that have a habit of becoming important later. Examples include the position of the driver's seat, the position(s) of the side windows, the position of the parking brake, and the position of the shifting lever.

FRAMES OF REFERENCE AND "SQUARING THE VEHICLE"

The usual frames of reference outlined in the preceding chapter are not particularly useful with moveable objects such as vehicles. Various fixed reference points on the vehicle itself can and should be used. These include trim, moldings, hood centerlines, and door edges. For example, an entry hole in the driver's door might be described as "Eight inches back (to the rear) from the front edge of the driver's door and 10 in. down from the lower edge of the driver's window sill." These natural reference lines in a vehicle should be integrated into any subsequent photographs depicting reconstructed projectile paths. At some point the vehicle's height above ground level should also be measured but it must be recognized that this can change with vehicle loading and movement. Such potential changes can be evaluated later. Nonetheless the location of a bullet hole or impact site described by identifiable and fixed reference points can be re-established at any future time on the repaired vehicle or a comparable vehicle of the same make, model, and year of manufacturer and equipped with the same size of tires.

Every vehicle can be divided into natural planes such as the front-to-rear axis, a vertical axis, and the plane across its width. These can be useful in describing the general path of a striking projectile with the vehicle viewed as a box or, more correctly, a three-dimensional rectangle. A technique called "boxing" or "squaring" the vehicle naturally arises out of the vehicle-as-a-3D rectangle. This technique is used when examining a bullet-struck vehicle in a subsequent controlled environment and on a smooth, level surface. The vehicle is first leveled on a flat, level surface that provides sufficient working space on all four sides. Housekeeping matters such as the height of certain features (e.g., the top of the roof, the driver's window sill, the driver's side rocker panel) can be measured at any time although they may not be particularly relevant. Likewise any height changes due to driver and/or passenger weights as well as any roll or pitch changes in the vehicle can be determined with the vehicle on this flat and level reference surface.

Let us imagine a vehicle that has projectile strikes entering the front and the driver's side of the car. Two sturdy tripods with a taught string line between them are positioned a short distance forward and aft of the car so that the string just touches and parallels the driver's side of the vehicle. Two more tripods with a taught string line are positioned across the front of the car and at a right angle to the previous string line. These tripods are also moved toward the front of the car until the string line just touches or is immediately over the forward-most feature of the car. It may be desirable to do the same thing at the rear of the vehicle if for no other reason than to establish the size of the "box" that will represent the car. We now have a coordinate system with which we can define and measure azimuth angles as well as describe impact sites on the vehicle. The reference point (RP) in this example will be the intersection of the two string lines at the left-front of the vehicle. The earlier example of a bullet hole that was found to be 8 in. back from the front edge of the driver's door and 10 in. down from the lower edge of the window sill now has an additional identity in this new coordinate system. It is determined to be 60 in. to the rear of the RP and 32 in. above the flat, level surface upon which the vehicle is standing. A properly inserted and positioned trajectory rod through the driver's door is photographed from directly above the intersection of the probe and the string line. From this the azimuth angle is determined. The vertical component is measured as described in the previous chapter using the photographic method with a plumb line or with an inclinometer, or both. This same probe with an attached and co-aligned laser can be used to evaluate possible, improbable, and even impossible gun heights and to re-check both azimuth and vertical angle components of the bullet's approach to the "box."

Shots into the front of the vehicle, into the hood, and into the windshield are referenced and measured relative to the line across the front of the vehicle. This "box" with the trajectories for each shot can now be integrated with diagrams and various views of the actual scene. The "box" can be pitched up or down, rolled to one side or the other, and the trajectories go with it. This is best done with a computer program and model of the scene into which the pertinent information regarding the vehicle size (the "box" size) and the trajectories associated with the vehicle can be illustrated in any view of interest.

PERFORATING PROJECTILE STRIKES TO VEHICLES

The general properties and behavior of bullets striking sheet metal, glass, rubber, and plastic have been presented in Chapters 3 and 6 and now we have all of these materials and more in motor vehicles. They are also arranged in varied shapes and ever-changing angles. It must be recognized that this can make the impact sites from two shots fired from the same position and along very similar flight paths look very different. For example, a grazing strike to the hood of a car and a perforating shot into the front fender could be the consequence of two shots fired from the same location that only differ by 6 in. in the heights of their pre-impact flight paths. These sudden changes in shape and contour along with variations in underlying support structures can also affect the amount and nature of any deflection experienced by the bullet. These complicating factors may once again create the need for some subsequent empirical testing on an exemplar vehicle.

The tracking of a perforating projectile can be difficult because of the various internal structures and components in vehicles. It may be necessary to cut viewing windows in sheet metal surfaces and ultimately disassemble door panels to make sure that there was not some internal structure hit that caused a change in direction of the bullet. Consider the situation illustrated in Figure 9.1a where two entry bullet holes are found in the exterior of the

Figure 9.1

(a) Cross-Sectional View of Probes Passed through Two Perforating Shots to the Driver's Door of a Vehicle. (b) Cross-Sectional View of the Actual Paths of Two Perforating Shots to the Driver's Door of a Vehicle

(a)

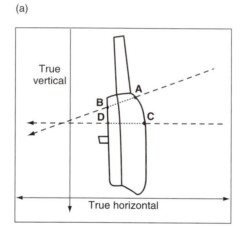

(b)

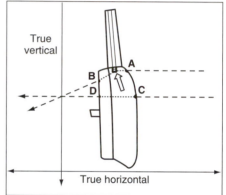

This drawing illustrates the danger of assuming an entrance and an exit hole represent a straight line. Probes passed through A-B and C-D may, at first, seem reasonable until a careful inspection of the internal components of the car door are undertaken. When this is done, it is found that the bullet that produced entry hole "A" struck the metal carrier for the driver's window after which it was defected downward and exited at "B" (see Figure 9.1b). Note the substantial differences in the back-extrapolated vertical angles for these two shots into the car door.

This is a cross section of the same door as shown in Figure 9.1a. The raised driver's window and carrier have been added to the figure after a careful inspection of the internal components revealed that the bullet that produced entry hole "A" struck the metal carrier for the driver's window (arrow) after which it was defected downward and exited at "B." The correct representation of the pre-impact flight path for strike 'A' is represented by the path from the entry hole ("A") to the impact site in the window carrier. The reader should also note the substantial differences in the plane of the sheet metal at entry points "A" and "C" relative to each bullet's flight path. This can easily produce entry holes of different shapes even though the multiple shots strike the vehicle along near-identical flight paths.

The nature of the impact damage to the bullets recovered from the driver's body should allow them to be associated with their respective impact sites in the driver's door.

driver's door and two exit holes are subsequently noted in the interior door panel. The roundness or out-of-roundness of bullet holes in automotive sheet only provides a very general notion of incident angle (see Figure 9.2), so passing trajectory rods through these holes and along the tracks illustrated by lines A-B and C-D might seem reasonable and appropriate. This results in widely divergent azimuth and vertical angles between these two shots yet the police officer-shooter claims that he fired two rapid shots from a single location and stance about 10 feet out from the side of the vehicle. The officer goes on to state that he did this when the driver made a rapid, non-compliant movement as he approached the stationary vehicle. If these divergent angles are true representations of the two bullets' flight paths, then the shooting officer either moved between these two shots or the vehicle moved or some combination of

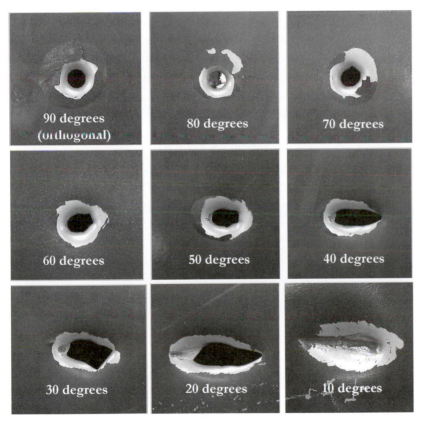

Figure 9.2
The Appearance of Entry Bullet Holes in Automotive Sheet Metal at Various Incident Angles

90 degrees (orthogonal)

80 degrees

70 degrees

60 degrees

50 degrees

40 degrees

30 degrees

20 degrees

10 degrees

These bullet holes in automotive sheet metal were produced with 147 gr. Federal Hydra-Shok bullets fired at the indicated incident angles with a Model 39 S&W 9 mm pistol. As can be seen, the shape of the holes between 70° and 90° are essentially round making the location of subsequent impact points within the vehicle of critical importance in defining the pre-impact flight path of the bullet. (Photos and test shots, courtesy of Michael Haag.)

both. A careful disassembly and inspection of the internal components of this door later reveals that a heavy cross member was struck by the bullet that entered at point A and the true path of this shot is as depicted in Figure 9.1b. The lessons here are several: two points do not necessarily represent a straight line and at some point the examiner needs to look inside complicated vehicle structures such as doors that have been penetrated or perforated by gunfire.

Sometimes the troublesome internal components in doors actually carry a benefit. The positions of important moveable components such as the operating rod or lever for the door handle, the door lock or the window carrier may be struck by the bullet. Careful inspection of this damage and its spatial relationship to the responsible bullet's path may answer questions such as: *was the driver in the process of opening the door when the shot was fired?* Or, *what position was the shattered window in when the shot occurred?*

Other moveable components such as the steering column, the steering wheel, tires, wheels, and drive shaft may provide useful and otherwise overlooked clues when struck by bullets. The task of trajectory reconstruction is relatively easy if one is so lucky to be dealing with a vehicle that was shot while at rest and there is ample time available to carefully reconstruct the trajectories of the bullet strikes to such a vehicle. This is accomplished by one or more of the techniques described in the previous chapter. These reconstructed trajectories can be related to the usual vertical and horizontal frames of reference and the associated scene. Included and excluded positions for the shots naturally flow from this situation. With vehicles that have been moved from the scene or that were moving when shot the work becomes more difficult. This is the more common situation for which the "squaring the car" technique was developed.

PENETRATING PROJECTILE STRIKES TO VEHICLES

It is often the case that bullets enter a motor vehicle and do not exit. Sheet metal or one panel of glass is relatively easy to perforate but there are numerous major heavy metal structures inside doors and other locations that will defeat many small arms projectiles, particularly pistol bullets. This is quite like the example in the previous chapter of a bullet that enters a wall but that does not exit. After the appropriate measurement and documentation of the location of this bullet hole, a borescope with an integral illumination system might seem useful in locating the next impact site or even the bullet itself. But we will still need to properly position a trajectory rod or laser and the bullet must ultimately be recovered if at all possible. The author prefers the "viewing window" approach for accomplishing both trajectory reconstruction and bullet recovery. This is somewhat labor-intensive but the pulling off of door panels and other similar components can disturb the geometry of certain bullet-struck sites within them so that one cannot be certain that they are in the same position when the panel is replaced and reconstruction efforts are renewed. A high speed electric Dremel tool is used to slowly and carefully cut a combined viewing window and access panel in sheet metal. Just as before, this can be done adjacent to the bullet hole or it can include the bullet hole itself. If the struck surface possesses significant curvature to the degree that crossed strings would *not* occupy the position of the bullet hole once the panel is cut, an alternative is recommended if the excised panel includes the bullet hole. The excised panel is placed on a copy machine with the area of the bullet hole in contact with the copier's platform. Clear plastic transparency film is used rather than paper. The result will be a plastic sheet with the outline of the bullet hole visible as well as the outline of the edges of the panel. This plastic sheet can now be matched back to the excised area and taped in place. An externally mounted laser is positioned and adjusted until the beam passes through the *copy* of the bullet hole and onto the internal impact site or the embedded bullet. A digital incli-nometer can be placed on the laser (if it is suitably constructed) or carefully shuttled in under the laser beam to read off the vertical component of the angle. The azimuth angle can likewise be determined in several ways, all of which are virtually the same as described in the previous chapter. The examiner need only to apply a little forethought and planning to measure or calculate the angular components of the bullet's pre-impact flight path relative to the appropriate axis of the "box" representing the car.

RICOCHET MARKS, GRAZE MARKS, AND NON-PENETRATING STRIKES

Ricochet and graze marks can occur on any surface of a vehicle and can be exceedingly difficult to locate in some instances. A soft lead bullet traversing the windshield can leave a "Chisum trail" without cracking the glass. A graze mark across the hood may only produce a one- to two-inch blemish in the paint. Further evaluation will require a test for copper and lead by the transfer tech-nique. Direct strikes into very heavy structures such as thick steel bumpers, wheels and axels may result in heavy fragmentation of the bullet. In these cases patterns of copper and lead transfers and especially lead splash can establish the mark as bullet-caused. The previous description of lead splash providing useful information as to direction must be approached very cautiously here. Any curvature in the struck surface stands to play a major role in the direction taken by spattered and vaporized lead.

SUMMARY

Vehicles present special challenges insofar as recognizing, identifying, and reconstructing projectile damage to them. The wide variety of materials and the complex shapes and surfaces represented in vehicles greatly increase the difficulty in reconstructing the direction or sequence of projectile strikes. Shots fired inside dwellings and buildings have the advantage that struck surfaces are usually static and often consist of relatively flat surfaces that make natural planes of reference. Vehicles, on the other hand, are often moving when struck by gunfire or are moved to some other location shortly after being struck. All of these factors require more diligence and effort during the examination of vehicles. The technique of squaring the vehicle while on a flat and level surface provides a useful frame of reference that can be used for back-extrapolation to the scene. The behavior of most of the materials represented in vehicles and their interaction with projectiles has been covered in Chapter 6. The pertinent ballistic phenomena described in that chapter were sheet metal behavior during bullet penetration and perforation, pinch points, lead-in marks, fracture lines in painted metal surfaces, laminated and tempered glass, sequence of shots through tempered glass, bullet holes in tires, and plastic bed lines in trucks.

REFERENCES AND FURTHER READING

Garrison, D.H., Jr., "Examining Auto Body Penetration in the Reconstruction of Vehicle Shootings," *AFTE Journal* 27:3 (July 1995), pp. 209–212.

Garrison, D.H., Jr., "Crown & Bank: Road Structure as it Affects Bullet Path Angles in Vehicle Shootings," *AFTE Journal* 30:1 (Winter 1998), pp. 89–93.

Haag, L.C., "Bullet Penetration and Perforation of Sheet Metal," *AFTE Journal* 29:4 (Fall 1997).

Hueske, E.E., "Lateral Angle Determination for Bullet Holes in Windshields," *SWAFS Journal* 27:1 (February 2005) pp. 39–42.

Lattig, K.N., "The Determination of the Point of Origin of Shots Fired into a Moving Vehicle," *AFTE Journal* 23:1 (January 1991).

Mitosinka, G.T., "A Technique for Determining and Illustrating the Trajectory of Bullets," *Jr. For. Sci. Society* 11:55 (1971).

Nennstiel, R., "Forensic Aspects of Bullet Penetration of Thin Metal Sheets," *AFTE Journal* 18:2 (April 1986).

Prendergast, J.M., "Determination of Bullet Impact Position from the Examination of Fractured Automobile Safety Glass," *AFTE Journal* 26:2 (April 1994), pp. 107–118.

Salziger, B., "Shots Fired at a Motor Vehicle in Motion," *AFTE Journal* 31:3 (Summer 1999).

CHAPTER 10

BULLET TRACK ("TRAJECTORY") DETERMINATIONS IN GUNSHOT VICTIMS

INTRODUCTORY COMMENTS

As previously pointed out, the term "trajectory" rightfully belongs to that aspect of exterior ballistics that deals with bullets following long distance flight paths through the atmosphere. Having said this, the reader will be regularly confronted with investigators, lawyers, and even pathologists who insist on referring to wound tracks and the paths taken by bullets in gunshot victims as "trajectories." The preferred phrases are *wound track* or *wound path* but no doubt we will continue to hear them described as trajectories.

Projectiles penetrating or perforating bodies travel over *very* short distances (compared to the previously described shots in the atmosphere) and *may* follow straight paths but since they are not moving through air, they may also deviate as they pass through tissue, organs, or other structures. Impacts with bone also increase the likelihood of deflection. This is due to the varying densities and strength of such tissues or organs and the fact that bullets cannot maintain their spin stabilization in media other than air. Elongated bullets such as most rifle bullets, because of their aft centers of gravity, are more susceptible to deviation from their initial path when penetrating tissue than short, nearly round pistol bullets. Pistol bullets tend to follow straight paths in tissue unless destabilized prior to impact with the victim or deformed in some asymmetrical manner during tissue penetration (see Figures 10.1 and 10.2).

It should also be realized that various structures and organs inside the body are not absolutely fixed in their positions. The course taken by a bullet through some of these organs may have been slightly different in life than they later appear during the autopsy. Additionally, a body at the moment it sustains a gunshot wound, may be in some twisted or bent position only to be straightened

Figure 10.1

Paths of Contemporary Bullets in a Tissue Simulant

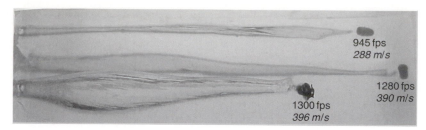

These three 40 gr., .22-caliber LRN bullets struck this block of MBM tissue simulant at the impact velocities shown in the photograph. They have all followed relatively straight paths in this homogeneous medium. It should be noted that the fastest bullet did *not* achieve the deepest penetration. This is because this bullet expanded as it entered the tissue simulant thereby presenting a greater cross-sectional area to the medium. Areas of bullet expansion and yaw are shown by diameter increases in the "wound" tracks.

Figure 10.2

Path of an Unstable Bullet in a Tissue Simulant

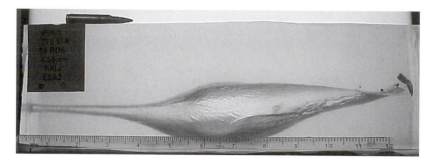

The 62 gr., .22-caliber (5.56 mm) SS109 bullet is a representative example of a bullet that quickly yaws and looses its pre-impact stability as it penetrates tissue or tissue simulants. This bullet penetrated the medium in a nose-first attitude for the first 3 in. then achieved maximum yaw after about 6.5 in. of penetration and came to rest after 13 in. of penetration. It also deviated from its initial course by about 10°. This is because of its long, pointed shape and aft center of gravity. This particular bullet was fired into this block of MBM tissue simulant from a distance of 100 yd (91 meters) and impacted the medium with an impact velocity of 2688 fps (819 m/s). Bullets previously destabilized due to ricochet or impact with some intermediate object usually achieve less penetration and often followed deviated paths in tissue and tissue simulants.

out later on the autopsy table where the tracking and description of wound paths takes place.

Many forensic pathologists routinely prepare anatomical diagrams that depict wound paths produced by projectiles. These diagrams, by convention, show the body in an erect, standing position with the hands fully open and the palms turned out or forward. As a consequence, many individuals attempt to apply the descriptions and illustrations of gunshot wounds in this position to the moment the gunshot wounds were sustained at the actual shooting scene.

This, of course, can give rise to some strange and erroneous body positions at the scene and at the moment the decedent was shot.

Gunshot wounds are customarily numbered with the top of the head used as a starting point. By this method gunshot wound number one (GSW-1) would be nearest to the top of the head. Although there will be a warning somewhere in the pathologist's report that these numbers are *not* an indication of wound sequence, they are occasionally taken as such. There are occasions where the pathologist can sequence certain gunshots but this finding will be specifically described in the autopsy report. An excellent example arises from the "crack rule" or "T" test described in Chapter 6 for shots through plate glass. Bone behaves in the same manner as single strength glass so multiple shots to the head can often be sequenced in the same way. Radial fractures from a second shot will be stopped when the intercept fractures from a previous shot. Because gunshot wounds must often be integrated with other "trajectory" information from the crime scene, there is a clear need to have some insight into the limitations and uncertainties associated with wound paths due to gunshots. The pathologist's information and observations regarding each gunshot wound is the starting point but not necessarily the ending point. It may be necessary to later follow up and discuss certain aspects of one or more gunshot wounds with the pathologist before his or her findings are incorporated into the reconstructive efforts at the scene. Finally, there is considerable variation among forensic pathologists regarding their knowledge of wound ballistics to the point that it may be desirable to confer with a pathologist having special training and knowledge in this area. Issues relating to penetration depths for a particular bullet, fragmentation characteristics, wound path deviation, and incapacitation due to specific gunshot wounds are all examples.

ENTRY AND RE-ENTRY WOUNDS

These wounds can be of immense help in determining what did and did not occur during a shooting incident. Consider a situation where a policeman's bullet entered the dorsal side of the left hand and exited the wrist of a decedent and then enters the sternum slightly in yaw where it follows an essentially straight path described as front to back and slightly downward. The officer claims that the subject was holding a rifle up to his right shoulder in a shooting configuration and pointed towards him when he fired multiple shots. The wife of the decedent claims that her husband had dropped the rifle and had raised his hands above his head (in a surrender position) when the officer fired the shots. The value of the autopsy findings is obvious in this case although examining the rifle for spattered bio-matter (from the hand/wrist wound) and any bullet damage might be of additional value in evaluating these two divergent accounts.

SHORED GUNSHOT WOUNDS

These wounds are of special reconstructive interest and value. The appearance of this term in an autopsy report should always attract the reader's attention. The term "shored" has been used to describe a situation where the skin at the wound site was reinforced in some way at the moment the projectile produced the wound. There are shored *entrance* wounds as well as shored *exit* wounds. A shored entrance wound is typically associated with perforating gunshot wounds to the upper arm with the arm in contact with the chest. Both the exit in the arm and the re-entry wound to the chest display atypical but characteristic features to the experienced pathologist. The skin around the margin of the exit site in the arm is shored by the skin of the chest. The skin around the re-entry site in the chest receives a slap from the shored skin at the exit site in this unique configuration. The position and alignment of these wounds have obvious reconstructive value particularly where the position of the struck arm is in dispute.

Shored *exit* wounds are more common and also arise in situations where the skin around the exit site is abraded at the moment the bullet stretches and breeches the skin. A subject lying on the ground or against a wall when shot and with the bullet's exit site supported by the surface will sustain a shored exit wound. A shored exit can even be produced by the shoring effect of a car's seatbelt across the exit wound. Shoring of an exit wound can also be the result of tight supportive garments and clothing such as a belt or girdle but these potential sources can be evaluated from a subsequent inspection of the decedent's clothing. Substantial shoring combined with a bullet that barely has the remaining energy to perforate the skin may result in the bullet being retained at the exit site. The author once received a bullet removed from a split in the skin at a shored exit site in the left cheek of a decedent. This bullet was found to contain embedded asphalt and mineral particles along with the expected blood and bone particles. The decedent was found on an asphalt parking lot next to an aluminum baseball bat that the shooter said was being swung at him when he defended himself by shooting the subject.

A witness described the incident quite differently. According to this witness, the shooter was confronted by the decedent who had threatened the shooter with the baseball bat. The shooter produced a gun whereupon the subject tried to run but fell instead at which time the shooter fired a shot into his head. This bullet, as recovered at autopsy, is shown in Figure 10.3. A large bone chip *and* the black asphaltic material can be seen embedded in this bullet. The former is no surprise after the bullet's journey through the decedent's skull but the asphalt and mineral inclusions in this bullet in combination with the shored "exit" wound show the witness's account to be correct and the shooter's account to be false.

Figure 10.3

Bullet Recovered from a Shored Gunshot Wound (see Color Plate 9)

This photograph shows the heavily damaged nose of a .38-caliber lead bullet resting on the medical examiner's "red tag" immediately following its removal from the partial exit wound in the decedent's head. The large white particle is a bone chip. The embedded black material at several locations in the gray lead of the bullet and on the bone chip is asphalt acquired when bullet attempted to exit the victim but was shored by the exit site resting against asphalt at the scene of this homicide.

PROJECTILE PATH DETERMINATIONS FOR GUNSHOT WOUNDS

A number of sources and methods are available to estimate a bullet's path in a gunshot victim. These are as follows:

MEDICAL EXAMINER'S DESCRIPTION

Many forensic pathologists, after tracking the course of a projectile, will provide their assessment of its angular components (in the vertical and horizontal planes) relative to the body in the anatomically erect position, e.g., "After entering the sternum, the bullet proceeded front to back, downward 20° and left to right 45° coming to rest just under the skin in the mid-back." The pathologist's tracking process can take several forms—visual assessment during the internal examination, insertion of a probe in the wound track followed by an estimate or even measurement of the track relative to the planes of the body.

Note: While one might reasonably assume that pathologists would look for evidence of projectile deviation or deflection, it may be prudent to contact the pathologist who performed the autopsy and inquire into his/her methodology where angular components are described in the autopsy report. Inquiry should also be made into the level of certainty that the pathologist is prepared to give to each such wound track particularly if numerical values have been stated.

Most forensic pathologists are loath to commit to numerical angles for good reason. Consequently their autopsy reports will often describe the wound track in much more general and subjective terms, "After entering the sternum, the bullet proceeded from front to back, slightly downward and left to right coming to rest just under the skin in the mid-back."

In this later situation and in the absence of a diagram illustrating the wound paths, it may be desirable to contact the pathologist and explain the need for a more detailed description or illustration of the wound track; that he or she is the proper source of this information and that it is needed to assist in the reconstruction of the shooting scene. Offer to provide the pathologist with one or more blank anatomical diagrams and ask him or her to draw in the best estimate of the path of the projectile for each gunshot wound.

It is important to recognize that the best of measurements is an estimate and has some uncertainty associated with it. Based on numerous test shots into tissue simulants with a wide variety of bullets, an uncertainty of $\pm 5°$ to $\pm 10°$ for wound paths is not unrealistic.

AUTOPSY PHOTOGRAPHS

Autopsy photographs showing probes inserted in the wound tracks, *if* photographed from the appropriate viewpoints, can be used to derive the nominal angular components of the bullet's track. Unfortunately, just as with many scene photographs, the appropriate camera positions are not always represented among such photographs. If this is the case, angular estimates derived from them will be compromised or be incapable of being made. One would be well advised after having personally made such estimates of the angular components from autopsy photographs to consult with the pathologist that performed the autopsy and obtain his or her concurrence with your estimates.

It should also be realized that such probed tracks *may* not be truly representative if the body was twisted or in some orientation substantially different than the usual supine position on the autopsy table. As just mentioned, it may be desirable to discuss the probing process with the pathologist who inserted the probe(s) and take advantage of his or her recollection, methodology, etc. Such consultation(s) and the information derived from them should be added to one's note package on the case.

LOCATION OF ENTRY/EXIT/BULLET RECOVERY SITE/STRUCTURES— ORGANS PERFORATED BY THE PROJECTILE

Nearly all pathologists will record the location of bullet entry wounds, exit wound, and/or recovery sites for projectiles in a coordinate system. These are typically referenced to a datum such as the top of the head or the planar surface of the foot on the same side of the body as the wound or projectile recovery site. Common anatomical reference points (such as the umbilicus) may also be noted. Photographs are invariably taken during the autopsy, copies of which should be obtained or studied in all cases where bullet path is of interest. A few pathologists will also provide the *length* of the bullet's path. This is very useful since a simple trigonometric calculation (the sine function in this case) can provide the angular components when combined with the measured locations of the entry and exit wound or the recovery site of a bullet in the body. However, another word of caution must be offered. A bullet that entered at point "A" and exited at point "B" may not necessarily have followed a straight line. This is particularly true with elongated rifle bullets. These often deviate from their initial path after they have penetrated a few inches. In these situations the initial path taken by the bullet is a much better indictor of the bullet's pre-impact flight path relative to the position of the body at the moment the wound was sustained (see Figure 10.2).

These locations (entry site, exit site) plotted on an anatomical diagram, a computer model, or a living model comparable in size, weight, and proportions to the decedent represent an additional means of *estimating* the path of the projectile in the decedent's body.

ANCILLARY-SUPPLEMENTAL INFORMATION

Radiographs (X-ray films) of the decedent represent potentially useful supplemental information since they may reveal a trail of small bullet (lead) fragments along the bullet's path. This phenomenon can provide additional information regarding the track of the projectile. Struck or fractured bones may also be visible as well as the location of any bullet(s) that failed to exit the body.

The *clothing* from a gunshot victim should *always* be sought and examined when available. This is especially true in cases where the victim has survived. In such cases the clothing is often discarded by the hospital unless early interdiction takes place on the part of investigators. On rare occasion such clothing has been found in the possession of the decedent's family.

The nature of a bullet hole in clothing can reveal much regarding the pre-impact stability of the bullet, its design (e.g., spitzer-pointed vs. hollow point), and whether it passes through some intermediate object prior to striking the victim. This information should be compared to the findings and observations of the pathologist regarding the entrance wound.

An examination of the recovered bullet from the body of a victim is also very useful in assessing the probable behavior of the bullet as it made its way to its final position of rest. The type of bullet, the nature and location of any damage to the bullet, symmetrical vs. non-symmetrical expansion or deformation are all useful properties in evaluating the probable behavior of the bullet during its passage into or through the body. For example, an evenly expanded hollow point bullet is likely to follow a straight course in soft tissue. The same bullet that only expands on one side of the hollow point will be prone to deviating during penetration into the body.

SURVIVORS OF GUNSHOT WOUNDS

These individuals make for interesting subjects particularly if they are willing to assist or participate in the investigation. Their entry and exit wounds (if present) can usually be seen as unpigmented skin or scar tissue. This holds true for many, many years. But with the passage of time and changes is the person's physique, the correctness of the locations of these sites may be difficult to establish. Many gunshot victims still carry the bullet in their bodies for the reason that it was deemed safer and less invasive to leave the bullet in place rather than subject the individual to surgery. The presence of the bullet provides a real advantage as long as it is still in its original position of rest. This can be determined with the help of a radiologist and the subject's cooperation. A contemporary X-ray film is prepared using the same view as one or more archival X-ray films taken immediately after the incident. These are compared and any relocation of the projectile noted. If the gunshot victim is willing to have new X-ray films prepared, one should contemplate and discuss several views with the radiologist. If the person has changed very little in height and body weight since the incident, the author would recommend the placing of a small lead or radio-opaque marker on the entry wound site. From a knowledge of this location and the location of the bullet, the subject should be oriented so that the wound track is in profile (or parallel) to the film plane. If there are remaining issues about the design (brand) of the bullet, lateral and A/P exposures should be taken. One of these views should reveal anything unique about the bullet such as the *Hydra-Shok* post, the talons of the *Black Talon* bullet or the separated jacket, and small pellets of the Glaser *Safety Slug*. Bullets and any other radio-opaque object will appear slightly, to substantially larger on an X-ray film depending on where the object was in relation to the film plane. This is because the X-rays emerge as a cone or radiation and thereby produce a magnifying effect at the film plane. Reasonable estimates of the bullet's caliber are possible if a suitable scale is positioned at the same height as the bullet above the film plane and the shank of the bullet can be seen. In this way an appropriate correction can be made and the diameter of the bullet estimated. Since bullet caliber is not an infinite variable (they come in steps such as .22, .25, .30 caliber, and so on), such a measurement may allow the elimination of certain choices and a reduction to one remaining candidate among a limited universe of contenders.

Survivors that sustained one or more *perforating* gunshot wounds present a special problem unless they were struck in a clothed area of the body and the clothing has been retained. In the absence of this we are often left with the victim's recollection and the opinion of the emergency room doctor regarding the direction of fire. The first can be in error and the latter is often in error. Emergency room doctors are in the business of saving lives, not documenting or interpreting the directionality of gunshot wounds. Even the sketches found in ER documents are of little or no use. Their knowledge and training in the interpretation of gunshot wounds are usually inadequate and the views they hold about wound ballistics appear to be derived from movies and television. It will be an especially fine day when the reader-investigator finds photographs of any kind of the entry and exit gunshot wounds amid the patient's medical records. With the popularity and simplicity of digital cameras, it is hoped that this will change so that an experienced forensic pathologist can study and interpret the wounds.

One might think that asking the gunshot victim would solve the dilemma but the author has seen more than one case where the subject's account was incorrect as evidenced by the obvious entry and exit holes in the retained clothing. These may have been failures of memory or intentional deception. Either way, the subject's recollection is not physical evidence.

The one remaining hope is the existence of any X-ray films taken before any surgical procedures were undertaken. The fracturing of any bones and the distribution of any resultant particles or fragments of bullet metal could answer the question of directionality.

SUMMARY AND CONCLUDING COMMENTS

It has been said elsewhere in this book but deserves repeating at this point—*All measurements are estimates.* There is some degree of uncertainty in every measurement we take. This is not a fatal flaw in our efforts. We simply must be prepared to provide reasonable and demonstrable uncertainty limits for the type of measurements we are making. Following this we must then assess what effect they might have on the determination we are trying to make. One of the very first questions to be addressed in any reconstruction is—*What is in dispute?* Consider a situation where the pathologist is able to say that a wound path was front to back and downward after the bullet entered the decedent's chest at a height of 52 in. When pressed for numbers, he concedes that he could be off on the height of the entry wound by ± 1 in. and the downward angle could be as low as $10°$ to as steep as $30°$, a threefold range in values for the angle.

A bullet graze mark on the edge of a doorway at the shooting scene shows the fatal bullet to have been traveling essentially parallel to the floor and at a height of 48 in. before it struck the decedent. The shooter says that the subject was approaching him in a slightly crouched and forward-leaning posture. Because the decedent was found on the floor on his back, it has been suggested that the victim was reeling back and leaning away from the shooter when he was shot. Scenario 1 is supported by the scene and autopsy findings; scenario 2 is excluded by the same findings despite the seemingly wide margin of error in the medical examiner's estimates of the wound path.

REFERENCES AND FURTHER READING

Di Maio, V.J.M., *Gunshot Wounds—Practical Aspects of Firearms, Ballistics and Forensic Techniques*, Elsevier Science Publishing Co., NY, 1985.

Fackler, M.L., "Ordnance Gelatin for Ballistic Studies," *AFTE Journal* 19:4 (October 1987).

Fackler, M.L., S.D. Woychesin, J.A. Malinowski, P.J. Dougherty, and T.L. Loveday "Determination of Shooting Distance from Deformation of the Recovered Bullet," *JFS* 32:4 (July 1987) pp. 1131–1135.

Haag, L.C., "Base Deformation as an Index of Impact Velocity for Full Metal Jacketed Rifle Bullets," *AFTE Journal* 33:1 (Winter 2001).

MacPherson, D., *Bullet Penetration—Modeling the Dynamics and Incapacitation Resulting from Wound Trauma*, Ballistic Publications, El Segundo, CA, 1994.

Sellier, K.G. and B.P. Kneubuehl, *Wound Ballistics and the Scientific Background*, Elsevier Publishers, Amsterdam 1994.

Spitz, W.U. and R.S. Fisher, *Medicolegal Investigation of Death—Guidelines for the Application of Pathology to Crime Investigation*, 2nd edn, Charles C. Thomas Publisher, Springfield, IL 1980.

CHAPTER 11

TRACE EVIDENCE CONSIDERATIONS ASSOCIATED WITH FIREARMS EVIDENCE: EVIDENTIARY AND RECONSTRUCTIVE ASPECTS

INTRODUCTION

There is a failing in some forensic laboratories that deserves to be rectified if the complete and reliable reconstruction of certain shooting incidents is to take place. It is regrettable that some firearm examiners consider themselves just that—exclusively examiners of firearms and ammunition components. This is shortsighted and contrary to good forensic science particularly when it comes to the various types of trace evidence that might be present on fired bullets or expended cartridge cases. The usual reasons given for such a narrow view of one's role are:

"It's not my job," or the corollary "That's someone else's job."

"I don't do that." (trace evidence analysis)

"I'm not qualified/trained in trace evidence analysis."

"No one asked me to look for such evidence."

"I didn't know it was of interest (important) to anyone," or the corollary to this—"I wasn't told anything about the case."

All of these reasons are those of an individual operating at the technician level. They are *not* satisfactory explanations for an individual who professes to be a forensic scientist.

While it indeed may not be within the personal skill of every person carrying out firearm examinations to analyze or identify trace evidence associated with shooting incidents, it *is* incumbent upon them to be diligent and observant, to note the presence of such evidence, to document and protect it for others to examine and to have some understanding of its potential value (i.e., to know what sort of tests can be performed on such evidence and what such tests can show). Failing to consider the presence of trace evidence on and in submitted firearms, fired cartridge cases, and recovered bullets will result in missed opportunities to answer critical questions in some future investigation or legal action. The presence of trace evidence on any one of these items can be more important than the matching of the evidence bullet or cartridge case to the submitted firearm. This then is the purpose of this chapter.

LOCARD'S PRINCIPLE REVISITED—SOME EXAMPLES OF TRACE EVIDENCE TRANSFERS AND DEPOSITS

The Locardian concept of mutual evidence transfer is the guiding theme in this chapter.

Trace evidence that may be *left by* a bullet or cartridge

- metal (copper, copper/zinc alloy, lead, alloyed lead, aluminum, nickel at impact sites and/or in the bore of the firearm)
- powder residues in the bore of a firearm from the last shot fired
- primer residues in the bore of the firearm
- lead splash
- bullet wipe
- bullet lubricants
- colored primer lacquer on the breechface of a firearm following discharge of the cartridge
- paint from the tips of color-coded bullets (tracer, AP, API) in the magazine or feed ramp of a firearm
- impressions left by decelerated bullets in wood, sheet rock, sheet metal
- impressions by bullets in soft, plastic surfaces such as polypropylene bed liners in pickup trucks
- cast-off blood from a bullet exiting a wound.

Trace evidence *on* a bullet

- powder particles/powder imprints in the base of a fired bullet
- primer residues in the base of a fired bullet
- powder particles from the bore/chamber of the gun embedded in the bearing surface of a fired bullet
- from an impacted surface (paint, asphalt, concrete, wood, soil grains, glass, etc.)
- from the perforated clothing worn by a gunshot victim (threads, fibers)
- from organs or structures in the human body (bone, hair, brain matter).

Trace evidence *on* or *from* a firearm (not previously mentioned)

- back-spattered blood from close proximity and contact gunshot wounds
- fibers from the garment or fabric in which the firearm was carried or in contact
- impact damage to the firearm with inclusions associated with the surface impacted
- impact damage to the surface struck by a dropped or thrown firearm
- unique or novel debris associated with sound suppressors (GSR-containing steel wool)
- metal transfers on skin or other surface from intimate contact with the firearm
- tissue from abrasions or lacerations sustained during the discharge and cycling of a semiautomatic firearm with the hand or fingers in an inappropriate location (such as the sharp corners at the rear of the slide)
- DNA from one or more persons that have handled the firearm.

COMMON EXAMPLES OF TRACE EVIDENCE ON OR EMBEDDED IN RECOVERED BULLETS AND APPLICABLE ANALYTICAL METHODS

- imprints of fabric patterns, wire screen, "facets" from striking failed tempered glass, etc.

    photography for this and all remaining examples
- propellant imprints/powder particles in the base of bullets

    removal and retention of any powder particles
- glass (powdery appearance; unique morphology and composition under the SEM)

    tests of refractive index, density, color, appearance under UV light, chemical properties by SEM/EDX analysis
- sheetrock-wall board (chalky appearance) composed of calcium sulfate (gypsum)

    particle morphology and chemical composition by SEM/EDX analysis
- soil (bullet has abraded surface with directionality; embedded sand-like particles)

    polarized light microscopy of soil grains; *in situ* analysis by SEM/EDX
- asphalt (heavy damage with white chalky mineral inclusions and brown/black tar-like deposits)

    asphaltic material is soluble in toluene, evaporate on filter paper/TLC plate is fluorescent under UV light; HPLC analysis; mineral material by SEM/EDX

- wood/plant material (fibrous appearance)
    polarized light microscopy examination; plant DNA tests
- hair/hair fragments (characteristic appearance in most cases)
    polarized light microscopy; *in situ* characterization by SEM; mitochondrial DNA
- paint transfers/paint chips (paint at the "pinch-point" for strikes to painted metal surfaces; painted bullet tips; paint chips in hollow point cavities; lacquer sealants)
    infrared spectroscopy; microspectrophotometry; elemental composition *in situ* by SEM/EDX
- plastic/rubber (smeared appearance)
    infrared spectroscopy; elemental composition *in situ* by SEM/EDX
- fibers (characteristic appearance)
    infrared spectroscopy; polarized light microscopy; morphology by SEM
- bone particles (waxy, translucent appearance)
    Ca and P (calcium phosphate) and morphology by SEM/EDX
- tissue (stringy, amber-colored uneven diameter material)
    serological tests; DNA tests
- blood (characteristic appearance in most cases)
    serological tests; DNA tests.

EXAMINATION PROTOCOL AND COMMON ANALYTICAL TECHNIQUES

- Stereomicroscopy
- Documentation
- Photography
- Polarized Light Microscopy
- SEM/EDX
- FTIR (infrared spectroscopy)
- Microspectrophotometry
- Chemical tests
- Thin layer chromatography/HPLC
- Serology/DNA.

SEQUENCE OF EVENTS THROUGH TRACE EVIDENCE—THREE CASE EXAMPLES

The *order* and/or *location* of multiple sources of trace evidence relates to the sequence of ballistic events experienced by the recovered bullet.

CASE 1

Consider the case where multiple pistol shots are fired at a motorcyclist traveling on a concrete highway. One of these shots perforates the rear tire of the motorcycle causing the rider to lose control, crash, and suffer fatal injuries. A grouping of fired cartridge cases effectively locates the shooter's position but the important issue is whether the motorcyclist was approaching or departing the shooter's location. One of several damaged bullets recovered from the roadway shows transfers of white paint in a flat, low angle ricochet mark. This site is partially overwritten by another ricochet event that occurred with the bullet in yaw and is later found to contain concrete particles. Overlaying both of these are smears of black rubber.

A careful inspection of the shape of the entry hole in the rear motorcycle tire shows it to have been caused by a destabilized and deformed bullet. A subsequent search of the scene reveals a low incident angle ricochet mark in the white paint of the highway's fog line. A few feet downrange of this mark is another mark containing bullet metal on the concrete highway divider.

Putting these together

There is only one solution to this arrangement of trace evidence. This is the bullet that perforated the rear tire of the motorcycle (rubber transfers). It did so in a destabilized orientation (irregular entry hole in the tire and the location of the rubber transfers on the bullet). The bullet first struck the white fog line at low angle and ricocheted into the concrete divider in a destabilized orientation (bare concrete damage partially overwriting the ricochet site containing white paint). It then ricocheted from the bare concrete divider and perforated the motorcycle tire (rubber transfers on top of both areas of ricochet damage). The question of *who* shot the tire was never in dispute so matching the bullet to the shooter's gun is hardly necessary. The critical issue was the location of the motorcycle relative to the shooter when this shot was fired. The trace evidence on this bullet combined with the locations of the two bullet impact marks on fixed surfaces and the behavior of ricocheted projectiles allow this question to be answered.

CASE 2

An alleged gun-wielding subject has been fatally shot by a police officer in a public park. It is not in dispute that the officer fired 3 shots with his 9 mm pistol. He later describes the distance between them as about 10 feet for all three shots. Two of these three shots struck the subject and produced the following wounds:

GSW-1 (fatal) entered the upper left chest passing between the 3rd and 4th ribs then entering the thoracic cavity where it perforated the heart after which it exited the back after passing just to the right of T4. The path was front to back, left to right, and slightly upward.

GSW-2 (non-fatal) entered the right mid-anterior thigh, passed through soft tissue, grazed and fractured the femur, then exited at the rear of the right thigh. The path was front to back and downward about 30°. This was a survivable wound according to the medical examiner. (*Note*: The numbering system used by forensic pathologists for multiple gunshot wounds is *not* an effort to assign shot sequence but merely a method for later reference.)

The first paramedics on the scene found the decedent on his back and as they moved the body they observed a fired bullet in some bare dirt in an area that was previously under the body. A search of the ground with a metal detector revealed no other bullets in the area where the decedent fell, nor were any bullets subsequently found in his clothing.

In a later interview the shooting officer and his partner describe the subject as falling over backward *after* the third shot whereupon the shooting officer rushed forward and secured what was later found to be a BB pistol next to the subject's body. Several citizen witnesses, on the other hand, later come forward and claim that the officer fired two quick shots whereupon the subject staggered back and fell onto his back. According to these witnesses, the officer then moved forward, stood directly over the body, and fired a shot into the subject's chest. This precipitates much political and public furor to include the formation of a special review committee and a march on city hall by outraged citizens.

A careful and thorough laboratory examination of the three fired cartridge cases, the one recovered bullet, the officer's 9 mm pistol, and the decedent's clothes revealed the following:

The three cartridge cases were of the same brand but from different lots or production periods in that one cartridge was plain brass and the other two were nickel plated. Additionally, the plain brass cartridge had residues of flattened ball powder in it as well as a ring of black sealant material inside the case mouth. The two nickel-plated cartridge cases contained a few particles of unperforated disk-flake powder and no sealant. A clean cotton patch pushed through the bore of the officer's pistol showed a few particles of unperforated disk-flake powder. No powder residues were found around the bullet holes in the front of the decedent's red wool pullover sweater or the upper portion of his blue cotton jeans. Examination of the bullet found under the body by paramedics showed loosely adhering soil grains on nearly all areas of the bullet but no impactively embedded soil grains. Traces of black sealant material were present on the shank of this bullet. Further inspection revealed a large bone chip in the partially expanded hollow point cavity. Underneath this bone chip was a small wad of blue cotton fibers.

Subsequent test-firing of the officer's 9 mm pistol with both types of ammunition produces easily discernable powder patterns out to two feet with cartridges loaded with the unperforated disk-flake powder and clear powder patterns out to three feet with cartridges loaded with the flattened ball powder.

Ejected cartridges of both types landed to the right and rear of the gun when held in the normal, grip down-barrel parallel to the terrain position. When pointed essentially straight down and fired, the ejected cartridge cases went to the right and slightly forward.

Putting these together

The bullet on the ground is associated with the non-fatal leg wound (blue cotton fibers and bone). Recall that *no* bones were struck by the bullet that produced the fatal chest wound and the clothing in this area was composed of red wool.

This bullet is associated with the plain brass cartridge (containing black sealant material and flattened ball powder).

This bullet can*not* be the last shot (the bore of the pistol contains unperforated disk-flake powder residues from the last shot and the plain brass cartridge contained flattened ball powder).

The soil adhering to the bullet is of no reconstructive value but the absence of *impact* damage with the ground is an important observation that suggests that this bullet simply fell out of the decedent's clothing as he struck the ground or as he was moved by the paramedics. This type of damage due to *impact* was previously illustrated in Figure 7.14 in Chapter 7.

When the shooting officer is asked to position himself over a suitable stand-in for the decedent and point his gun at the stand-in's chest (as described by the citizen witnesses), the separation distance measures 8 in. ±2 in. Since there is no suggestion or likelihood of an intervening object, the absence of any GSR pattern on the decedent's sweater effectively excludes the citizens' account.

The absence of the bullet from the fatal wound being under the body also refutes the coup-de-grâce to the chest account.

A subsequent review of the scene photographs and diagram show the three fired cartridge cases to be in a relatively small area to the right and rear of the position where the shooting officer places himself. This is in agreement with the normal ejection characteristics of his pistol and ammunition and does not support the citizens' account. It does support the shooter's account.

Aftermath and further work

Approximately one year later a civil lawsuit is filed despite these findings. This prompts a renewed and extended search for the fatal bullet. Approximately 70 yd from the decedent's position and what would be downrange according to the police account is found a 9 mm bullet embedded approximately 1 in. in the trunk of a large elm tree. This bullet shows little evidence of expansion but appears to have struck the tree somewhat in yaw. Its hollow point cavity is plugged with bark and wood fibers. It is ultimately matched to the officer's pistol but DNA tests for traces of biological material associated with the decedent are negative.

Note: This is another example of *the absence of evidence is not necessarily evidence of absence*. And the reason is that some bullets make it through gunshot victims without picking up identifiable DNA. Moreover, the passage of time combined with microbial activity could degrade any DNA traces that might have been present if this is indeed the fatal bullet and not the missed shot. But how can this final question be resolved? Penetration tests (to be discussed more in Chapter 13) could be carried out with the idea that direct shots over the 70 yd distance with the gun-ammunition combination might provide a means of discriminating a direct strike from a decelerated strike. But a careful removal of the bark and wood from the hollow point cavity might be more productive and certainly much easier. If this is the fatal bullet, it passed through the red wool sweater first, then the decedent's body followed by the rear of the sweater and on to embed itself in the tree. This question is answered upon the discovery of several red wool fibers underneath the plug of bark and wood in this bullet's hollow point cavity.

CASE 3

Arson investigators at a suspicious warehouse fire note what they believe to be a bullet hole in a smoke-stained but surviving window. Before the window is removed for laboratory examination they also find a perforating pair of holes through opposite sides of a metal drum of solvent. Using these three points of reference, they go on to find an irregular hole in the wall opposite the window from which a heavily damaged rifle bullet is ultimately recovered. This projectile turns out to be a WWII vintage .30-'06 U.S. military M1 tracer bullet fired from a rifle with 6 land and groove rifling with the land widths 1/3rd the width of the grooves. It also shows conspicuous "checking" of its ogive from having perforated glass. (*Note*: This sort of damage was illustrated in Figure 6.19 in Chapter 6. Glass particles have a unique appearance under the SEM and can be easily distinguished from soil grains by an experienced microscopist.) The investigators now believe this fire was started by a shot fired from a distant knoll on the warehouse owner's property. He denies any involvement but does volunteer his .30-'06 caliber hunting rifle when asked if he owns any firearms in this caliber. He also tells investigators that he only shoots one brand of commercial hunting ammunition in this rifle and does not own any tracer cartridges of any kind. A box of this hunting ammunition is also turned over to investigators.

Evidence Examination and Findings

Careful laboratory examination of the rifle reveals that it has not been cleaned since it was last fired as evidenced by sooty residues on a previously clean cotton patch pushed through the bore. The general rifling characteristics of the rifle (6-right, $L_{\text{wdt}} = 1/3G_{\text{wdt}}$) do not allow it to be excluded but the evidence bullet has no surviving individual characteristics that would allow the responsible rifle to be identified. Prior to any test-firing of the rifle, a few particles of what appears to be red paint are noted adhering to the feed ramp leading to the chamber. A faint ring of clear red lacquer particles are also observed on the bolt face around the firing pin aperture. Infrared spectroscopy, microspectrophotometry, and non-destructive elemental analysis by SEM/EDX show these materials to compare favorably to the red primer annulus lacquer on WWII .30-'06 cartridges and the red identification paint used on the tips of M1 tracer bullets of the same period. This revelation leads to a testing of the bore residues collected on the cotton patch by SEM/EDX. Residues of corrosive perchlorate primer mixtures used in WWII .30-'06 ammunition and traces of the elements found in the igniter and tracer compositions are identified in these residues. *None* of these materials are used in contemporary .30-'06 ammunition. Likewise, the commercial hunting ammunition volunteered by the warehouse owner *lacks* any red sealant around the primers, possesses standard non-corrosive lead styphnate primers and is loaded with jacketed soft point bullets containing lead cores.

All of these findings will be very useful to investigators when they next question and confront the warehouse owner.

SUMMARY AND CONCLUDING COMMENTS

It was neither the author's purpose nor the expectation to convert the reader to a trace evidence analyst. Rather it was to acquaint the reader with the great variety of trace evidence types that may be found on firearms and associated firearms evidence and the reconstructive implications of such evidence. The hypothetical examples presented in this chapter were all constructed from real cases. A fiber, a few particles of glass or bone, a paint chip in the hollow point cavity of a bullet recovered at a shooting scene can be as important as, sometimes more important than, the matching of the bullet to an impounded firearm. Yet these materials may go unnoticed and even get lost during the subsequent comparison process if the firearm examiner does not have an appreciation for their potential value and does not take measures to note and protect such minute evidence. The same is true for fired cartridge cases that may bear fragile transfers of paint or mineral material on their rims or mouths from an impact with an adjacent surface. Trace evidence on a firearm or in the bore of a submitted firearm is quickly lost or destroyed if the examiner's first action is to test-fire the gun. Checklists and written examination protocols alone are not enough. Individuals who collect and impound firearms evidence and those who first receive and examine firearms evidence in the laboratory must think beyond their speciality. While it is true that the analysis of trace evidence may be someone else's job, recognizing, documenting, and protecting such trace evidence falls squarely on the individuals who collect firearms evidence and on the forensic scientist who first examines this evidence. As was illustrated in the hypothetical cases examples, trace evidence can exonerate the innocent as well as implicate the guilty. It is neither prosecution evidence nor defense evidence. It is evidence that we must protect and hold in trust until it can be fully and properly analyzed at which time its story will be told.

REFERENCES AND FURTHER READING

Laible, R.C., ed., *Ballistic Materials and Penetration Mechanics*, Elsevier Scientific Publishing Co. (1980).

McCrone, W.C., L.B. McCrone and J.G. Delly, *Polarized Light Microscopy*, Ann Arbor Science Publishers, Ann Arbor, MI (1978).

CHAPTER 12

CARTRIDGE CASE EJECTION AND EJECTION PATTERNS

OVERVIEW

With automatic and semiautomatic firearms, the location of expended cartridge cases can be important in establishing the approximate location of a shooter's firearm when one or more shots have been fired. It may also be possible to determine whether a shooter moved from one location to another, or alternatively failed to move in any significant way, during the discharge of multiple shots. The value of cartridge case locations improves if one also has an azimuth line or direction of fire for the associated shots. Ejected cartridge cases may also acquire trace evidence deposits if they strike nearby surfaces with sufficient force and the surface is of a type that generates transfer evidence. The paint on most interior walls of houses and office buildings is a good example of such evidence. In some instances, where the impact force is sufficient, the ejected cartridge will leave a faint, crescent-shaped indentation in the struck surface. These are very subtle marks and no doubt go unnoticed by many scene investigators. Their presence combined with the ejection characteristics of the responsible gun can refine the positioning of the shooter's firearm. Figure 12.1a,b shows an expended cartridge case with adhering paint on its mouth and the corresponding crescent-shaped mark impressed in the sheetrock wall within a few feet or less of where the pistol was fired. From a brief study of the first photograph it should be apparent how delicate this trace evidence is and how easy it would be to lose it by careless handling or improper packaging.

SCENE WORK—TERRAIN/SUBSTRATE CONSIDERATIONS

The issue of cartridge case location at a shooting scene might seem straightforward. The customary matter of each casing's location in a suitable coordinate system followed by individual packaging in appropriate containers is usually accomplished in contemporary police work. Except for photographs, the *nature*

Figure 12.1

(a) Fired Cartridge Case with Adhering Paint on Its Mouth. (b) Shallow Impact Mark in a Painted Surface from an Ejected Cartridge Case

(a)

This 9 mm cartridge case was found on the floor of a shooting scene. Only through careful handling and preservation was the adhering white paint preserved. This trace evidence is visible on the mouth of the cartridge at the 6 o'clock position in this photograph. A corresponding crescent-shaped impact site was found many months later when the author examined the scene. This mark, combined with the correspondence in the color and composition of the paint and the ejection characteristics of the gun, allowed the approximate position of the shooter to be established.

(b)

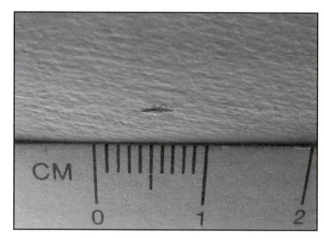

This crescent-shaped indentation (just above the 0.5 cm mark) in a painted wooden surface at shooting scene was caused by the impact of an ejected cartridge case. Given their small size, such marks can be difficult to locate and are often overlooked but can be very useful in refining the position of the shooter.

of the surface on which such cartridge cases were found is seldom recorded however. Examples: "uncut grass, victim's front yard; yard slopes towards the street about −2°", "level asphalt parking lot with much exposed aggregate" or "thick green shag rug-living room."

Photo-documentation and a description of the surface is needed because it is of considerable interest and importance in assessing:

- how an expended cartridge came to be in a particular location (e.g., uneven terrain with significant elevation changes in the area of deposition vs. level terrain)
- how much additional movement, if any, the ejected cartridge case could experience after its initial impact with the substrate (e.g., essentially no further movement would be expected with soft soil or sand; however, a casing may continue to bounce several feet or even yards over smooth concrete)
- how certain trace evidence, physical damage, or changes to the cartridge case might have been acquired or have taken place (e.g., an ejection port of a semiautomatic pistol within a few inches of a brick wall at the moment of discharge leaving abrasive damage on the casing with corresponding red mineral inclusions).

Cartridges ejected over hard surfaces such as asphalt and concrete will usually show several small impactive sites on the case rim and/or case mouth when examined under a stereomicroscope. When fired from a fast-moving vehicle driving over the same type of surface, these "dings" in the cartridge case will be numerous. These minute damage sites often contain small amount of embedded trace evidence characteristic of the surface upon which they fell. With a little experience and test-firing over the same type of surface, the examiner can come to recognize this type of "damage" and differentiate it from later post-impact damage such as described in the next section.

RELOCATED CARTRIDGE CASES

Post-ejection events *must* be considered before any interpretation is attached to the location of a recovered cartridge case. These include such things as being stepped on or run over by vehicles that might relocate a fired cartridge. Fortunately, the generally soft nature of cartridge cases make them quite susceptible to impact damage, abrasion, and/or deformation if stepped on, run over, or kicked. These events *usually* (but not always) leave evidence of such an event on/in the cartridge, e.g., multiple, deep abrasive gouges in the case wall and/or crushing of the cartridge mouth. These marks, once recognized, evaluated, and properly interpreted, can indicate whether the questioned cartridge case has gone through a normal extraction–ejection cycle or not. These same marks and the trace evidence embedded in them may also provide clues to the type of surface upon which the cartridge case fell or the surface against which it was ejected. The laboratory examiner *must* study the marking left on fired cartridges during the extraction/ejection process of the responsible gun. This should be done using cartridges of the *brand* and *load* involved in the case and after inspecting them to make sure they are free from any manufacturing marks or other marks that might be confused with firearm-generated marks. The isolation and elucidation of other firearm-generated marks such as magazine lip or latch marks, chamber marks, and/or slide scuff marks will also be necessary so as not to confuse such marks with those actually associated with the extraction/ejection process. Laboratory familiarity with ejection port dings is also of critical importance because they can crush the mouth of a cartridge casing. These types of marks are certainly post-firing, but are not the consequence of relocation damage.

FIREARM—AMMUNITION PERFORMANCE

In semiautomatics and machine guns, certain requirements of impulse to the bolt or slide followed by adequate impactive forces to the cartridge from the ejector must be met to extract and eject it from the gun properly. These events are directly related to the condition of the gun, as well as the performance of the ammunition during discharge. All of these events and their corresponding marks on the casings leave a variety of clues regarding the normal/proper performance of both the ammunition and the gun. Examples of marks and characteristics left on casings that can indicate whether or not a casing was fired in a normal manner include: cartridge case expansion, firing pin drag, breechface impressions, extractor marks, ejector marks, chamber marks, and/or ejection port dings. Cartridges experiencing abnormally high or low pressure excursions in performance during discharge often reflect such events in the markings on the casing and/or the degree of expansion of the fired cartridge case.

The comparison of these various markings on evidence cartridge cases with those produced on test-fired cartridges of the same brand, type, and load fired in the same gun provides the means for assessing which variable or firearm component may be responsible for a particular event under investigation. Varying selected parameters such as bullet weight or powder charge will allow one to evaluate the role or contribution to ejection pattern of any particular ammunition parameter of interest or concern to the examiner. This statement is not an invitation for someone inexperienced in ammunition loading to assemble cartridges of unreasonable or uncertain pressures because there can be a high degree of risk associated with such endeavors.

REVIEW OF MARKINGS ON FIRED CARTRIDGE CASES

Figure 12.2 summarizes nearly all of the markings that might occur on a fired cartridge case. Some of these are, or may be, generated *prior* to discharge and are included in the general category of chambering marks. These may occur while the live cartridges are retained in the magazine of the firearm (e.g., latch marks on the head of a shotgun shell when it becomes the next cartridge to be chambered). Chambering marks *may* occur as the bolt or slide of a semiautomatic pistol is *retracted* and the underside of the slide drags along the length of the top cartridge in the magazine (slide drag mark) or when the cartridge is stripped from the magazine (magazine lip marks). Scuffs on the cartridge case body and/or its head (12 o'clock slide scuff mark) may occur as the breechface impacts the 12 o'clock position of the head of the top cartridge in the magazine, and is forced up the feed ramp and into the chamber.

Depending on how the gun is designed and how it has been loaded, the extractor may strike the head/rim area of the cartridge case, ride up and over

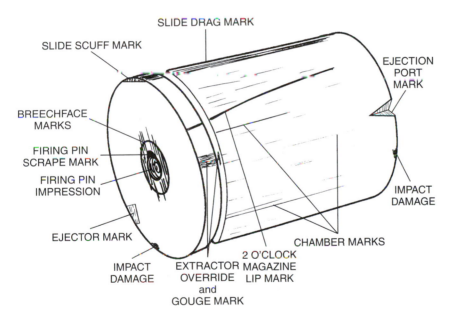

Figure 12.2

Marks on Fired Cartridge Cases

This figure summarized the various firearm-induced marks that can be found on cartridge cases that have been chambered, fired, and ejected from a semiautomatic pistol. Many of these marks possess individuality and can be associated with the responsible firearm. Others have reconstructive value and allow one to determine how the cartridge came to be loaded in the firearm, e.g., manually inserted in the chamber or stripped from a magazine after slide retraction as shown in this figure. Any impact damage as a result of ejection over hard surfaces such as concrete and asphalt usually takes the form of small "ding" marks around the mouth and/or rim of the cartridge case. Two such areas of impact damage are illustrated in this figure. These often contain minute particles of mineral material that can be further characterized under a suitable microscope.

the rim, then snap down into the extraction groove leaving as many as three marks in the process. A cartridge dropped in the chamber followed by the release of the slide into battery stands the greatest likelihood of creating these types of marks on pistol cartridges. Such a chambering process could be repeated multiple times, creating numerous marks of interest, *none* of which are the consequence of cartridge case extraction. If the gun's design calls for the cartridge to slip under and behind the extractor during the loading process, then the only mark produced by the extractor at this point might be a small nick or gouge in the bottom of the extraction groove or immediately next to the rim in cartridges lacking an extraction groove. Finally, during the discharge phase the extractor *may* produce an indentation on the forward facing side of the rim as it forcibly pulls the cartridge case from the chamber. Creation of this type of mark is particularly likely with locked breech guns as opposed to straight blowback guns. The extractor in a straight blowback design is not functional because the cartridge case is literally blown out of the chamber. An extractor is present in blowback guns to allow for extraction of live cartridges from the chamber during the unloading process. Chamber markings (or additional chamber markings) may be acquired, particularly in blowback guns, during this phase as the swollen cartridge is expelled rearward out of the chamber.

Finally, the head of the cartridge casing is ultimately struck by the ejector, typically causing the cartridge case to pivot around the extractor, or simply be knocked clear of the breechblock. With manually operated guns like bolt action or lever action rifles, an ejector mark may not be visible simply because of the greatly reduced force with which cartridges typically strike the ejector, as compared with semi or fully automatic firearms. During their final exit, the cartridge cases may strike one or more areas around the margin of the ejection port and acquire additional marks as a consequence. With some firearms this occurs with great regularity and reproducibility, and leaves an outstanding mark on the body of the cartridge case or adjacent to the mouth of the cartridge case. For these reasons, these ejection port dings are often identifiable as having been created by a specific firearm. The fact that the casing is striking areas on the gun during the ejection process greatly affects the ejection characteristics of the fired cartridges. These types of marks typically cannot be created except in an actual firing process.

In studying the normally generated marks on cartridges, the examiner should design a testing protocol that will isolate the source and production of the various marks on fired cartridges. This is important for a number of reasons. First, the examiner should be prepared to state whether the fired cartridge appears to be a *normally* chambered, fired, extracted, and ejected cartridge. Secondly, any atypical or foreign marks deserve to be explained since they may be important and may relate to some non-standard use of the firearm or some sort of misadventure such as restricted or retarded slide travel. A situation where this might be important is that of a purported struggle over the gun. A weak hold with some recoil-operated guns, for example, may reduce the slide travel and result in a weak strike by the ejector, with a corresponding reduction in ejection distance and/or direction. This *may* occur in awkward holds on a pistol with self-inflicted shootings or accidental discharges during the unsafe handling of a firearm.

A suggested laboratory protocol is as follows:

- Obtain cartridges of the same brand and type as those involved in the incident. Disassemble an evidence cartridge and one of the reference cartridges to ensure powder morphology and quantity are in agreement. If available and of sufficient quantity, use evidence ammunition for testing.
- Examine and select cartridges free of any pre-existing marks that might later be confused as having been produced during the loading–firing–extraction– ejection cycle.
- To examine magazine lip marks, and if the firearm uses a detachable magazine, load some of these cartridges in the magazine and then strip them out of the magazine by pushing against the head at the 12 o'clock position, with a non-defacing tool such as the eraser end of a pencil. As each cartridge comes up to the top of the magazine, place a suitable index mark on it so the source and position of any magazine lip marks can be ascertained.
- To examine chambering marks from the feeding process, chamber a few cartridges using the full forward force of the slide, but remove them gently from the chamber so that feed ramp marks, slide scuff marks, chamber marks from the loading process and extractor marks from the initial phase of this process can be isolated. It may be useful to do this by inserting the magazine with the slide locked back, as opposed to retracting the slide over a previously inserted magazine containing cartridges. A slide drag mark can only occur with the latter method. The presence or absence of a slide drag mark may be important when inter-comparing cartridge cases from a multi-shot incident.
- Insert a cartridge in the chamber by hand and close the action. In this manner, the examiner can look for any extractor override marks.
- It may be desirable to vigorously cycle several live cartridges through the gun's action to see if a visible ejector mark can be left in the head of a live cartridge by this means. This should only be done with utmost care, with the gun pointed downrange. It is also advisable to keep one's hand clear of the ejection port in case some malfunction or alteration allows the ejector to strike the primer and possibly discharge the cartridge. A live cartridge, with its projectile and propellant charge, is much heavier and possesses more momentum than a fired cartridge casing. For this reason, live cartridges

cycled through guns at shooting scenes often acquire an ejector mark. This may be of value in situations where live cartridges are found at a scene or in verifying that an ejector mark in a fired cartridge is indeed the consequence of the discharge process and not some previous cycling of the live cartridge through the gun's mechanism.

- Obtain test-fired cartridge casings using a normal or firm hold, making provisions to capture these cartridges without them impacting hard surfaces.
- Obtain test-fired cartridges with any alternate holds deemed appropriate or germane to the incident under investigation. Capture the cartridges in the same manner and compare the markings with the previous cartridges.
- If appropriate, obtain test-fired cartridges from semiautomatic pistols with the slide delayed or restricted. This can be accomplished by wrapping multiple rubber bands around the slide and frame. Quantification of the additional force for slide retraction can be measured with a force gauge.

LABORATORY EXAMINATION OF EJECTED CARTRIDGE CASES

Insofar as questions and issues regarding cartridge case location and ejection patterns are concerned, the laboratory examination of fired cartridge cases should include the following types of examinations:

- an inspection of the interior of the cartridge case for foreign debris (dirt, sand, water spots, spider webs, etc.) in order to address the issue of "freshness" or the alternative, signs of exposure to environmental conditions;
- an inspection and description of the exterior of the cartridge case for adhering foreign debris (dirt, mud, sand, water spots, stains, etc.), brightness, shininess vs. tarnishing of the surface;
- a check of any out-of-roundness of the case mouth or case body which might be indicative of post-firing crushing;
- an examination for any abrasive or impactive effects with non–firearm related surfaces (with impacts on concrete and asphalt, the rim and/or the mouth of the cartridge case typically receives one or more rough "dings" or indentations, often with small particles of embedded mineral material in these areas);
- an examination of any gouged or striated areas on the case wall that are non-firearm related (kicking a cartridge case while resting on a hard surface or running over a cartridge case with an automobile tends to leave characteristic damage to the casing in this area);
- an assessment of any other form of adhering trace evidence such as paint, plaster, stucco that might be associated with an area with which the cartridge case impacted after discharge.

Documentation of any of these phenomena by sketches and/or photography is important particularly in situations where trace evidence is seen adhering to the cartridge case since this could easily be dislodged during subsequent handling and examination.

GENERAL PROTOCOL FOR EJECTION PATTERN TESTING

Test-firing the actual firearm with the same type of ammunition in an open area with some sort of prearranged coordinate system will be necessary. Both circular coordinates and Cartesian (rectangular) coordinates have been used for this purpose. The author has routinely used the Cartesian coordinates system, with the intersection of the X and Y axis directly below the ejection port of the firearm and the Y axis aligned with the barrel. Multiple metal tape measures are laid out on the surface to form the X/Y axis and to provide a grid for later measurements. The proper positioning of the gun relative to the coordinates system can be accomplished by dropping a plumb line from the trigger guard or other suitable location to the designated spot on the surface. Alternatively, a tripod can be set up so that the gun can be held at the appropriate location. A preselected aiming point should be established to achieve the desired vertical angle. For example, if the shooting event in question is believed to have involved a 45° downward shot(s) then an aim point that yields a downward 45° barrel angle should be employed. For most shooting events, an aim point yielding a horizontal (0°) barrel angle would be used. Gun height or the shooter's use of prone, kneeling, or standing positions, gun rotation, and possibly even the nature of the hold on the gun may need to be considered and varied to properly evaluate the implications of cartridge case location(s) at a shooting scene.

The nature and behavior of the surface on to which the evidence cartridge case(s) fell is yet another consideration. A cartridge falling on smooth sand will remain at the impact site but the same cartridge landing on concrete will seldom, if ever, remain at its initial impact site. The cylindrical shape of cartridge cases, their displaced centers of gravity, and their propensity to rebound from hard surfaces all combine to send them off in various directions from the initial impact site for most surfaces. The fired cartridge casings themselves emerge from the ejection port of the gun with a tumbling motion, and at departure angles and velocities that can vary slightly from shot to shot. The net effect is that a series of fired cartridge casings will land in an *area*, but not on a fixed spot. The size, shape, and location of this area become the subject of ejection pattern testing. The reproducibility of the ejection pattern for a particular gun and ammunition combination is illustrated in Figure 12.3. This figure shows an aerial view for three consecutive

Figure 12.3

The Reproducibility of Cartridge Case Ejection Pattern

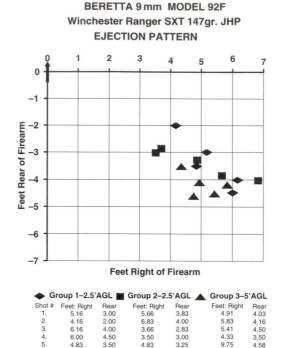

BERETTA 9 mm MODEL 92F
Winchester Ranger SXT 147gr. JHP
EJECTION PATTERN

Shot #	Group 1–2.5'AGL		Group 2–2.5'AGL		Group 3–5'AGL	
	Feet: Right	Rear	Feet: Right	Rear	Feet: Right	Rear
1.	5.16	3.00	5.66	3.83	4.91	4.03
2.	4.16	2.00	6.83	4.00	5.83	4.16
3.	6.16	4.00	3.66	2.83	5.41	4.50
4.	6.00	4.50	3.50	3.00	4.33	3.50
5.	4.83	3.50	4.83	3.25	9.75	4.58

This plan view shows three consecutive cartridge case ejection patterns for a 9 mm Beretta pistol fired over a non-resilient surface (uncut grass). The gun position is at the 0/0 axis intercept point with the arrow indicating the direction of fire. Note the height change for Group 3 in this figure, and its minimal effect on ejection pattern for this particular gun–ammunition combination.

cartridge case ejection patterns for a 9 mm Beretta pistol fired over a non-resilient surface (uncut grass). Note the height change for Group 3 in this figure, and its minimal effect on ejection pattern for this particular gun–ammunition combination.

Figure 12.4 shows a more complicated pattern for a .40 S&W pistol fired over level asphalt. The dashed lines represent the path from the initial impact point with the asphalt, represented by triangles, to their final positions of rest, which are represented with bold dots. In most investigations it is the final position of rest that represents the pattern in which we are interested, but a study of this type is useful to appreciate and understand the post-impact behavior of the ejected cartridge cases. To conduct such a study, an assistant will be needed to promptly identify and mark the initial impact site after each shot. This must be followed by a means of associating the particular cartridge case with the specific impact site. This same assistant can also provide information on the typical heights (if any) the ejected cartridges rise above the level of the gun. This can

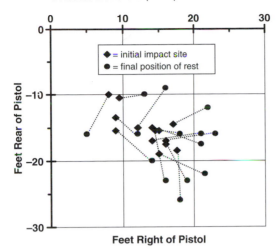

EJECTION PATTERN OF A .40 S&W BERETTA
Remington 155gr. JHP
14 rounds over level asphalt –pistol 48-in. AGL

Figure 12.4

Cartridge Case Ejection Pattern over a Hard Surface

This diagram shows the pattern for a .40 S&W pistol fired over level asphalt and the effect of impact with a hard and somewhat uneven surface. The gun position is at the 0/0 axis intercept point. All of the fired cartridges ejected to the right and rear of the pistol when fired in the normal configuration and with the barrel parallel to the substrate. The dashed lines represent the path from the initial impact point with the asphalt (represented by diamonds) to their final positions of rest (represented by circles).

be important if there is the possibility or question of a fired cartridge passing over a fence or the roof of a car before it landed on the ground. Setting up a video camera at the same height of the gun, and at an appropriate position and distance behind the gun, can be useful in several ways. When properly positioned, the camera can record the flight characteristics, as well as the behavior of the cartridges after ground impact. If photography of the final pattern of ejected cartridges is contemplated, their positions should be identified with suitable markers. Once the markers are in place, at least one photograph should be taken from behind the pattern, with the shooting position in the field of view, and another with a profile view of the pattern and shooting position (see Figure 12.5a and 12.5b).

Figure 12.6 shows the effect of pointing the particular test firearm downward at a 45° angle compared to the same pistol fired parallel to the terrain. The general ejection characteristics of most semiautomatic pistols are either right, or right and rear. In these situations, when such a pistol is pointed down, the rearward component is shortened with the result that the fired cartridges impact the surface further and further forward as the downward angle is steepened.

Figure 12.5a

Photography of a
Cartridge Case Ejection
Pattern

This photograph shows a profile view of the pattern for a 9 mm pistol fired over level ground (coarse sand). Numbered markers have been placed over each of the fired cartridge cases. The tape measures provide a coordinate system and means of measuring the position of each cartridge case relative to the point immediately below the ejection port of the pistol. The tripod is used as a reference point for the positioning and re-positioning of the gun for each shot. The shooter's hands are barely touching the top of the tripod as each shot is fired. The author has also used a plumb bob and line of a selected length suspended from the shooting hand and oriented over the 0/0 axis point for this purpose.

Figure 12.5b

Photography of a
Cartridge Case Ejection
Pattern

This photograph shows the same pattern with the camera positioned in front of and slightly to the right of the shooter. The numbered markers for these ten shots have been turned to face the camera. A view from behind the gun and shooter accomplishes the same end. The views shown in these two figures allow one to better understand and visualize a cartridge case ejection pattern.

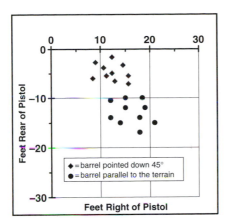

Figure 12.6

Cartridge Case Ejection Pattern: The Effect of Vertical Angle Changes .40-Caliber Glock Fired over Coarse, Level Sand Winchester 180 gr. JHP Ammunition

This pistol was fired ten times with a conventional grasp with the barrel 48 in. above and parallel to the terrain after which an additional ten rounds were discharged with the pistol pointed down at a 45° angle and a height above ground level of 32 in. These results are typical for most right-rear ejecting guns in that a downward shooting angle moves the pattern of ejected cartridge cases forward.

CARTRIDGES EJECTED FROM MOVING VEHICLES

The behavior of various cartridge cases ejected from moving vehicles traveling over common asphalt at speeds ranging from 10 to 40 mph has been studied. As one might expect, the increasing speed of the vehicle increases the scatter of the cartridges and continues to move their final resting place further and further away from the launch or ejection point in the direction of vehicle travel. Microscopical examination of the brass cartridge cases also showed increasing damage with increasing vehicle speed in the form of multiple "dings." Increased vehicle speed also corresponded to an increasing number of sites of embedded mineral material from repeated impacts with exposed stones in the asphaltic concrete. This impact damage quickly exceeded what was found with an ejection over the same surface from a stationary position.

With adequate testing it may be possible to set some limits on the speed of moving vehicles from which shots are known to have been fired based on the distribution of such fired cartridges relative to the trajectories of the shots and the damage the cartridge cases have sustained.

ADDITIONAL CONSIDERATIONS

To cycle properly, recoil-operated guns need to be held with a reasonably firm grasp. During the discharge process, the slide and barrel recoil and move rearward together for a short distance, and subsequently unlock while the frame is secured in the hand(s) of the shooter. A poor or loose grip on some recoil-operated pistols can affect the extraction and ejection process to the point of a stoppage or jam. A very loose grip can altogether prevent the casing from being extracted from the chamber in some pistols. Although it might seem unlikely that anyone would hold and fire a gun with such a hold, it may be necessary to evaluate this potential variable in certain cases.

In some semiautomatic firearms the position and/or force exerted on the extracted cartridge by the top cartridge in the magazine can have an effect on the subsequent ejection of the cartridge. This can often be seen in AK-47 and SKS-type firearms. The cartridges in these magazines are staggered so that the top cartridge in the magazine is alternating between left and right. The result may be seen in alternating paths of the ejected cartridges. Likewise, the force acting on the top cartridge in the magazine may play a role in the ejection of a fired cartridge. When a magazine is nearly full, the upward force on the top cartridge is much greater than when it is the last cartridge in the magazine. This *may* have an effect on the ejection of the fired cartridge. This can easily be evaluated during test-firing but the easier approach is to design the ejection pattern tests around the facts of the case. If the shooter is known to have fired four shots starting with a fully loaded magazine, then conduct the ejection tests in the same manner. Whether the firearm was canted or not can be another issue. Largely as a result of movies, the rotation of a firearm either right or left has become a popular cross-examination question at trial. In general, if a normally right ejecting pistol is rotated 90° to the left, pointing the ejection port skyward, casings will tend to travel upward and to the left of the pistol. Rotating the pistol to the right, pointing the ejection port to the ground, will tend to leave casings almost at the shooter's feet.

Proper training in gun-handling and shooting position require the grip to be in a vertical orientation for proper sight alignment. Just because it is improper to cant a firearm to the right or to the left does prevent a shooter from doing so. If this issue is raised and is deemed to be of concern, then it too must be added to the list of variables to evaluate in any ejection pattern testing. If the shooter is cooperative, then the question of canting the gun needs to be put to him or her in a neutral fashion. Interview questions such as: *Could you please demonstrate your best recollection of how you were holding the pistol when you fired? [Photograph this position] Could you demonstrate how you believe you were standing at the moment you fired? What is the highest you believe you could have been holding the gun at the time of the shot?—and the lowest? [Get measurements of the height of the gun above the surface] How were you trained to hold and shoot this pistol?* There are, no doubt, additional questions that deserve to be asked of the shooter depending on the facts and possible issues in the particular case, but these can serve as a starting point.

MANUALLY OPERATED FIREARMS

The techniques described in this chapter are *not* applicable to mechanically operated firearms such as bolt-action and lever-action rifles, pump shotguns, and single shot pistols. With these and like firearms, the location of the fired cartridge case only provides some insight into the location where the expended cartridge was removed from the gun by the handler. This location may or may not relate to the shooting position.

SUMMARY AND CONCLUDING COMMENTS

It must be said that of all the matters investigated and evaluated in shooting incidents, the interpretation of cartridge case location and ejection, is fraught with the greatest uncertainty. This is meant as a strong word of caution and not as a statement of futility. Despite the numerous variables, and recognizing that some of these are beyond our ability to know, we can often state with reasonable certainty where the semiautomatic pistol, rifle, or shotgun were *not* fired. The same would hold true for fully automatic firearms. This is fully in keeping with the approach employed in the *Scientific Method*; i.e., ruling out certain hypotheses. Consider the following:

- A cartridge case can only be ejected so far by the gun that fired it.
- The design of the particular gun sets some limits on the direction that the fired cartridge must emerge from it.
- While it is true that the performance of the ammunition may play a significant role in the extraction and ejection of fired cartridges, commercial ammunition is typically a highly refined and consistent product. Subsequent laboratory inspection of the evidence and test-fired cartridge cases can provide considerable insight into the actual performance of the evidence cartridges during the discharge, extraction, and ejection process.

With semiautomatic and fully automatic firearms, cartridge case location, as an expression of the shooter's approximate location, can be important in evaluating various accounts and theories related to a shooting incident. The accuracy of such determinations increases with an increase in the number of shots fired and/or trajectory information as a consequence of perforating bullet strikes to relatively fixed objects at the scene. For example, two compact groups of pristine, undamaged cartridge cases, all fired by the same semiautomatic pistol, at two very different locations clearly show a change in the shooter's position. If some of these shots have struck and perforated fixed objects, such as a wooden fence followed by a nearby wall of a building, considerable improvement in establishing the shooter's location will be realized by integrating the trajectory information with a cartridge case ejection pattern.

Trace evidence adhering to, or embedded in, recovered cartridge cases can provide very useful, sometimes critical information in certain instances, yet it is often overlooked. Shots fired from moving vehicles may have special value or importance. With sufficient testing over the same or comparable surface, some reasonable inferences as to the speed of a vehicle from which one or more shots is believed to have been fired can be drawn from the location of the cartridge case and the degree of impact damage sustained by the casing prior to its coming to rest. Trajectory information from the scene, if available, will greatly enhance the accuracy of such an evaluation.

Post-shooting incident damage to expended cartridge cases as a result of being stepped on or run over by vehicles is usually apparent and distinguishable from the small sites of abrasive damage around the rim or case mouth acquired by impact with hard surfaces such as concrete or asphalt.

Markings on cartridge cases produced by the firearm and/or the loading practices of the shooter may allow the *first* shot to be discriminated from subsequent shots. This would be very significant in a situation where the fired cartridges showed obvious movement by the shooter, for example the shooter was either advancing or retreating as he fired.

REFERENCES AND FURTHER READING

Garrison, D.H., Jr., "Reconstructing Drive-By Shootings from Ejected Cartridge Case Location," *AFTE Journal* 25:1 (January 1993).

Haag, L.C., "Cartridge Case Ejection Patterns," *AFTE Journal* 30:2 (Spring 1998).

CHAPTER 13

TRUE BALLISTICS: LONG RANGE SHOOTINGS AND FALLING BULLETS

INTRODUCTION

For several centuries there has been a special fascination with exterior ballistics and long range shooting. The fact that bullets travel faster than the eye can see immediately instills a degree of mystery and awe to the process of a bullet's flight from muzzle to target. Add to this the ability to strike very small targets at very long distances, the wonder intensifies. Long range shooting contests were once the sort of events that kings, queens, and presidents opened and often attended. Participating in the long range rifle matches at Camp Perry, Ohio, is the dream of almost every American rifleman. The seemingly incredible long range feats with muzzle-loading black powder rifles equipped with metallic sights in the late 1800s can be read about in books such as that by Ned Roberts and F.W. Mann. Such matches are still held every year in Raton, New Mexico, at a special range design for shooters of black powder firearms. The German *Schutzenfest* is another example of the use of beautifully crafted muzzle-loading rifles fired over long distances. These black powder rifles fire lead bullets at relatively low muzzle velocities (on the order of 1200–1400 fps). The invention of smokeless powder, stronger steels, and jacketed bullets moved muzzle velocities into the 2000–3000 fps regime. Modern breech-loading rifles using high velocity smokeless cartridges with jacketed bullets and optical sights have advanced the marksman's capabilities during the last century so that targets as small as an 8 in. pie plate can be consistently struck from a range of 1000 meters and a high quality benchrest rifle with a skilled marksman can regularly put 10 bullets in the same hole at 100 yd. The Summer and Winter Olympics contain several types of rifle competition with participants from many countries. None of these examples of long range marksmanship could take place if the flight of projectiles did not obey certain laws of physics and if scientists and engineers had not developed the necessary mathematics to both describe and predict their flight though the atmosphere. It is this same knowledge and the application of relatively complex equations through the use of modern computer programs that will be applied to shooting incidents involving shots from substantial distances.

THE BASICS OF EXTERIOR BALLISTICS AND THEIR FORENSIC APPLICATION

Whether deliberate or unintentional, modern bullets even from handguns can easily travel a mile or more if launched at a relatively high departure angle. Bullets from centerfire rifles are capable of several miles of flight before impacting the ground or other object. Although greatly slowed by air resistance, such bullets usually retain sufficient velocity and energy to produce serious and even fatal wounds.

A basic understanding of exterior ballistics only requires one to grasp a few fundamental concepts. While it is true that there are a number of forces acting on a projectile in flight that will never be accurately known, I would point out that there are typically more and greater uncertainties associated with an actual incident than those associated with any subsequent ballistic calculations we might carry out.

This is an important distinction between the forensic scientist dealing with a shot from long range and a ballistician intent on delivering a particular type of bullet or artillery projectile to a distant target with extreme accuracy. The military and research ballistician knows the muzzle velocity of his projectile and the existing meteorological conditions with exactitude. The forensic scientist, on the other hand, will never know the actual muzzle velocity associated with a long distance shot. The environmental conditions at a shooting scene might be known with a little more certainty depending on the scene location and knowledge of the time of the incident. However, the uncertainties that confront the forensic scientist are not fatal to many reconstructive efforts involving exterior ballistics. The substantial analytical power of contemporary exterior ballistics programs and speed of modern computers allow the effects of these variables on a bullet's flight to be assessed. Crosswinds can be introduced, and changes in air temperature, barometric pressure, relative humidity, and, of course, variations in muzzle velocity can all be isolated with these programs and the effect on bullet path and flight time calculated for a set of conditions.

The military ballistician or high-tech target shooter knows much more about his firearm, the performance of his ammunition, and the meteorological conditions under which the shot will be fired. But their purpose is to hit a specific target. In a forensic investigation a bullet has arrived at some known location after traveling a considerable distance. In this situation exterior ballistic calculations are used to locate a search *area* from which a long range shot could have originated and to rule out other areas. A subsequent search of this area may lead to the recovery of pertinent physical evidence or the location of important witnesses. Either or both of these outcomes can result in the solution of the incident and the apprehension of the shooter.

There are a number of parameters associated with a bullet's flight that are either essentially constant or readily measurable with modern instrumentation. The foremost of these is gravitational attraction. This is, for all practical purposes, constant and its effect on a bullet's flight is quite predictable. The average sea level value for the earth's gravitational acceleration is 32.174 f/s/s or 9.807 m/s/s. A fired bullet regardless of its muzzle velocity will be acted upon immediately after it leaves the muzzle and will ultimately fall to the ground. If, for example, a shot were taken with the centerline of the bore pointed directly at a distant target, a fired bullet would always strike low. It is therefore necessary to elevate the gun's bore *above* the shooter's *line of sight* (LOS) sufficiently to strike the intended target at a particular distance. The LOS is, of course, a straight line through the particular sighting system to the intended target or *point of aim* (POA). Since the sights on a firearm are typically above the centerline of the bore, the bullet's path (for a properly sighted-in rifle) will pass through the LOS (usually about 20–30 yd beyond the muzzle for most rifles), rise above it, and re-intercept it at the distance for which the gun and sight system have been set (the POA). Once a bullet has exceeded this distance it will continue to fall below the POA.

These properties and parameters of a bullet's flight from gun to target are depicted in an exaggerated form in Figure 13.1. An inspection of this figure

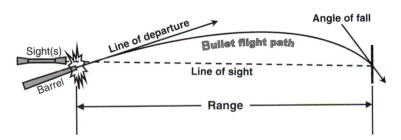

Figure 13.1

The Basic Parameters and Components of a Long Range Trajectory

This drawing shows a view of a long range trajectory with x-axis (range) greatly compressed and the y-axis expanded. The relationship of the sight to the axis of the barrel is especially exaggerated to illustrate the relationship between the line of sight and the path of the bullet. The effect of gravity acts on the bullet immediately upon its emergence from the barrel causing it to follow a curved path that is always below the line or departure. It should also be noted that the angle of fall is always greater than the angle of departure and the highest point in a bullet's trajectory is beyond the true midpoint.

reveals that a bullet will either strike high or low on a target or object at any point other than the two points where it intercepts the LOS. These two points can be referred to the *near-zero* and the *far-zero*. The distances to these two points are dependant on the nature of the sighting system (it's height above the centerline of the bore), the angle between the axis of the bore and the LOS, the muzzle velocity of the bullet, the bullet's exterior ballistic characteristics, and the properties of the atmosphere through which the bullet moves.

The atmosphere is the other major force acting on a projectile in the form of air resistance. This is the same force we feel on our hand when we hold it out the window of a fast-moving car. While there are some weaker and much more subtle forces acting on a bullet in flight, gravity and drag are the two most important. Those readers who desire a more thorough description of these other forces are directed to the excellent article in the *AFTE Journal* by Ruprecht Nennstiel. The extensive section on exterior ballistics in the *5th Edition of the Sierra Reloading Manual* or the very cerebral book by Robert McCoy are two additional choices. The primary forces of gravitation attraction and air resistance (drag) are depicted in Figure 13.2.

Air resistance slows the bullet down and gravity pulls it back to earth. At high velocity, especially at supersonic velocities, the deceleration due to air resistance is substantial. At subsonic velocities it is much less. Figure 13.3 shows a typical plot of velocity vs. distance for a 7.62NATO bullet. This plot is quite useful in showing the initial steep drop in velocity when this bullet is substantially supersonic. The inflection point in this otherwise smooth curve at about 340 m/s (ca. 1115 fps) is the nominal speed of sound below which one can see that the bullet's velocity loss per unit of flight distance nearly flattens out. The plotting of velocity vs. time gives a very similar graph.

Figure 13.2

The Primary Forces Acting on a Bullet in Flight

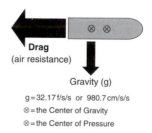

Drag
(air resistance)

Gravity (g)

g = 32.17 f/s/s or 980.7 cm/s/s
⊗ = the Center of Gravity
⊗ = the Center of Pressure

The primary forces acting on a bullet are aerodynamic drag and gravitational attraction. Gravitational attraction is a constant accelerating force whereas drag is constantly changing. Drag forces are many times that of gravity at high velocities, particularly when the projectile is supersonic. There are other more subtle and much smaller forces acting on projectiles in flight that are not illustrated here.

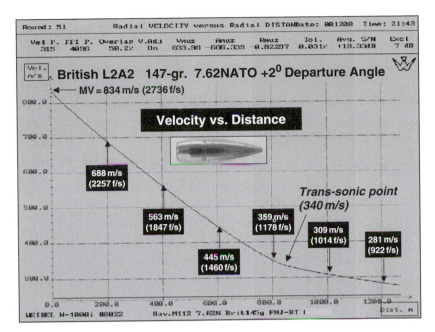

Figure 13.3

A Plot of Velocity vs. Distance for a Supersonic Rifle Bullet

This plot illustrates many interesting properties of a bullet in flight. Note the very rapid loss of velocity over distance while the bullet is supersonic followed by a sudden change in slope at the transonic point followed by a much reduced loss of velocity in the subsonic region.

The slowing of a projectile in flight can be expressed by the equation:

$$F = -\tfrac{1}{2}\rho V^2 A C_D$$

where F is the decelerative force or drag, ρ (rho) is the density of the atmosphere (which is dependent on barometric pressure, altitude, temperature, and humidity). The ICAO standard sea level value for the density of air is 0.076474 pounds per cubic foot (0.01226 grams per cubic centimeter). The standard atmosphere is 59 °F, has 78% relative humidity and a pressure of 29.53 in. of Hg (750 mm Hg). V in the equation is the velocity, A is the cross-sectional area of the projectile, and C_D is the drag coefficient. C_D is an experimentally derived factor that makes the equation fit the data. The drag coefficient is *not* a constant and varies with velocity as well as the properties of the projectile including its stability in flight. Figure 13.4 shows the relationship between drag coefficient (C_D) and bullet velocity expressed as Mach number for the same 7.62NATO bullet. This figure was produced from a Doppler radar track of an actual shot fired from one of the author's rifles. The Mach number is determined by dividing the velocity of the projectile by the speed of sound for the atmospheric conditions present at the time of the shot. Note the radical drop in drag coefficient as the bullet passes

Figure 13.4

A Plot of Drag Coefficient vs. Mach Number for a Supersonic Rifle Bullet

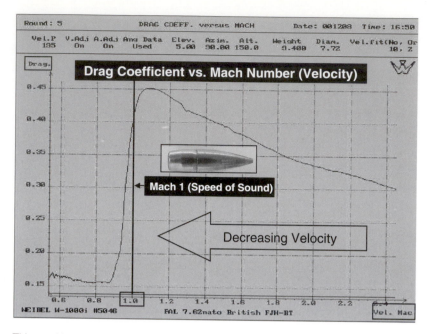

This graphic generated from a Doppler radar track of a 7.62NATO bullet shows the gradual increase in drag coefficient as the supersonic bullet approaches the speed of sound (Mach 1 under the site conditions) followed by a rapid drop to a much lower and near-constant value in the subsonic region.

through Mach 1. The V^2 parameter in the equation should help one understand the rapid increase in decelerative force with increasing velocity. For example, a velocity difference between 100 and 3000 fps represents a 30-fold increase in velocity but amounts to a 900-fold increase in the decelerative force acting on the bullet based on the square of velocity in the drag equation.

It is not so much that this formula will be of use in casework. Rather it is presented for those readers interested in the scientific basis of exterior ballistics. The decelerative drag force (F) can be thought of as an ever-changing (because of changing velocity and changing drag coefficient) braking force applied to the nose of the bullet. These forces are substantial at high velocities such as 2700 and 3200 fps (the nominal muzzle velocities of the 7.62NATO and M193 5.56 mm bullets). By way of an example, the drag equation was used along with the C_D values from Figure 13.4 to calculate the drag forces acting on the nose of the 7.62NATO bullet in the standard atmosphere and at velocities of 2700 and 900 fps (Mach 2.4 and 0.8 respectively). These computations yield drag forces of 1.39 and 0.080 lb. for the two selected velocities acting on a bullet that weighs 0.021 lb. These, in turn, translate to decelerations of 2130 and 123 f/s/s (feet per second per second). These decelerative forces are many times that of gravity (32.174 f/s/s) and should help the reader understand and appreciate the primary forces acting on a bullet in flight. An inspection of the drag equation also shows that those factors that increase or decrease the density of the atmosphere (ρ) directly affect the decelerative force (F). The variations in atmospheric conditions are easily handled by all of the currently available exterior ballistics programs for PCs.

The *Ballistic coefficient* is another type of form-fitting factor that can be related to the performance of a standard bullet. The ballistic coefficient (BC) adjusts or scales the standard drag deceleration of a standard bullet to fit the non-standard bullet. This non-standard bullet is our evidence bullet. There have been a number of "standard bullets" developed and thoroughly studied over the last century but the one that has become the gold standard is the G_1 bullet and the exterior ballistic tables that go with it. No one is likely to be shot with, or shot at by, a standard bullet because it is a 1 in. diameter, 1 lb. bullet with a form or shape factor of 1. The ballistic coefficient (BC) of the standard G_1 bullet is 1.00 by definition. The BCs of actual bullets are derived from actual test-firings and performance measurements. Another way to view the numerical values for ballistic coefficients is as a performance indicator. Bullets with high BCs will retain their velocities better than bullets with lower BCs. An example should improve the reader's understanding of BC and the effect of bullet shape on exterior ballistic performance. The standard bullet fired under standard atmospheric conditions at sea level with a muzzle velocity of 3000 fps will have a velocity of 2902 fps at 100 yd. The *Sierra Bullet Company* makes two, 150 gr., .30-caliber jacketed rifle bullets that have very different nose shapes. Their sharply pointed FMJ-BT bullet (very similar to the 7.62NATO bullet) has a nominal BC of 0.40 for velocities in the area of 3000 fps. This bullet will have a 100 yd velocity of about 2759 fps. Their round nose bullet of the same weight has a BC of about 0.20. With a muzzle velocity of 3000 fps, this bullet will fall to 2531 fps after 100 yd of flight. An inspection of the velocity *loss* values should bring the concept of ballistic coefficient into sharper focus. The standard bullet lost 98 fps over the 100 yd flight. The .30-caliber FMJ-BT bullet lost 235 fps and the RN bullet lost 469 fps over the same distance. Dividing 98 by 241 and 448 gives 0.41 and 0.21 respectively. It should also be noted that the velocity loss for the round nose bullet with the nominal BC of 0.20 was *twice* that of the pointed bullet with the higher BC of 0.40. This is not coincidence.

The ballistic coefficient for a particular bullet typically does not perfectly parallel the performance of the standard bullet at all velocities and consequently varies with velocity. As a result, some databases for exterior ballistic programs provide adjusted BCs for selected velocity zones. Such programs will improve the accuracy of any computations but are not absolutely critical for most forensic applications.

Not knowing the exact ballistic coefficient of a particular brand, weight, and design of a bullet is not critical so long as we have a reliable or measured BC close to the expected muzzle velocity and over the range of fire involved in the case under investigation. BC values for virtually all common military and commercially available bullets are readily obtainable from a number of sources. They can also be measured with the Oehler M43 PBL chronograph system or derived from actual performance data. This equipment is easily within the budgets of nearly any laboratory and is capable of providing reliable and useful measures of the G_1 ballistic coefficients of any projectile that can be accurately fired over a 50–100 yd distance. Whether taken from the bullet manufacturer's literature, ballistic reference sources, Doppler radar data, or derived with instrumentation such as the Oehler M43 system, it is the G_1 ballistic coefficient that is needed and utilized by all currently available exterior ballistics programs for PCs to carry out any number of exterior ballistic calculations.

CASE SITUATIONS—AN OVERVIEW

The shooting incidents considered in the previous chapters have all involved relatively close range events where the distances are sufficiently short that the flight time of the bullet amounts to a few thousandths of a second and the flight path of the bullet is, for all practical purposes, a straight line. In these close range situations any objects or victims that were in motion at the time of a shot can be considered stationary insofar as the bullet's flight is concerned. Think of these situations like a flash photograph where the resultant picture shows people known to have been in motion frozen in time and space. A 115 gr. 9 mm bullet with a muzzle velocity of 1250 fps fired from the sidewalk 10 feet away from a passing car will traverse this distance, strike and perforate the sheet metal of a car door, and then strike the driver in a time frame of about 0.01 sec. At 30 mph, the car will have only moved 5 in. between the discharge of the gun and the impact of the bullet in the driver. In this situation, the pistol would have been pointed at the driver's door when it was discharged. Now consider this same bullet traveling over a distance of 500 yd requiring a flight time of about 1.7 sec. It arrives with a residual velocity of about 665 fps-sufficient to perforate the sheet metal door and interior door panel and wound the driver. Questions that immediately come to mind are: Where was the vehicle when the shot was fired? Where was the gun pointed when the shot was fired? What were the shooter's view of the car and the ultimate impact area of the bullet when the shot was fired?

Long distance shootings (100 yd and beyond) are more complicated and more interesting from a scientific and reconstruction standpoint. In these situations the curved flight path of the bullet and the bullet's flight time may become important issues. Likewise, back-extrapolations in an effort to locate the shooter's probable position must now consider fundamental terminal and exterior ballistic phenomena. Figure 13.1 shows the principle aspects of long distance shooting. In this diagram, the positioning of an optical sight and the associated LOS have been exaggerated to illustrate the relationship between the LOS and the path of the bullet. If a cardstock witness panel were to be located at any point other than the previously defined near-zero and far-zero positions, we would have a bullet hole above or below the LOS. This is referred to as the *point of impact* (POI). If one were looking through the rifle's optical sight for any one of these shots, the position of the crosshairs on the witness panel would be referred to as the POA. LOS and POA are, in effect, the same thing. It is only a matter of the view. In a profile view such as Figure 13.1, LOS would be applicable. Looking through the sights of the gun, POA would apply. The vertical distance between the LOS and the path of the bullet is one of the many items of information provided by contemporary exterior ballistics programs that is of value to the shooting reconstructionist. If one has the necessary information regarding the gun, ammunition, range of fire, scene topography, and meteorological conditions, an appropriate exterior ballistics program can provide useful information regarding the relationship between the sight picture (POA) and the entry wound position (POI). These same programs provide bullet drop data, time of flight (TOF) data, downrange velocity, energy and momentum of the bullet, as well as the amount of wind drift for any particular

Range (yd)	Velocity (fps)	Bullet Path (in.)	Bullet Drop (in.)	Time of Flight (sec.)
0	2800	−1.50	0.00	0.00000
20	2769	−0.48	−0.09	0.02155
40	2738	0.35	−0.36	0.04334
60	2708	1.00	−0.82	0.06538
80	2677	1.46	−1.46	0.08766
100	2647	1.73	−2.30	0.11020
120	2617	1.80	−3.34	0.13300
140	2587	1.67	−4.58	0.15606
160	2556	1.33	−6.02	0.17939
180	2526	0.77	−7.68	0.20300
200	2496	0.00	−9.56	0.22689

Table 13.1

Trajectory for .308 diameter 168 gr. JHP-BT at 2800 fps Muzzle Velocity

G1 Ballistic Coefficients of 0.462 @ 2600 fps and above and 0.447 @ 2100 to 2600 fps
Site Altitude of 7000 ft. MSL
Barometric pressure = 29.92 in. Hg
Temperature = 55 °F
Humidity = 30%
Crosswind = 0 mph
Far Zero = 200 yd
Rifle Sight Height above Bore = 1.5 in.

crosswind. Any of the meteorological parameters previously described can be varied and the effect on the bullet's flight calculated. Table 13.1 is an abbreviated ballistics table for a 168 gr., .30-caliber bullet used by many S.W.A.T. marksmen in the United States.

A table very similar to this was prepared for a high-profile case involving a relatively long distance shot by a law enforcement marksman who was positioned approximately 200 yd from his target located across a small canyon. The shooter was located on a mountain side at an elevation of about 7000 ft. MSL. The nominal muzzle velocity of this bullet–rifle combination was 2800 fps. The rifle was sighted in for 200 yd (POA = POI at 200 yd). The BC values for the bullet were 0.462 for the velocity region of 2600 fps and above and 0.447 for the region of 2100–2600 fps. An inspection of this and other similar tables provides further insight into the flight of bullets through the atmosphere. For example, the time of flight for this bullet to cover 200 yd is a little less than $1/4$ sec. This is a potentially useful piece of information since it bears upon how much movement a target or adversary could undergo while the bullet is in route. If the target was actually at a distance of 160 yd, the bullet would have struck 1.33 in. above the point of aim as viewed through the telescopic sight. If one could have directed a laser through the bore of the rifle just as the bullet was fired, the laser beam would have illuminated a spot 9.56 in. above the bullet point of impact 200 yd away. A simple arctangent calculation using the Bullet Drop value of 9.56 in. and the Range of 200 yd (7200 in.) can provide an estimate of the Departure Angle (Line of Departure) for the particular shot; in this case $+0.076°$ if the target was at the same elevation as the shooter.

A more familiar plot is shown in Figure 13.5. This shows a profile view of a maximum range shot for 00 buckshot tracked by Doppler radar. The classic asymmetrical shape of a trajectory in air is clearly evident in this graph. It also shows that the *Angle of Fall* is significantly steeper than the *Angle of Departure*.

For a particular projectile, muzzle velocity, and environmental conditions, there is a relationship between the angle of fall at the impact site and the range from which the shot occurred. Sophisticated programs such as Nennstiel's EB4.1 allow the user to reconstruct and back-extrapolate a bullet's trajectory based, in part, on terminal ballistic information from the scene. With a little work, the trial-and-error method can be used with simpler programs to accomplish the same thing. Table 13.2 provides a computation with *Sierra's Infinity* program for a standard, 230 gr., FMJ-RN .45 Automatic bullet fired at sea level out to a distance of 1000 yd over level terrain. The *Angle of Fall* can be calculated by taking the *change in height* for the Bullet Path over a relatively short distance at the end of the trajectory (1000 yd in this example). For this purpose, the Range Increment value was reset in this computation to give 5 yd intervals (see the lower bold values in Table 13.2). A height change of 36.44 in.

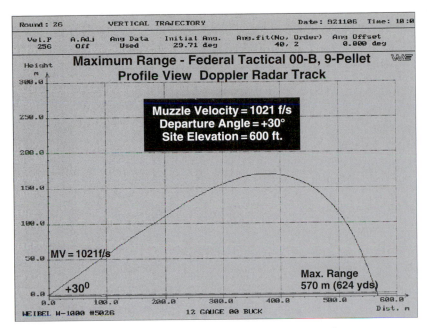

This profile view of a maximum range shot for 00-buckshot was derived from a Doppler radar track and is similar in many respects to the drawing in Figure 13.1.

over a distance of 5 yd (180 in.) forms a right triangle and the tangent of the *Angle of Fall* is equal to 36.44 divided by 180. The use of a pocket calculator with $\tan^{-1}$ capabilities will show the *Angle of Fall* to be about $-11.4°$. Inserting this value in Nennstiel's EB4.1 program along with the estimated muzzle velocity of a standard, 230 gr., .45 Automatic bullet of 850 fps gave a calculated range of fire of 986 yd, a flight time of 5.32 sec., and an *Angle of Departure* of $+6.91°$; good agreement with the 1000 yd and $+6.95°$ values derived from the *Sierra Infinity* program.

The starting point at a shooting scene for any retrograde trajectory extrapolation is either angle of fall information or penetration information or a combination of the two. Consider the following hypothetical case. A bullet, later determined to be a standard, 230 gr., FMJ-RN .45 Automatic bullet, enters a residence through a screen window, nicks the edge of a table after which it just has sufficient energy to perforate a sheetrock wall and fall to the bottom of the space behind this perforated surface. No recognizable mark is left on the next surface about 4 in. downrange of this bullet hole. A check of the alignment of these three points with a laser positioned outside the screen window shows them to form a straight line over a distance of 10 feet. The vertical component of the bullet's entry into the residence measures $-12° \pm 0.5°$. In this example, we have both trajectory information and terminal ballistic information in the form of penetration/perforation behavior. While *two* reference points *may*

Table 13.2

Trajectory for .452 in. diameter 230 gr. FMJ-RN .45 Automatic Bullet

Range (yd)	Velocity (fps)	Bullet Path (in.)	Bullet Drop (in.)	Time of Flight (sec.)
0	850	−0.50	0.00	0.000
50	814	212.77	−6.19	0.180
100	781	412.95	−25.48	0.368
150	751	598.89	−59.00	0.564
200	722	769.42	−107.93	0.768
250	695	923.27	−173.54	0.980
300	669	1059.11	−257.17	1.200
350	645	1175.49	−360.25	1.429
400	621	1270.88	−484.32	1.666
450	598	1343.61	−631.06	1.913
500	577	1391.87	−802.26	2.170
550	556	1413.72	−999.87	2.436
600	536	1407.04	−1226.01	2.714
650	516	1369.53	−1482.99	3.002
700	498	1298.66	−1773.32	3.301
750	480	1191.71	−2099.73	3.613
800	463	1045.68	−2465.23	3.938
850	447	857.28	−2873.09	4.276
900	432	622.91	−3326.93	4.628
950	417	338.60	−3830.69	4.995
1000	403	0.00	−4388.76	5.378
995	**405**	**36.44**	**−4330.38**	**5.339**
1000	**403**	**0.00**	**−4388.76**	**5.378**

850 fps Muzzle Velocity
G_1 Ballistic Coefficient of 0.166
Sea Level Site Elevation
Barometric pressure = 29.92 in. Hg
Temperature = 59 °F
Humidity = 60%
Crosswind = 0 mph
Far Zero = 1000 yd
Rifle Sight Height above Bore = 0.5 in.

represent a straight line, *three* points in alignment *do* define a straight line. This obviates any concern regarding deflection in this hypothetical case, therefore the three points provide a very reliable expression of this bullet's final flight path. The normal muzzle velocity of .45 Automatic ammunition loaded with this bullet is well known (ca. 825–850 fps). The observable facts fit a long range shot. Conversely, a shot fired from somewhere immediately outside the window with a downward angle of 12° can be effectively eliminated because of the minimal penetration of the evidence bullet. Occupants of the house upon being interviewed only report hearing what they thought was the impact of a rock or golf ball inside the house. No loud sound like a gunshot was heard nor was any fired cartridge case found on the property along a back-extrapolation of the laser path. Controlled tests could also be carried out in the laboratory to establish and refine the penetrative capabilities of this type of ammunition striking sheetrock at various impact velocities. Ultimately, one of the many contemporary exterior ballistic programs can be used to calculate various angles of fall from selected ranges until a nominal value of $-12°$ is obtained. This may require multiple computations, but once the approximate range is found, then other parameters such as flight time, terminal velocity, and angle of departure can be extracted from the ballistic data. Aerial views, topographical maps, city or county maps, in conjunction with the range-of-fire estimates and the azimuth of the bullet's approach, are then used to establish the search area. Let us further assume in this example that a teenage subject is found and later admits to taking his father's pistol and firing a shot at a bird on top of a telephone pole located in his backyard. The astute investigator asks the shooter to show him where he stood and how he held and sighted the pistol. This angle is later measured and found to be $+35°$. There is little need to even run the calculations in this example because the angle of fall ($-12°$ in this case) for projectiles fired in the atmosphere will always be greater than the angle of departure. In this example, the angle of fall is *less* than the angle of departure. The final exercise should involve returning to the shooter's location and photo-documenting any view of the victim's residence from the shooter's location and some means of illustrating the calculated $+7°$ departure angle for the actual shot. This parameter was calculated from the arctangent relationship of the 1000 yd drop value of 4389 in. (see Table 13.2) divided by the distance of 1000 yd (36,000 in.).

MAXIMUM RANGE TRAJECTORIES

The explanation given by the shooter in the hypothetical case raises the issue of maximum range for small arms projectiles. There is published data on this subject but several of the contemporary exterior ballistics programs will provide reasonably accurate answers to this question.

The *maximum* range of typical small arms projectiles discharged in the atmosphere is achieved at departure angles of 30–35°, not 45° as one might expect from the elementary physics of such an event. Table 13.3 provides some sea level examples for a number of common bullets calculated with the *Sierra Infinity* program. This program also allows other site elevations and environmental conditions to be entered and the maximum range recalculated. As one would expect, the thinner air at higher altitudes and/or higher temperatures will result in a slightly longer flight in both time and distance. The ability to set a maximum range value for a particular bullet is useful in casework because such a calculation sets the *maximum distance* from which a

Table 13.3

Some Maximum Range Calculations for Some Common Bullets (Standard Sea Level Conditions*—Sierra Infinity Program)

Bullet Description	Muzzle Velocity (fps)	Departure Angle	Maximum Range (yds)
Pistols			
50 gr. FMJ 25 Auto	760f/s	+30°	1212yds
71 gr. FMJ 32 Auto	905f/s	+30°	1630yds
95 gr. FMJ 380 Auto	955f/s	+27°	1176yds
115 gr. FMJ 9 mmL	1190f/s	+30°	1885yds
Note: YPG 1996 Shot # 127 9mm 115-gr. FMJ @ +35°/600 ft. MSL/21°C/MV = 1167 fps traveled 1982 yd, $V_{REM.}$ = 281 fps, Angle of fall = −67°			
124 gr. FMJ 9 mmL	1120f/s	+30°	1920yds
158 gr. LRN 38Spl	850f/s	+31°	1900yds
158 gr. JSP 357Mag	1235f/s	+30°	1955yds
180 gr. JHP 40S&W	1015f/s	+30°	2094yds
Note: YPG 1996 Shot # 135 40 S&W 180-gr. FMJ @ +35°/600 ft. MSL/21°C/MV = 1025 fps traveled 1878 yd, $V_{REM.}$ = 290 fps, Angle of fall = −67°			
240 gr. JSP 44 Mag	1180f/s	+29°	2112yds
185 gr. JHP 45 Auto	1000f/s	+30°	1862yds
230 gr. FMJ 45 Auto	850f/s	+31°	1806yds
Note: YPG 1996 Shot # 140 45 Auto 230-gr. FMJ @ +35°/600 ft. MSL/21°C/MV = 880 fps traveled 1780 yd, $V_{REM.}$ = 288 fps, Angle of fall = −67°			
Rifles			
55 gr. FMJ 5.56 mm	3200f/s	+28°	3414yds
110 gr. FMJ 30Carb	2400f/s	+28°	2577yds
123 gr. FMJ 7.62 mm	2300f/s	+29°	3417yds
168 gr. JHPBT 30-cal	2800f/s	+30°	4905yds
123 gr. FMJ 7.62 mm	2300f/s	+29°	3417yds
250 gr. SPJ 338Win	2600f/s	+31°	4954yds
300 gr. JHP 45–70	2000f/s	+30°	2920yds

* Sea level 59 °F 78%RH 29.53 in. Hg.
YPG = U.S. Army Yuma Proving Grounds.

recovered bullet could have come. It is also useful in the previous case of the youthful shooter. Had he truly fired a shot at a departure angle of +35°, the bullet would have traveled about 1800 yd and returned to earth at an angle of fall of −67°.

A number of firings for maximum range have been carried out with some of these bullets at the U.S. Army Yuma Proving Grounds at Yuma, Arizona, during the period of 1992 through 2005. The Doppler radar tracks and tabular data from these shots provide a very useful means of comparison with the calculated values. Several of these high angle, maximum range shots are included in the data in Table 13.3.

LAGTIME

There is another interesting property associated with long range gunfire that has forensic implications and value. With the blast of the gun, sound marches forward and toward the target at a constant speed. A supersonic bullet will be forward of the sound front and arrive at the target or some downrange point ahead of the sound of the gunshot. A bullet with a muzzle velocity below that of sound will arrive at the target after the arrival of the sound of the gunshot. But it can, and often is, more complicated than these two examples. The forward speed of a bullet is constantly changing due to air resistance but the speed of sound remains the same at the particular site location and meteorological conditions. This means that a supersonic bullet that is initially out in front of the sound of the shot can rapidly become subsonic and continue to lose velocity. In this situation the sound of the shot will eventually overtake and pass the decelerating bullet and arrive at the target before the bullet. The relationship between the bullet's flight time and the time for the sound of the gunshot to arrive at some selected downrange location is called *Lagtime*. If, in some future case, we have a serendipitous audio recording of the arrival or passage of the bullet at a known location and the arrival of the sound of the distant gunshot, we have a very useful piece of information that relates to the range of fire. Such cases are not unheard of and the author has used this information in several cases to assess the distance from which a shot could and could not have come. Let us return to the residence where the .45 Automatic bullet came through the screen window and imagine that the homeowner was leaving a voicemail message on a friend's telephone. We learn this because an investigator asked the homeowner what he was doing when he heard the impact of the bullet. The flight time for the .45 Automatic bullet over a distance of 1000 yd was calculated to be 5.38 sec. The speed of sound at the site elevation and conditions is later determined to be 1125 fps. At this speed, it will take sound 2.67 sec. to cover the 3000-feet (1000 yd) distance. Subtracting 2.67 sec. from 5.38 sec. gives 2.71 sec. An analysis of a copy of the friend's telephone message tape shows a faint "pop" 2.70 sec. before the much louder impact of the bullet and the expletive of the homeowner.

This same concept can be used to test people's accounts of a shooting incident. It is not uncommon that in non-injury cases a complaining witness/victim will describe in vivid detail the sound of a shot and the passage of the bullet just inches away from his head and ear. No bullet or bullet impact site is found, so investigators are left with this account. The account, of course, may be true and correct in some circumstances but it is relatively easy to evaluate if we know something about the gun and ammunition purportedly involved. In this example, the complainant states that while standing in his garden, he saw a hunter with a rifle just beyond a fence at the far end of his property (later determined to be 300 yd away). He heard a shot coming from the location of the hunter then heard the bullet pass by his ear with a hissing sound. A fired .308 Winchester caliber cartridge is later found at this location. The hunter is also located and admits to firing a shot with his .308-caliber rifle but in a totally safe direction. The gun and ammunition are impounded and subsequently tested in the laboratory. The average muzzle velocity of the ammunition is 2800 fps ± 25 fps and the ballistic coefficient of the bullet is 0.35. Even the most basic exterior ballistic program will reveal that this bullet is still supersonic at 300 yd downrange. Such a bullet cannot and will not produce a hissing sound as it passes by a listener. Instead it will produce a loud, sharp "crack" like that of a bullwhip. This is due to the shockwave generated by supersonic bullets (see Figure 13.6).

Figure 13.6

Shadowgraph of a Supersonic Rifle Bullet in Flight

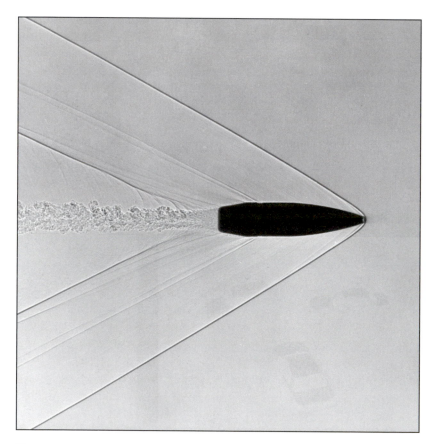

This shadowgraph provided by Dr Beat Kneubuehl of Switzerland shows the multiple shockwaves generated by a supersonic rifle bullet. It is these shockwaves that are responsible for the loud "crack" heard *after* such a bullet passes one's location. Wake turbulence can also be seen behind this bullet. Close inspection of this photograph will also reveal the reduced drag benefits of the boattail shape. Wake turbulence makes a relatively small contribution to the total aerodynamic drag on a supersonic bullet. Once this bullet becomes subsonic, the shock waves disappear and the aerodynamic drag becomes substantially smaller.

A subsonic bullet, on the other hand, will produce a hissing sound as it forces its way through the atmosphere. Moreover, under the conditions in this case the bullet will arrive at the listener's location 0.4 sec. *before* the sound of the gunshot. The complainant's account is mistaken at best.

PENETRATION AND PROJECTILE DEFORMATION AS AN EXPRESSION OF RANGE OF FIRE

The depth of penetration in gunshot victims and various inanimate materials bears a relationship to impact velocity as does any deformation or damage suffered by the projectile as it produces the wound in the victim. Both are range-dependant parameters and can be evaluated through test-firings into the appropriate media using downloaded cartridges and a suitable chronograph to produce selected impact velocities in the particular medium. The recovered projectile must also be carefully examined for any evidence of impact with intermediate objects. Any bullet hole in clothing as well as the appearance of the entry wound is also of critical importance in addressing the possibility of ricochet, deflection, or the perforation of an intermediate object. Penetration depth *may* obey a linear relationship with impact velocity but this will have to be evaluated through test-firings in the appropriate medium. Non-linearity usually arises from varying amounts of bullet yaw in the test medium (e.g., ordnance gelatin or ballistic soap) or varying degrees of projectile deformation. For example, double aught (00B) buckshot pellets will penetrate more deeply in tissue and tissue simulants at 900 fps where they retain their spherical shape than at 1300 fps where they flatten upon impact and have a much larger area of presentation as they advance into tissue or tissue simulants. In the region from 300 fps to about 1000 fps their penetration depth will obey a linear relationship with impact velocity. The ultimate object in any evaluation of this sort is to prepare a series of fired projectiles into a suitable medium that show a reproducible relationship between impact velocity, penetration depth and projectile deformation. The hypothetical case of the residence struck by the .45 Automatic bullet is a relatively easy example. A section of the actual sheetrock wall could ultimately be used for ballistic testing after some preliminary shots at high and low velocities are fired into sheetrock of the same thickness. The objective would be to find that velocity where the bullet is just able to perforate the sheetrock and then to compare this value with the calculated residual velocities of long range shots.

BULLETS FROM THE SKY

Falling bullets from the reckless discharge of small arms in populated areas become a matter of concern every New Year's Eve and 4th of July in the United States. This concern is with some justification since occasional injuries and even deaths have occurred from apparent high-angle shots. Misinformation and misconceptions abound, however, regarding this subject. The most common is the claim (usually by a police chief or county attorney) that a bullet fired straight up returns to earth with the same velocity that it left the muzzle of the gun. This, of course, would only be correct if the earth had no atmosphere. A truly *falling* bullet from a vertical to near-vertical discharge will return to the surface at its terminal or free-fall velocity where the retarding drag force (air resistance) of the existing atmosphere equals the weight of the projectile. A bullet's terminal velocity will depend not only on its weight and shape but also on the *way* the bullet returns; base-first, nose-first, or tumbling. Although not at all likely, a nose-first return would result in the highest free-fall velocity. If the spinning bullet maintained its stability as it reached its apogee, it would very likely return base-first. If it lost its gyroscopic stability, a tumbling return would likely occur. Of the three choices, the tumbling mode would result in the lowest terminal velocity.

The matter of free-falling bullets has been of interest for many years. Numerous calculations and practical tests around the time of WWI and for the decade afterward addressed military rifle bullets of the day—the .30-'06, the .303 British, and the German 7.92 × 57 mm Mauser. The 174 gr., Mark VII, .303 bullet with a muzzle velocity of 2440 fps was calculated to rise to an altitude of 9000 feet in 19 sec. then return in 36 sec. for a round trip time of 55 sec. Actual firings produced round trip times of 48–51 sec. Computations for the 150 gr. flat base .30-'06 bullet with a muzzle velocity of 2700 fps carried out around 1920 gave a calculated round trip time of 49.2 sec. and a free-fall velocity of 300 fps. Out of 500 carefully fired vertical shots, Julian Hatcher was able to document four returning bullets all four of which were found to have impacted base-first. One of these struck a soft pine plank and left a 1/8 in. impression of the *base* of the bullet.

There is at least one program for PCs that will calculate vertical ballistics and the answers obtained appear to be in good agreement with empirical tests such as those cited above and tests carried out by the author with spherical projectiles. The program is called *Baltec1*. It requires the bullet's weight, diameter, length, the length of the ogive, the diameter of any meplat, the bullet's G_1 ballistic coefficient, muzzle velocity, and the elevation of the site for the vertical discharge. Nose-first, base-first, and tumbling modes can be selected for the bullet's return to the surface. The *Baltec1* program was used to calculate the vertical firing results shown in Table 13.4. It was also used to calculate values for the historic examples previously cited. These calculations were in good agreement with each other and provided a useful check of the program. An inspection of the terminal velocity column of Table 13.4 reveals that most common bullets fired vertically return to earth with velocities on the order of 150–250 fps. This is of considerable interest since the threshold velocity of the perforation of human

Table 13.4

Vertical Ballistics for Some Common Cartridges and Projectiles

Cartridge Name	Bullet Wt. (gr.)	Muzzle Vel. (fps)	Ballistic Coeff.	Max. Alt. (ft.)	Ascent Time (sec.)	Terminal Velocity	Descent Time (sec.)	Round Trip Time (sec.)
.22 Short	29 LRN	1095	0.098	3014	10.0	168-BF	21.5	31.5
						134-TU	25.0	35.0
.22 Long Rifle	40 LRN	1255	0.132	3867	12.5	198-BF	23.5	36.0
						142-TU	30.0	42.5
.25 Auto	50 FMJ	760	0.090	2288	9.4	191-BF	15.8	25.2
						146-TU	18.6	28.0
.32 Auto	71 FMJ	905	0.132	3342	11.7	187-BF	21.6	33.3
						158-TU	24.4	36.1
.380 Auto	95 FMJ	955	0.079	2450	9.4	187-BF	16.9	26.3
9 mmL (Win)	115 JHP	1255	0.142	4034	12.7	210-BF	23.4	36.1
9 mmL	124 FMJ	1110	0.172	4415	13.3	219-BF	24.6	37.9
.38Spl. (Rem)	158 LRN	755	0.142	3004	11.4	237-BF	17.4	28.8
.38Spl. Speer pdt. 4647	158 LRN	950	0.170	4040	13.2	241-BF	22.0	35.2
						182-TU	26.0	39.2
.38Spl. Speer pdt. 4623	158 LSWC	950	0.123	3296	11.6	238-BF	19.0	30.6
						167-TU	23.0	34.6
.40 S&W	180 FMJ	990	0.170	4142	12.9	231-BF	22.6	35.5
						180-TU	26.7	39.6
.41 Mag.	210 JSP	1300	0.165	4537	13.6	247-BF	23.3	36.9
.44 Mag.	240 JHP	1180	0.172	4519	13.6	249-BF	23.1	36.7
.45 Auto	230 FMJ	850	0.166	3661	12.7	230-BF	20.6	33.3
						192-TU	23.0	35.7
5.56 mm (.233Rem.)	55 FMJBT	3240	0.250	8024	17.0	244-BF	38.0	55.0
						141-TU	60.0	77.0
.30Carb.	110 FMJ	1990	0.166	5129	13.7	239-BF	26.0	39.7
7.62 × 39 mm	123 FMJ	2400	0.320	8556	19.0	264-BF	38.0	57.0
						158-TU	57.0	76.0
30–30Win	150 JSP	2390	0.217	6539	15.6	282-BF	28.7	44.3
.30-'06	180 JSP	2700	0.382	10,103	20.6	323-BF	37.5	58.1
#4 Buck	19.4 sph.	1350	0.025	994	5.3	124	10.6	15.9
00 Buck	53.5 sph.	1350	0.028	1341	6.4	149	12.1	18.5*

BF = Base First
TU = Tumbling
* Actual test-firings at 3000 ft. MSL gave 19–20 sec.

skin is on the order of 200–330 fps for typical small arms projectiles striking skin in their normal, nose-forward orientation. This substantial range of impact velocities for skin penetration arises from a number of sources: skin thickness and age, a shored or unshored impact site, and projectile nose shape. Regardless of these reasons and the relative wide range of values, the fact emerges that being struck by any of the most common bullets truly *falling* from a vertical to near-vertical discharge stands to be a very unpleasant experience but not one likely to produce a deep penetrating or fatal wound. This should, in no way, be seen as an endorsement or approval of the practice of discharging firearms into the air as a means of celebration or self-amusement. The most important fact to emerge from a study of the data in these tables is that relatively deep penetrating and fatal wounds, if the result of distant shots, are *not* vertical firings but rather high angle discharges (e.g., 20°–45° departure angles). In these situations the fired bullets typically arc over in flight and return to earth in a nose-forward orientation. Because of this, such bullets retain their velocity much better than a bullet tumbling from the sky or falling back to earth base-first.

There are other interesting observations to be made from the data in Table 13.4. All of these round trip flight times are such that the sound of the shot, if even heard by witnesses at a subsequent victim's location, will have typically occurred 30–45 sec. prior to the arrival of the falling bullet.

SUMMARY

The most important first step in any effort to identify a point, or more correctly, an area of origin for any long distance shot is that of exclusion. Determining from where the shot could *not* have come allows us to focus on more productive matters.

Aside from pure luck and/or informants, any hope of solving these long distance shootings should include:

- information regarding the appearance of the entry wound, a detailed description of the wound track, the structures involved, and the depth of penetration;
- the recovery and subsequent examination of any clothing perforated by the bullet, the careful examination of the bullet for evidence of ricochet or deflection prior to producing the wound, and any evidence of the bullet's orientation at the moment of impact with the victim (or nearby intervening object(s));
- the careful examination and evaluation of the scene to isolate those zones or corridors from which the injury bullet could and could not approach the victim;
- a careful search of the scene for any bullet-caused damage that could provide directional information regarding the path of the incoming bullet;
- any recollections of witnesses as to any distant sounds shortly prior to or subsequent to the victim's receipt of the injury (its character, loudness, and timing); and
- any serendipitous video or audio recordings that might include the sound of the shot and the arrival of the bullet.

REFERENCES AND FURTHER READING

Barnes, F.C., *Cartridges of the World*, 10th edn, Krause Publications, Iola, WI, 2003.

Di Maio, V.J.M., *Gunshot Wounds—Practical Aspects of Firearms, Ballistics and Forensic Techniques*, Elsevier Science Publishing Co., NY, 1985.

Fackler, M.L., "Ordnance Gelatin for Ballistic Studies," *AFTE Journal* 19:4 (October 1987).

Fackler, M.L., S.D. Woychesin, J.A. Malinowski, P.J. Dougherty and T.L. Loveday, "Determination of Shooting Distance from Deformation of the Recovered Bullet," *JFS* 32:4 (July 1987) pp. 1131–1135.

Garrison, D.H., Jr., "Reconstructing Bullet Paths with Unfixed Intermediate Targets," *AFTE Journal* 27:1 (January 1995) pp. 45–48.

Garrison, D.H., Jr., "The Effective Use of Bullet Hole Probes in Crime Scene Reconstruction," *AFTE Journal* 28:1 (January 1996) pp. 57–63.

Garrison, D.H., Jr., "Crown & Bank: Road Structure as it Affects Bullet Path Angles in Vehicle Shootings," *AFTE Journal* 30:1 (Winter 1998) pp. 89–93.

Haag, L.C., "Suggested Method for Calibration of Gelatin Blocks," *AFTE Journal* 21:3 (July 1989).

Haag, L.C., "Vertical Ballistics," *AFTE Journal* 22:1 (January 1990).

Haag, L.C., "Falling Bullets: Terminal Velocities and Penetration Studies," *IWBA Journal* 2:1 (1995).

Haag, L.C., "Extended Ballistic Properties of Some Law Enforcement 9 mmP Cartridges," *AFTE Journal* 29:3 (Summer 1997).

Haag, L.C., "Base Deformation as an Index of Impact Velocity for Full Metal Jacketed Rifle Bullets," *AFTE Journal* 33:1 (Winter 2001).

Haag, L.C., "Design, Exterior and Terminal Ballistic Performance of 5.56 × 45 mm SS109/M855 Bullets," *AFTE Journal* 33:1 (Winter 2001).

Haag, L.C., "The Forensic Uses of the Oehler Model 43 Personal Ballistics Laboratory System," *AFTE Journal* 34:1 (Winter 2002).

Haag, L.C., "The Exterior and Terminal Ballistics of 00 Buckshot," *AFTE Journal* 34:2 (Spring 2002).

Haag, L.C., "The Sound of Bullets," *SWAFS Journal* 24:1 (May 2002) also *AFTE Journal* 34:3 (Summer 2002).

Haag, L.C., "Skin Perforation and Skin Simulants," *AFTE Journal* 34:3 (Summer 2002).

Haag, L.C., "Sound as Physical Evidence in a Shooting Incident," *SWAFS Journal* 25:1 (January 2003).

Haag, L.C., "Light and Sound as Physical Evidence in Shooting Incidents," *AFTE Journal* 35:3 (Summer 2003).

Hatcher, J.S., *Hatcher's Notebook*, 3rd edn, The Stackpole Co., Harrisburg, PA, 1966.

LaGarde, L.A., *Gunshot Injuries*, 2nd edn, Lancer Militaria, Mt. Ida, AR, 1991.

Laible, R.C., ed., *Ballistic Materials and Penetration Mechanics*, Elsevier Scientific Publishing Co., Amsterdam, Oxford, NY, 1980.

MacPherson, D., *Bullet Penetration—Modeling the Dynamics and Incapacitation Resulting from Wound Trauma*, Ballistic Publications, El Segundo, CA, 1994.

Mann, F.W. and H.M. Pope, *The Bullet's Flight from Powder to Target*, Wolfe Publishing Co., Prescott, AZ, 1980.

McCoy, R.L., *Modern Exterior Ballistics—The Launch and Flight Dynamics of Symmetric Projectiles*, Schiffer Military History, Atglen, PA, 1999.

Moss, G.M., D.W. Leeming and C.L. Farrar, *Military Ballistics*, Brassey's, London-Washington, 1995.

Nennstiel, R., "Accuracy in Determining Long Range Firing Position of a Gunman," *AFTE Journal* 17:1 (January 1985).

Nennstiel, R., "Forensic Aspects of Bullet Penetration of Thin Metal Sheets," *AFTE Journal* 18:2 (April 1986).

Nennstiel, R., "Doppler Radar Applications in Forensic Ballistics," *AFTE Journal* 24:2 (April 1992).

Nennstiel, R., "How Do Bullets Fly?," *AFTE Journal* 28:2 (April 1996).

NRA Firearms Fact Book, 3rd edn, The National Rifle Association of America, Washington DC, 1989.

Rathman, G.A., "The Effects of Material Hardness on the Appearance of Bullet Impact Damage," *AFTE Journal* 20:3 (July 1988) pp. 300–305.

Roberts, N.H., *The Muzzle-Loading Cap Lock Rifle*, Wolfe Publishing Co., Prescott, AZ, 1991.

Sawyer, C.W., *Our Rifles, Volume III, Firearms in American History Series*, Pilgrim Press, Boston, MA, 1920.

Sellier, K.G. and B.P. Kneubuehl, *Wound Ballistics and the Scientific Background*, Elsevier Publishers, Amsterdam, 1994.

Stone, R.S., "Calculation of Trajectory Angles using a Line Level," *AFTE Journal* 25:1 (January 1993) pp. 21–24.

Trahin, J.L., "Bullet Trajectory Analysis," *AFTE Journal* 19:2 (April 1987).

CHAPTER 14

SHOTGUN SHOOTINGS AND EVIDENCE

INTRODUCTION

Shotguns present special challenges to crime scene investigators, medical examiners, and firearms examiners. This is largely due to the great variety and complexity of this type of ammunition compared to bulleted cartridges.

There are numerous types of wads, shot collars, shot cups, buffering materials, as well as many sizes and compositions of shot available in this type of ammunition. Figures 14.1, 14.2, and 14.3 provide three examples. All of the items shown in these photographs have important evidentiary and reconstructive value. They also have their own special ballistic properties that can play a vital role in establishing the distance from which the shot was fired and the location of the shooter at the moment of discharge.

One of the frustrating aspects of shotgun shootings is that it is very seldom possible to identify shot pellets as having been fired from a particular shotgun. There are rare exceptions to this statement when large shot (buckshot) loaded in old-style shotshells is used and the sides of these pellets rub against the bore of the gun during discharge (see Figures 14.3 and 14.4). Expended shotshells, on the other hand, can often be matched back to the responsible gun using the same principles employed for bulleted cartridges, i.e., firing pin impression, breechface signature, etc. The printed information on the size of an expended shotshell can often provide information regarding the size of the shot, the amount of shot, and the purpose of the load. This information can lead to other facts about the ammunition. For example, a trap load and a field load with the same size of shot and made by the same manufacturer will have different types of wadding or shotcups in them. Some American shotshells contain fine plastic granules mixed with the shot to act as buffering material. All of these components (wads, shot, shotcups, and plastic buffer) emerge from the muzzle of a shotgun upon discharge and all have useful exterior and terminal ballistic properties.

Figure 14.1

20-ga. Shotshell and Fired Components (see Color Plate 10)

Figure 14.2

Sectioned 12-ga. Shotshell Containing #8 Shot (see Color Plate 11)

These figures show several common types of wadding in contemporary shotshells. Multiple cardboard or fibrous wads may be in a single shotshell such as the 20-gauge shell depicted in Figure 14.1. The upper wad will take up impressions of the pellets during discharge as will the plastic shot collar unique to certain Winchester brand shells. The sectioned shell in Figure 14.2 contains a one-piece plastic shotcup in which the pellets are nested. Two of the four petals of this shotcup can be seen in this figure.

Figure 14.3

Sectioned 12-ga. Shotshell Containing 00-Buckshot

Figure 14.4

Fired 00-Buckshot Pellets from a Buffered Shell

Figure 14.3 shows a typical American buckshot load with plastic buffering material filling the spaces between the individual pellets. This particular shell contains both plastic and fiber wads. All of these components emerge from the muzzle of the gun and have exterior ballistic properties of reconstructive value. The three fired 00-buckshot pellets in Figure 14.4 show the characteristic "orange peel" imprint of the plastic buffering material acquired during discharge. One of these pellets has been oriented to show the area where it contacted the bore of the gun. This mark establishes the pellet as having come from a shell like that of Figure 14.3 as opposed to a shell employing a shotcup (such as shown in Figure 14.2) where the pellets are prevented from contacting the bore of the gun.

SHOTGUN DESIGN AND NOMENCLATURE

Shotguns may be semiautomatic, pump action, lever action, bolt action, or of break-open design. Many of the semiautomatic and pump action guns allow for quick removal and replacement of the barrels for different hunting and sporting purposes, so this should be kept in mind. Break-open designs are used in single shot and double-barreled guns. Double-barreled guns are manufactured with their barrels side-by-side or over-and-under. The bores of shotgun barrels are normally smooth (unrifled) but special purpose barrels with rifling are available. When they are encountered they are usually associated with deer hunting in the Eastern United States or in special law enforcement applications. The performance of these guns with shot cartridges is radically different than the same gun with a conventional smooth bore. The muzzles of the more common smooth bore guns may possess a certain amount of constriction. This is called "choke" and its function is to tighten the flight pattern of shot for certain hunting applications. It should be noted that the amount of choke is often different between the two barrels of double-barreled shotguns. Moreover, some shotguns have a replaceable choke that is threaded into the muzzle of the barrel, while other guns may have adjustable chokes. Since the type and amount of choke affects pattern size, it is of critical importance that no changes be made to any adjustable choke present on an impounded gun. The author would recommend marking the position of any adjustable choke in some non-invasive manner upon the seizure of such shotguns.

Shotguns were developed for small game hunting (ducks, quail, rabbits, etc.) although special purpose loads are available for deer hunting (buckshot and solid slug loads) as well as for personal protection, law enforcement, and military purposes. Less-lethal munitions using shotshells containing "bean-bags," rubber shot, and chemical munitions are also a relatively recent development in shotgun ammunition.

For traditional small game hunting the shells are loaded with selected sizes of spherical lead shot hardened with antimony. The large caliber lead spheres used in buckshot loads are usually composed of plain lead although they may possess an exterior plating of copper or nickel. The various sizes of shot were developed for different types of game. The various sizes and weights of shot were discussed and illustrated in Chapter 2 in the section entitled *the Worth of Weight*. High-flying geese with their thick layer of feathers, for example, require a larger size of shot than a rabbit that breaks and runs from under foot. Traditional shotgun pellets were made of lead but presently such pellets may be made of steel, tungsten, or bismuth, as well as composites of steel and tungsten. These developments are the consequence of concerns regarding lead contamination of lakes and waterfowl areas.

The size of the bore in shotguns is expressed in terms of *gauge*. The actual bore diameters for shotguns are given in Table 14.1. The four most common gauges are shown in italics.

Table 14.1

Shotgun Gauges and Bore Diameters

10 gauge 0.775 in. (19.6 mm)	*20 gauge* 0.615 in. (15.6 mm)
12 gauge 0.730 in. (18.5 mm)	28 gauge 0.550 in. (14.0 mm)
16 gauge 0.670 in. (17.0 mm)	*410 gauge** 0.410 in. (10.4 mm)

* *Not* a true gauge.

CHOKE AND PATTERNING

As previously pointed out, many shotgun barrels will have some amount of choke present at the muzzle. This constriction is usually described as *Full, Modified, Improved Cylinder, or Cylinder* choke with the last one being somewhat of a misnomer since it involves no constriction. The term *Cylinder Bore* may also be used for this situation. The type of choke machined into a shotgun barrel is usually marked somewhere on the barrel but this may exist in some codified form and not be immediately understandable. There are special, inexpensive gauges made for gunsmiths that will quickly determine the type of choke present if any but careful laboratory measurement of the inside diameter at the muzzle will resolve the question.

An inspection of Table 14.2 for a 12-ga. shotgun with a bore diameter of 0.730 in. should give the reader some idea of the amount of constriction involved for the various amounts of choke.

Table 14.2

Choke Constriction

Choke [12-ga.]	Muzzle Diameter (in.)	Constriction (in.)
Cylinder	0.730	0.000
Improved cylinder	0.721	0.009
Modified	0.712	0.018
Full	0.694	0.036

SHOT CHARGES AND DRAM-EQUIVALENTS

The numbering system for shot pellets, their diameters, and weights for individual pellets have been previously discussed in Chapter 2 but the *number* of pellets in a given shotshell is a separate matter. The total weight of a shot charge in a

Table 14.3

Lead Shot Pellets per shell

Shot Charge	American Shot Size							
	#2	#4	#5	#6	#7$^{1}/_{2}$	#8	#8$^{1}/_{2}$	#9
$^{1}/_{2}$ oz.	45	67	85	112	175	205	242	292
$^{3}/_{4}$	67	101	127	168	262	308	363	439
7/8	79	118	149	197	306	359	425	512
1	90	135	170	225	350	410	485	585
1$^{1}/_{8}$	101	152	191	253	393	461	545	658
1$^{1}/_{4}$	112	169	213	281	437	513	605	731
1$^{3}/_{8}$	124	186	234	309	481	564	665	804
1$^{1}/_{2}$	135	202	255	337	525	615	730	877
1$^{5}/_{8}$	146	220	276	366	569	666	790	951
1$^{7}/_{8}$	169	253	319	422	656	769	850	1097
2	180	270	340	450	700	820	910	1170

shotshell is given in ounces for American shotshells. In addition to the weight of the shot charge, the manufacturers' markings on the sides of most shotshells usually include the size of shot within the shell and a dram-equivalent value that relates to the velocity of the particular load.

The total number of pellets in a particular loading of shell will be determined by the size of the pellets and the charge weight. The tabulation in Table 14.3 gives the approximate number of pellets per shell for lead shot sizes #2 through #9. The exact number will vary slightly depending on exact alloy content and slight variations in pellet diameter. Different charges of shot are available in different product lines in the same gauge and from the same manufacturer. A 12-ga. target shell may contain only 1 oz. of #8 shot but a heavy field load by the same manufacturer may contain 1$^{1}/_{4}$ oz. of the same size of shot. An inspection of the table in Figure 14.3 reveals that there are approximately 410 pellets in the 1 oz. load and 513 pellets in the 1$^{1}/_{4}$ oz. load. This example illustrates the importance of counting the total number of pellet impacts when it is reasonably certain that the entire shot charge is represented at the scene of a shotgun shooting.

WADS AND SHOTCUPS

Examples of several styles of contemporary shotshells were previously illustrated in Figures 14.1, 14.2, and 14.3. The several types of wads and one shotcup shown in these figures all serve the same purpose; namely to seal off the powder gases from the shot charge and to drive the shot out of the barrel. The class characteristics of these components are also of special importance in determining the gauge of the responsible gun and the brand of the ammunition, but in addition they have useful exterior ballistic properties.

At very close range (inches to perhaps several feet) they will follow the shot charge into any gunshot wound. The pellets in the shot charge will still be very close together resulting in a single, large entry hole or wound followed by "billiard-balling" of the shot pellets inside the body. As the range increases, these wads or shotcups typically fall out of alignment with the shot charge and will strike to one side of the entry wound. They still have sufficient velocity and energy to produce satellite wounds on the body and defects in any clothing. Another very interesting and useful phenomenon associated with shotcups (see Figure 14.3) is the fact that the petals open out during the very early phase of flight as they encounter air resistance. Shotcups of this general type are made by all U.S. manufacturers of shotshells and they all go through a cycle of opening, then reversing ends (due to air resistance and an aft center of gravity) and then quickly lose velocity because of their light weight. The important phases of this cycle are illustrated in Figure 14.5. This opening and reversal cycle typically occurs in the first 3–4 feet of travel and has very important reconstructive value. At such close distances the shot charge is still en masse and will produce a single entry defect. In this situation, muzzle-to-wound distance determinations based on the entry wound appearance cannot usually distinguish between 12 in., 18 in., and 24 in., for example. The impact marks and/or impact

Figure 14.5

In-Flight Reversal of Shotcups

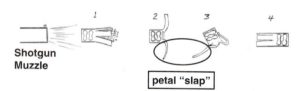

This drawing illustrates the four major phases of the exterior ballistic behavior of plastic shotcups fired from smoothbore guns. Within a few inches of the muzzle, the petals begin to open up as they encounter air resistance. At this stage the pellets are essentially en masse and still within the shotcup. Phase 2 shows the petals fully extended. Air resistance has slowed the shotcup sufficiently so that the pellets are fully separated from the shotcup. Because the center of gravity is toward the rear, the shotcup will consistently rotate as shown in phase 3 of this illustration. The velocity of a shotcup at phase 2 and 3 is on the order of 1000 fps, consequently the extended petals will typically produce an injury in living skin and leave a mark on most inanimate objects. This is called "petal slap" and is very useful in range-of-fire determinations because these events all occur within a few feet of the muzzle of the gun and in a very reproducible manner for the specific gun–ammunition combination. Phase 4 shows the final orientation of a shotcup in flight. This position may require 3–5 feet to achieve. However, the shotcup still has sufficient velocity and energy to produce an injury and sustain impact damage to its forward-flying base.

orientation of this type of one-piece plastic wad can often resolve distance determination questions over these separation distances. Being able to reliably distinguish a 12 in. standoff distance from a 24 in. standoff could mean the exclusion of a self-inflicted wound where the muzzle-to-wound distance combined with the wound path and gun dimensions preclude the victim from reaching the trigger.

One of the most useful and interesting marks produced by these shotcups occurs very early in their flight when one or more of the extended petals strike the skin. This leaves a visible rectangular mark for which the author has given the name "petal slap." Figure 14.6 shows an example of petal slap for three shots fired from a distance of 20 in. with a cylinder bore 12-ga. shotgun. When present, this mark will usually occur immediately adjacent to the entry wound produced by the associated shot charge. Petal slap relates to the range

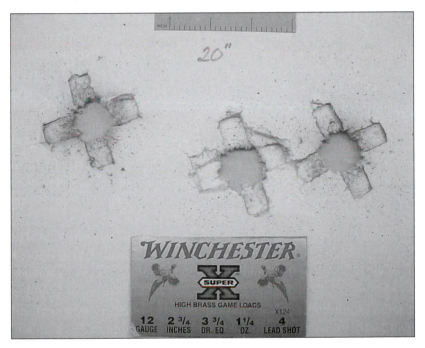

Figure 14.6
Shotcup Petal Slap

This photograph shows the results and reproducibility for three shots fired into foam board from a distance of 20 in. using a Remington Model 870 12-gauge shotgun with a 20 in. cylinder bore. At this distance the four petals of the plastic shotcups are fully extended for this particular gun–ammunition combination. The large holes produced by the $1\frac{1}{4}$ oz. of #4 shot (approximately 170 pellets) clearly show that the shot charge is still en masse at this distance. This would also be true for all distances less than 20 in. The behavior of the shotcups and particularly the shotcup petals would be significantly different at lesser distances and thereby provide a useful means of estimating the separation distance between muzzle and entry wound or site.

of fire for a particular gun–ammunition combination. Petal slap marks can also be seen or detected in clothing. If lead shot was contained in the wad's cup, the sodium rhodizonate test for lead will often raise these slap marks. The center-to-center distances of the imprints of the shot pellets in the interior of recovered wads or shotcups can provide useful information on the shot size.

There is a notable exception to this in-flight reversal behavior for shotcups. When fired from *rifled* barrels, they have repeatedly been observed to remain in a forward orientation with the petals extended. This is the consequence of spin stabilization from the rifling sufficient to overcome the overturning moment caused by the aft center of gravity in these components. This not only results in an increase in the distances over which they can produce petal slap but also reduces their overall range of flight. Careful examination of recovered shotcups fired from rifled barrels will reveal the presence of rifling marks.

POWDER, GUNSHOT RESIDUE, AND BUFFER MATERIAL

Traditional GSR in the form of soot and partially consumed powder particles are also often seen at close ranges of inches to a few feet. These residual materials have the same evidentiary and reconstructive value as previously described in Chapter 2. All of the physical forms of smokeless powder described in that chapter with the exception of tubular powder have been used in shotshells.

Many American shotshells also contain a granulated plastic material that fills the air spaces between the individual pellets in the shell (see Figure 14.3). This material is usually white in color and composed of either polypropylene or polyethylene. Its purpose is to reduce deformation of the spherical lead pellets that normally occurs during the very high accelerative forces experienced during discharge. The shape, size, and composition of this buffering material are related to the brand of ammunition. This material behaves very much like unburned powder and will produce very conspicuous stippling of the skin around close range entry wounds. Just as with powder stippling patterns, the patterns from this buffering material can be used to estimate range of fire in close proximity shotgun wounds involving this type of ammunition. This buffer material also produces an orange peel texture on fired buckshot pellets. Buckshot from unbuffered shells will have obvious flat spots (acquired during the discharge process) on their surfaces but *none* of the orange peel effect. The two effects can be seen in Figure 14.7. Consider the implications if the medical examiner removed pellets of both types from a purported victim of an accidental shooting. The presence of pellets from a buffered and unbuffered shotshell means that at least two shots were fired.

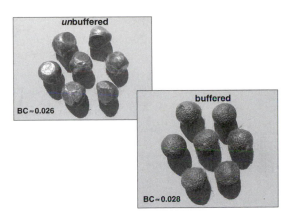

Figure 14.7

00-Buckshot from Buffered and Unbuffered 12-ga. Shotshells

The flat spots on the lead pellets fired from an unbuffered shell occurred during the discharge process and *not* as a consequence of any downrange impact. Without the particles of plastic buffering material (see Figure 14.3) between the pellets in the unfired shell, the very high accelerative forces during discharge deform the soft lead pellets. The "orange peel" texture on the relatively undeformed buckshot pellets at the lower right is a consequence of the same high accelerative forces of discharge that, in this situation, have produced impressions of the buffering material in the surface of the pellets and minimized their deformation.

The approximate ballistic coefficients for these .33-caliber, 53 gr. lead projectiles derived from Doppler radar tracks and tests with an Oehler M43 system are also shown in this figure.

THE EXTERIOR BALLISTICS OF SHOTGUN PELLETS

As the muzzle-to-victim separation distance continues to increase, the diameter of the pellet pattern will increase and the margins of the entry wound will now show scalloping. With a little more distance, satellite injuries from individual pellets appear outside the main entry wound. Finally, distances will be reached where the pellets have separated sufficiently that they create individual injuries or impact sites on inanimate objects. If the incident angle of a shotgun discharge is orthogonal, the pellet pattern will usually occupy a circular area the diameter of which is range dependent. This is illustrated in Figure 14.8. The size of the pattern at any point along the path of the discharge is largely controlled by choke. Gauge and barrel length has very little effect on pattern until one is dealing with a very short, sawed-off barrel. In this situation, the pellet pattern is often "blown" or disrupted by excessively high pressures at the muzzle as the pellets emerge. Another useful phenomenon occurs with substantially sawed-off barrels when the fired shell contains a shotcup rather than cardboard wadding. The bottom skirt of the shotcup will either be everted or blown away by these high pressure gases as it exits the muzzle.

The most common and accepted practice for a distance determination is that of carrying out a series of shots with the gun and ammunition at selected

Figure 14.8

Pellet Pattern vs. Range

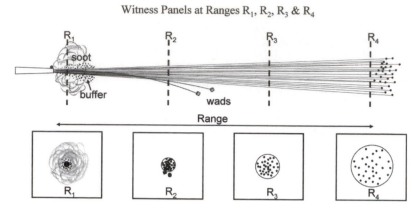

Witness Panels at Ranges R_1, R_2, R_3 & R_4

This illustration depicts the general behavior of the pellets, wads, buffering material, and GSR for a shotgun discharge. The vertical dashed lines represent witness panels placed at four selected distances from the muzzle of the shotgun. The lower squares show idealized results at these four ranges. At distance R_1, soot, powder residues, and buffering material are deposited on the witness panel. Stippling of skin and other surfaces can be expected at such close range (inches to perhaps 2 feet). The pellets are en masse at these standoff distances and produce a single, large entry hole through which the wads or shotcup also pass. Distance R_2 is beyond the reach of GSR and the buffering material. The pellets have now separated sufficiently to create individual holes in the witness panel. The two wads in this illustration have deviated to the point where they create satellite holes outside the pellet pattern. The pellet pattern in this and in the remaining witness panels (R_3 and R_4) can be enclosed within a circle of minimum diameter. A plot of range vs. the average minimum pellet pattern diameter for a series of shots at selected distances will typically produce a straight line for buckshot and the larger sizes of shot over the sorts of distances normally encountered in casework. The very small sizes of shot will often produce a plot of pattern diameter vs. range with a slight upward curvature.

distances followed by measurements of the minimum diameters of the circles that will include all of the pellets in each shot. A graph is then prepared from these test-firings that shows the relationship between average pattern diameter and range of fire. The validity and reliability of this graph and the sort of uncertainty limits associated with it depends on the number of shots fired at each distance. Although there is no set requirement for the number of shots at each distance, the person preparing and interpreting the results must be able to describe his procedure and provide confidence limits based on these tests. The other consideration is the evidence pattern itself. If one has an evidence pattern that fits nicely within a 3 in. circle, there is no need to fire more than one shot at 25 feet where the pattern is many times this size. For most sizes of shot over relatively short distances such as 30–40 feet or less, the relationship between average pattern diameter and range is a linear one. With the smaller sizes of shot such as #8 and #9 and/or substantial increases in distance, the plot of pattern diameter vs. distance may curve slightly upward. This is due to a combination of two factors. The spread of the pellets away from the center point of the pattern is occurring at an essentially constant rate based on *time*. This lateral spread has been measured by the author on several occasions and is on the order of 10–12 fps. Because of the very poor ballistic coefficients of shotgun pellets, the *forward* movement of the same group of pellets is decelerating *very rapidly* from its initial muzzle velocity of about 1300 fps. The plot of average pellet pattern diameter is based on *distance*, not on the time of flight. With a constant rate of lateral spread, the pellets will move farther apart during the second half of a select distance than during the first half of that distance. This is because it has taken a little more time to cover the distance between 25 and 50 feet than it took to cover the 0–25 feet distance.

Another consideration and potential difficulty is the presence of "flyers" in the evidence or test pattern. These are pellets that were probably deformed in some way and consequently strayed from the main group of pellets. This creates a quandary. Do we exclude such errant pellets or not? If so, how do we make the decision that a pellet impact site is a flyer? One approach is simply to be consistent. If flyers are included in the determination of pattern diameter, this should be noted and done for all test shots and the evidence pattern. The converse is also true. An alternate solution to the problem of flyers is the use of the *equivalent circle* method for pattern diameter determination. In this method a multi-sided polygon is drawn by connecting all of the peripheral pellet strikes with lines. The lengths of each line segment are measured, added together then divided by π to obtain the equivalent diameter of the circle having a circumference represented by sum of the line segments. This can be time consuming but removes the subjective element associated with including or excluding flyers. It also minimizes the skewing effect of flyers when they are included in the minimum circle method.

The final result derived from a proper patterning of the evidence shotgun and ammunition involved in the shooting incident will be a range of distances. A typical report might read: "Patterning of the evidence shotgun with the submitted ammunition and measurements of the pattern at the scene established a range of fire between 10 feet and 15 feet."

Non-orthogonal patterns on essentially flat surfaces produce an elliptical pattern whose minor axis is related to range of fire. The arcsine ($\sin^{-1}$) function on a pocket calculator can be used to determine the approximate incident angle following the division of the diameter of the minor axis by the diameter of the major axis. This concept is illustrated in Figure 14.9.

From classical maximum range computations and some actual test-firings, it can be seen that shotguns, particularly when loaded with the smaller sizes of shot, are very short-range firearms compared to bullets from rifles that may travel more than a mile when fired at departure angles of 20–30°. This is due to the very low ballistic coefficients for these small, spherical projectiles. The author measured the ballistic coefficient for 00 Buckshot with both Doppler radar and the Oehler

Figure 14.9

Non-Orthogonal Shot Patterns

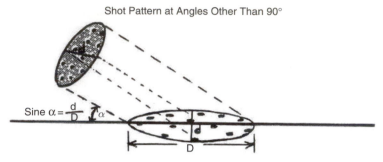

The discharge of a shotgun against a flat surface at incident angles other than orthogonal will produce an elliptical pattern much like that illustrated in the figure above. The minimum diameter of this ellipse is related to the range of fire (as previously shown in Figure 14.8). The approximate angle of incidence can be calculated by dividing the minimum diameter by the maximum diameter of the elliptical pattern and then using the arcsine function to obtain the incident angle.

M43 PBL system and obtained values on the order of 0.026–0.028. Steel air rifle BBs of 0.173 in. diameter yielded G_1 ballistic coefficient values of 0.009–0.010. By way of comparison, typical medium caliber pistol bullets have BCs on the order of 0.16–0.20 and modern rifle bullets have BCs as high as 0.50.

Some maximum range values of several common shot sizes fired at sea level are as follows: #8—240 yd, #6—275 yd, #4—300 yd, 4Buck—480 yd, 00Buck—610 yd. Actual firings and tracking with Doppler radar for several 00 Buckshot loads gave maximum range values of 618 and 639 yd at a site elevation of about 400-ft. MSL and required departure angles of 25°–30°.

Such maximum range determinations for shotguns are usually associated with issues of potential injury causation when someone is claimed to have discharged a shotgun in someone's direction but the pellets failed to strike them.

SUMMARY AND CONCLUDING COMMENTS

Although shot itself cannot normally be matched back to the gun that fired it, various class characteristic comparisons can be made between seized ammunition and fired components. These components can be numerous and varied and include the shot size, wads, shot collars, shotcups, and buffer material if present. In some instances, recovered shotcups and plastic over powder wads can be matched to the responsible gun when they have been sufficiently striated by the bore of the gun. In "no gun" cases, the gauge of the responsible gun and the brand of the shotshell can often be determined from recovered wads or shotcups. In some instances, the amount and size of shot can also be determined from these components. Recovered pellets are of obvious importance and a total pellet count in a complete pellet pattern can establish the weight of the shot charge.

It is, however, the relatively short range ballistics of the shot, the wads, shot collars, shot cups, and any buffer material that is of great interest and value to the forensic scientist or crime scene examiner attempting to reconstruct a shotgun shooting. The chemical composition and physical form of any plastic buffering material loaded in certain American shotshells have both brand identification and reconstructive value. This material can produce close range stippling in skin much like unburned propellant particles and thereby provide a means of estimating the muzzle-to-wound distance.

The exterior ballistic behavior of all of these components can be related to range of fire as well as the azimuth angle from which the particular shot originated. As with previous examples, this information can be used to locate possible points of origin for the shot, additional physical evidence, and witnesses. If this physical evidence includes one or more fired shotshells, these can usually be matched to the responsible gun. Carefully conducted and evaluated test-firings of the responsible gun with the appropriate ammunition will permit the determination of the range of fire. This is useful in testing various explanations, theories, or accounts of the incident.

REFERENCES AND FURTHER READING

Breitenecker, R., "Shotgun Wound Patterns," *Am Jour Clinical Path* 52 (1969) pp. 269–285.

Breitenecker, R. and W. Senior, "Shotgun Patterns I: An Experimental Study on the Influence of Intermediate Targets," *JFS* 12:2 (April 1967) pp. 193–204.

Di Maio, V.J.M., *Gunshot Wounds—Practical Aspects of Firearms, Ballistics and Forensic Techniques*, Elsevier Science Publishing Co., NY, 1985.

Dillon, J.H., "Graphic Analysis of the Shotgun/Shotshell Performance Envelope in Distance Determination Cases," *AFTE Journal* 21:4 (October 1989) pp. 593–594.

Ernest, Richard N., "A Reassembled Buckshot Load," *AFTE Journal* 23:3 (July 1991) pp. 792–798.

Ernest, Richard N., "Exploring the Possibility of Matching Fired Shotgun Ammunition Components to Unaltered Shotguns," *AFTE Journal* 24:1 (January 1992) pp. 28–36.

Ernest, Richard N., "A Study of Buckshot Patterning Variation and Measurement Using the Equivalent Circle Diameter Method," *AFTE Journal* 30:3 (Summer 1998) pp. 455–461.

Fann, C.H., W.A. Ritter, R.H. Watts and W.F. Rowe, "Regression Analysis Applied to Shotgun Range-of-Fire Estimations: Results of a Blind Study," *JFS* 31:3 (1986) pp. 840–854.

Garrison, D.H., Jr., "Field Recording and Reconstruction of Angled Shot Pellet Patterns," *AFTE Journal* 27:3 (July 1995) pp. 204–208.

Haag, L.C., "Double 00-Buck," *AFTE Journal* 13:3 (1981).

Haag, L.C., "Some Forensic Aspects of Spherical Projectiles," *AFTE Journal* 30:1 (Winter 1998).

Haag, L.C., "Once Fired, Twice Fired, Thrice Fired, More? A Novel Method for Assessing the Number of Firings of Shotshells with Plastic Bodies," *AFTE Journal* 34:1 (Winter 2002).

Haag, L.C., "Average Pellet-to-Pellet Distance for Estimating Range of Fire in Cases Involving Partial Pellet Patterns," *AFTE Journal* 34:2 (Spring 2002).

Haag, L.C., "The Exterior and Terminal Ballistics of 00 Buckshot," Part 1— *AFTE Journal* 34:2 (Spring 2002), *AFTE Journal* 35:1 (Winter 2003)—Part 2.

Haag, L.C. and M.G. Haag, "The Analysis and Comparison of Shotshell Buffers," *AFTE Journal* 32:3 (Summer 2000).

Haag, L.C. and E. Wolberg, "Shotgun Barrel Shortening Effects on Pellet Pattern, Velocity and Penetration," *First IWBA Conference*, Sacramento, CA (March 1994).

Heaney, K.D. and W.F. Rowe, "The Application of Linear Regression to Range-of-Fire Estimates Based on the Spread of Shotgun Pellet Patterns," *JFS* 28:2 (1983) pp. 433–436.

Lattig, K.N., "The Angle of Intersection of a Shot Pellet Charge with a Flat Surface," *AFTE Journal* 14:3 (July 1983).

MacPherson, D., *Bullet Penetration—Modeling the Dynamics and Incapacitation Resulting from Wound Trauma*, Ballistic Publications, El Segundo, CA, 1994.

McJunkins, S., "Identification of Plastic Shotgun Waddings," *AFTE Newsletter* (June 1970) p. 24.

NRA Firearms Fact Book, 3rd edn, The National Rifle Association of America, Washington DC, 1989.

Royse, D., "Identification Made on a Fired 00 Buckshot Pellet," *AFTE Journal* (October 1996) pp. 252–253.

Sellier, K.G. and B.P. Kneubuehl, *Wound Ballistics and the Scientific Background*, Elsevier Publishers, Amsterdam, 1994.

Speak, R.D., F.C. Kerr and W.F. Rowe, "Effects of Range, Caliber, Barrel Length and Rifling on Pellet Patterns Produced by Shotshell Ammunition," *JFS* 30:2 (1985) pp. 412–419.

Stone, I.C. and P.E. Besant-Matthews, "Effect of Barrel Length and Ammunition on Shotgun Range Patterns," *SWAFS Journal* (January 1985) pp. 10–12.

Watkins, R.L. and L.C. Haag, "Shotgun Evidence," *AFTE Journal* 10:3 (July 1978).

Wray, J.L., J.E. McNeil and W.F. Rowe, "Comparison of Methods for Estimating Range-of-Fire Based on the Spread of Buckshot Patterns," *JFS* 28:4 (1983) pp. 846–857.

CHAPTER 15

ULTIMATE OBJECTIVES, REPORTS, AND COURT PRESENTATION

A number of techniques for the documentation and processing of shooting scenes have been presented in the preceding chapters of this book. The reader has been exposed to a variety of exterior and terminal ballistic phenomena related to the behavior of projectiles in stable and unstable flight and their ultimate effects upon various target materials. The importance of trace evidence transfers associated with ballistic events and types of laboratory examinations available to interpret these materials has been demonstrated through case examples taken from actual shooting incidents. In one way or another they are all directed toward an effort to reconstruct what did and what did not occur in a shooting incident.

Virtually every discussion and example was directed toward one or more of the 12 objectives listed at the end of Chapter 1.

EXPLAINING WHAT WE DO

The examination of shooting scenes and subsequent reconstruction of the shooting incident is very much like any other event under investigation involving collisions, impact damage, trace evidence exchanges, pattern recognition, and the application of certain analytical tests. In some respects, the techniques employed during the processing of shooting scene can be viewed as an applied physical science, and as such the various methods and techniques are well known, routine, readily verified, and repeatable. The results of a properly conducted shooting incident investigation are documented and preserved allowing the accuracy of any findings to be evaluated at a later time. The reconstructive process must also involve unbiased analytical thinking and the application of *the Scientific Method* to evaluate various explanations of one or more events in a shooting incident. Just as with the broader fields of Forensic Science and

Criminalistics, the methods and techniques employed in the reconstruction of shooting incidents are drawn from other scientific disciplines such as chemistry, physics, mathematics, ballistics, trigonometry, and microscopy. For example, the sodium rhodizonate test for lead comes from the broad field of chemistry and the subspeciality of chemical spot tests for trace amounts of metals. The measuring of the angular components of a bullet's flight path utilizes very basic trigonometric calculations taught in every high school trigonometry course. The calculation of a bullet's flight time, angle of departure, and angle of fall involves the use of computers and exterior ballistics programs, the results of which can be checked for accuracy multiple ways including actual test-firings. The recognition of embedded bone particles and fibers in a bullet involves the use of a basic laboratory microscope. The identity of these and other trace materials can be established, rechecked, and verified in multiple ways. The analytical methods employed in a modern forensic laboratory are non-destructive. This is of great value because it means that other specialists and forensic scientists can re-examine the evidence if requested to do so by a litigant or ordered to do so by a court.

LEGAL CHALLENGES AND THE ROLE OF SHOOTING INCIDENT RECONSTRUCTION IN LITIGATION

The foregoing measures will not necessarily thwart critics. One litigant or the other will not be pleased with even the most thorough and well-documented reconstruction. *Any* physical evidence that implicates a particular person or interest in some form of wrong-doing is prejudicial. Legal challenges may be raised in several forms beyond that of any alleged prejudicial effect. The foreseeable claims might be that Shooting Incident or Shooting Scene Reconstruction is a *new science or technique* or that it is based on *untried and unproven scientific principles*. This claim, put to the author many years ago in a courtroom in Iowa, was the very genesis of this book. It should be apparent from the most basic reading of this book and an inspection of the many references that this is untrue. But the person on the stand in an evidentiary hearing or during a voir dire examination at trial needs to be prepared for such challenges. Bad testimony leads to bad case law. Both the attorney soliciting expert testimony regarding a shooting reconstruction and the witness called by this attorney needs to conversant regarding *the Scientific Method* and how objective, testable measures and techniques were utilized in the reconstruction of the particular incident. These matters are of great importance because there are rules of evidence and case law at both the state and federal level that govern the admissibility of evidence and expert testimony. A review of the important elements and language from these sources should be helpful.

The *Frye* decision by the 1923 DC Court of Appeals addressed the matter of expert testimony. This case is still the standard for admissibility in a number of states in the United States. *Frye* requires that the proffered technique and expert testimony have "—gained general acceptance in the particular field in which it belongs." Some states have added to this either by statute or case law. If the challenge is based on a claim that shooting reconstruction is a "new, untried and unproven science or technique," the challenge should clearly fail under *Frye*. The book by G.G. Kelly written over 40 years ago and referred to in Chapter 1 describing the reconstruction of numerous cases over the period of 1929–1958 should be noted. The many articles dealing with shooting scene reconstruction that have appeared in the Journal of the Association of Firearm and Tool Mark Examiners (*AFTE Journal*) over the last 30+ years is probably the most important asset to the proponent of such testimony. This is a peer-reviewed journal; reviewed by the relevant scientific community. AFTE contains the largest membership of forensic firearm examiners and criminalists dealing with shooting investigations in the world. Its membership is composed of both private examiners and scientists and members employed in government crime laboratories. Articles appearing in this journal have often been given previously at one of AFTE's annual international seminars. Upon submission to the journal editor, they are sent to one or more reviewers who have special knowledge or experience regarding the particular subject matter. Upon publication, the article is disseminated to the relevant scientific community. This allows others working in the same field to evaluate and test any new technique. If it cannot be replicated or flaws are found in the technique, a letter to the editor or a counter article should follow. There are, of course, other scholarly scientific journals in which articles dealing with shooting scene reconstruction have appeared with similar peer-review processes. Many of these have been cited in the references at the end of each chapter.

Finally, the witness should also be prepared to give common examples of shooting reconstructions that are carried out daily in forensic laboratories throughout the world. Examples would include distance determinations based on powder patterns and shotgun pellet patterns or the common cone fracture produced by a BB strike to a plate glass window allowing the direction of impact to be determined.

In 1975 new Federal Rules of Evidence were passed by the U.S. Congress. Federal Rule 702 addresses and controls expert witness testimony in all federal courts. Many state courts have essentially adopted this rule. Rule 702 states: *If scientific, technical or other specialized knowledge will assist the trier of fact to understand the evidence or to determine a fact in issue, a witness qualified as an expert by knowledge, skill, experience, training or education may testify thereto in the form of an opinion or otherwise.* The last part of this rule is particularly important. The use of the little word "or" means that the witness need not necessarily be a scientist with a degree in one of the physical sciences. The witness may presumably qualify as an expert based on experience alone. In actuality, most witnesses will have training (shooting reconstruction courses), knowledge (a study of the relevant literature), skill (practice in carrying out the various techniques associated with shooting reconstruction), and experience (prior casework). This should be brought out by the proponent attorney. Education (a degree in chemistry, physics, or mathematics) would be an asset but not a requirement. The use of *the Scientific Method* or scientific methodology does require the individual to possess a degree in one of the physical sciences.

Rule 702 goes on to set conditions regarding the opinion testimony as follows: (1) the testimony must be based on sufficient facts or data, (2) the testimony must be the product of reliable principles and methods, and (3) the witness has applied the principles and methods reliably to the facts of the case. It should be fairly easy to envision how the questioning might go in a motion hearing.

- Was there sufficient data for you to form an opinion in this case?
- What data did you rely upon in forming your opinion?
- What principles are involved in carrying out the tests you performed?
- What method(s) did you use?
- Are these recognized principles and methods among your peers? Please explain.
- Did you apply them in a manner that is designed to produce reliable results? Please explain.

It should be noted that infallibility is neither required nor possible in any human endeavor and the witness should be prepared to concede this. But he or she should also be prepared to explain those measures that were taken to minimize the chance for error and that the method used allows an independent review of the evidence and tests.

There are three other Federal Rules of Evidence that should be of interest to the proponent of expert testimony. These are Rule 104, 401, 402, and 403. Rule 104 provides that the preliminary questions concerning the qualifications of a person to be an expert witness and the admissibility of the evidence are in the province of the court. A jury will never hear the results of any reconstructive efforts if the court is not satisfied regarding the witness's qualification or the admissibility of the evidence examined by the witness.

Rule 401 defines relevant evidence as: *evidence having any tendency to make the existence of any fact that is of consequence to the determination of the action more probable or less probable than it would be without the evidence.* Rule 402 provides that nearly all relevant evidence is admissible.

However, Rule 403 excludes relevant *evidence if its probative value is substantially outweighed by the danger of unfair prejudice, confusion of the issues, or it is misleading the jury, or by considerations of undue delay, waste of time or needless presentation of cumulative evidence.*

In 1993 the U.S. Supreme Court changed the *Frye* standard when they handed down their decision in the *Daubert* case. The Court dismissed the "general acceptability" standard and replaced it with the Federal Rules of Evidence and added four criteria by which scientific testimony must be evaluated before it can be admitted. These four criteria are:

1. Testability of the Scientific Principle
2. Known or Potential Error Rate
3. Peer Review and Publication
4. General Acceptance in a *Particular* Scientific Community.

The Court provided guidance on how these criteria were to be weighed, so one might conclude that they have equal weight. These criteria should be relatively easy to meet if the witness has done the various things described in this text. We are dealing with the physical world and tangible evidence that can be photographed, measured, and tested in various ways with those tests being documented and preserved. These tests can be shown to be repeatable and reliable with their origins based on other established and accepted bodies of scientific knowledge. The author has regularly discussed confidence limits (uncertainty of measurements) and offered suggestions on how to evaluate "error rate" in actual case work. Point 3 (Peer Review and Publications) has already been discussed and the General Acceptance listed in point 4 naturally flows from the many articles regarding shooting scene reconstruction appearing in multiple journals. The existence of various training courses and workshops in Shooting Incident Reconstruction adds additional support to point 4. However, the witness needs to be cognizant of these criteria and rules of evidence if they are applicable in the jurisdiction where reconstruction testimony is contemplated.

There is an alternate view that holds that the requirements of *Daubert* do not apply if the witness is functioning in technical, rather than a scientific, role. This view arises out of the opening language of Federal Rule 702: *If scientific,* technical *or* other specialized knowledge *will assist the trier of fact—*.

It takes the position that the reconstruction of shooting scenes and shooting incidents is an applied science in which the witness utilized well-known and well-established scientific techniques and objective-measuring methods to arrive at his or her opinions. In this role the witness has acted purely as a technician by following a checklist or "cookbook" procedure at a shooting scene. This role does not involve any analytical thinking or use of *the Scientific Method*. In some respects this argument has some validity. A mark on a brick wall is observed at a shooting scene and the investigator tests it for copper and lead. The examiner in the laboratory sees some embedded material on a bullet from a gunshot victim, places it in the SEM and observes mineral grains. In these examples, the crime scene investigator and the laboratory analyst simply carried out the tests because it is a part of their job. Most, if not all, of the techniques, tools, reagents, and analytical instruments used in shooting incident reconstruction at this "technician level" are well known in the field of Criminalistics and its sub-specialty of Forensic Ballistics. But it would be dangerous to assume that every court will take the view that *Daubert* or Daubert-like requirements do not apply to the "technician." The American reader should remain mindful that *Daubert* and its progeny have removed the responsibility of determining the validity of a scientific procedure or test from scientists and transferred it to jurists. Anyone carrying out reconstructive efforts at shooting scenes, or even in a laboratory, should be familiar with the pertinent Federal Rules of Evidence and four requirements of *Daubert*. This same individual should also be prepared to address these rules and requirements as they might relate to the tests and procedures that were utilized in these reconstructive efforts.

REPORTS AND REPORT WRITING

Individuals employed by governmental agencies will no doubt have specific formats for the writing of reports in routine matters. There has been considerable variety in the style and amount of detail in reports coming out of American crime laboratories. The matching of a fired bullet to a submitted firearm is a common example. After the usual inventory and description of the various submitted items, the examiner's report might read, *The 9 mm Luger pistol, serial number 12345 submitted by Detective Smith on May 2, 2005, was determined to have fired the full metal jacketed bullet in the packet marked Item 6 from the medical examiner's office* or simply *Item 6 was fired by item 2*. In this latter situation, which happens all too frequently, the reader of such a report had to go back through the documents to figure out what item 6 and item 2 were and where they came from. Reports from the private sector and organizations show an even wider range of styles and formats depending on the complexity of the case.

There are some rules at the federal level in the United States that address reports by experts.

In criminal cases, Rule 16(G) of the Federal Rules of Criminal Procedure states, *At the defendant's request, the government must give to the defendant a written summary of any testimony that the government intends to use under Rules 702, 703, or 705 of the Federal Rules of Evidence during its case-in-chief at trial.* This would seem very general but at the same time would also seem to call for a more narrative report. This would be particularly true if the report related to a reconstruction of a shooting incident.

The rule regarding reports in civil cases is much more involved and explicit. Rule 26, section 2 subpart (B) of the Federal Rules of Civil Procedure reads, in part:

> *Except as otherwise stipulated or directed by the court, this disclosure shall, with respect to a witness who is retained or specially employed to provide expert testimony in the case or whose duties as an employee of the party regularly involve giving expert testimony, be accompanied by a written report prepared and signed by the witness.*
>
> *The report shall contain a complete statement of all opinions to be expressed and the basis and reasons therefore; the data or other information considered by the witness in forming the opinions; any exhibits to be used as a summary of or support for the opinions; the qualifications of the witness, including a list of all publications authored by the witness within the preceding ten years; the compensation to be paid for the study and testimony; and a listing of any other cases in which the witness has testified as an expert at trial or by deposition within the preceding four years.*

The author has prepared a sample report for a hypothetical shooting incident in which certain reconstructive efforts were carried out. This report includes all the required Rule 26 elements for a *civil* case at the *Federal level* plus the format and style preferred by the author. The hypothetical case used as the basis for this sample report involves an undercover federal narcotics officer who later states that he approached a parked vehicle for the purpose of a drug purchase. Upon reaching the driver's door, the driver suddenly produces a pistol and fires one shot through the tempered glass side window striking him in the left shoulder. As he staggered back and toward the rear of the car, he drew his pistol and fired three rapid shots. One of these shots struck the driver in the left forehead and killed him. A passenger in the vehicle survived with minor injuries from flying glass. The undercover officer also survived his shoulder wound and provided the foregoing account. By the time the vehicle is examined no visible glass remains in the driver's side window. Three cartridge cases from the roadway and one from inside the vehicle are recovered and impounded as evidence. These items, the bullets from the deceased driver and the wounded officer along with the two guns and remaining ammunition are submitted to the examiner.

The passenger's account is quite different. He states that he and his friend were simply sitting in the car listening to a music CD when this individual with a gun in his hand came up to the driver's window and beat on it with his gun. His terrified friend tried to start the car to escape the situation but could not. At this point the side window shattered and he saw the gun starting to come into the car. His friend happened to be armed and produced a pistol from somewhere in the vehicle and fired one shot toward his assailant at the same time as a fusillade of shots entered the car.

CASE #05/12345

REPORT

of
July 15, 2005
in the matter of

the Death of John Johnson
and
the Wounding of Officer Sam Smith

prepared for

The Office of the United States Attorney
123 North Main Street
Any City, USA

by

Lucien C. Haag
Criminalist
Carefree, AZ 85377

CASE #05/12345

INTRODUCTION

This investigation was initiated on July 8, 2005 upon the receipt of a request from AUSA Sterling Allgood of the Officer Involved Shooting (OIS) Review Detail of the U.S. Attorney's Office. A number of documents and photographs were also submitted to this examiner as follows:

- The Arizona State Police Case Report related to this July 6, 2005 incident
- The Arizona State Police Property Inventory of items impounded
- The ASP Scene Diagrams and a CD of digital scene photographs (TIFF images)
- The Autopsy Report of John Johnson (to include photographs and diagrams of the wound)
- A Witness Statement from passenger Jordan Jones dated July 6, 2005
- A transcript of the Interview of Officer Sam Smith dated July 7, 2005.

CASE OVERVIEW

This incident involved a confrontation between Officer Sam Smith and John Johnson that occurred on July 6, 2005 at approximately 8 pm in front of the residence at 2909 South Third Street.

According to Jordan Jones, an exchange of gunshots occurred between the deceased driver, John Johnson, and an armed male subject later identified as Sam Smith. The subject Smith approached the driver's side of the Johnson vehicle with a gun in his hand, yelled, and beat on the driver's side window with this gun until it shattered. The driver (Johnson) attempted to start the car but was unsuccessful. At this point Jones describes the subject's gun as entering the vehicle whereupon the driver produced a gun. Both guns fired at the same time. Jones believes that the driver only got off one shot when he slumped across the center console of the vehicle.

According to Officer Sam Smith, he approached the Johnson vehicle (a blue, 2002 Camarro) believing that it was occupied by a drug dealer with whom he planned to make an undercover buy. Upon reaching the driver's window he was surprised to see that it was not the individual he expected. What is more, this subject produced a gun and fired a shot at him through the driver's side window. He staggered back away from the window, drew his pistol, and fired 3–4 shots as he tried to position the "B" pillar of the vehicle between himself and the driver. It was only then that he realized that there was another occupant in the vehicle. About this time a marked patrol car that had been stationed around the corner arrived on the scene.

MATTER NOT IN DISPUTE

It is *not* in dispute that Officer Sam Smith fired multiple shots (3) at the decedent from his Glock pistol and that one of his shots resulted in the driver John Johnson's death from a single gunshot wound to the head. The entry wound was somewhat irregular and was located just above the left eye. The path was front to back and left to right.

It is also *not* in dispute that the driver fired one shot from a Colt. 45 Automatic pistol recovered from the floorboard of the decedent's vehicle.

RECONSTRUCTIVE ISSUES

From a review of the two witness's statements, the sequence of shots and the position of Officer Smith when he was shot and when he fired his pistol are at issue in this shooting incident.

SCENE AND VEHICLE EXAMINATION

The *scene* at 2909 South Third Street was inspected on July 11, 2005. Three circles of fluorescent orange paint were still quite visible on the pavement where three fired cartridge cases could be observed in the scene photographs. Four lines or stripes of the same type of spray paint corresponding to the positions of the four tires of the Johnson vehicle could also be seen. The distances between these locations were measured and related to the reference point shown in the ASP scene diagram. The character and appearance of the pavement was noted and photographed and the slope of the pavement in this area measured in both the N/S and E/W orientations with a digital inclinometer. All notes, measurements, and photographs taken during this scene inspection were placed in the laboratory file on this case.

The *Johnson vehicle* was examined at the ASP impound yard on this same date.

The general dimensions of the blue 2002 Chevrolet Camaro were taken to include the size of the opening of the missing driver's side window, the height of the window sill above ground level, and numerous other measurements all of which were recorded and placed in the laboratory file on this matter.

Inspection of the molding around the driver's door revealed multiple particles of diced tempered glass trapped in the upper molding. These were photographed in place, then marked as to their exterior surface, and secured in a small evidence collection box marked "05/12345—tempered glass from top edge of driver's side window." Following this the driver's door panel was carefully removed and the interior components examined and photographed. Remaining tempered glass was observed, photographed, marked as to interior/exterior surface then samples collected from the window carrier. Considerable shattered tempered glass was also observed and photographed in the bottom of the driver's door. This glass included sections with obvious radial fractures. These pie-shaped sections were reinforced with clear plastic tape, marked, and impounded with the previous items.

A probable bullet graze mark was observed on the front edge of the driver's side "B" pillar. This was photographed and measurements taken of its location relative to fixed reference points on the vehicle and relative to its height above ground level. A piece of copper bullet jacket was observed trapped in a chrome molding just forward of this apparent graze mark. This was photographed in place then removed and impounded for later laboratory examination. Chemical tests for copper and lead were carried out on this mark with positive results for both metals. The direction of this graze mark (based on the presence of a pinch point) was from back to front (relative to the long axis of the vehicle) and in to the area normally occupied by the driver.

No subsequent downrange impact point was found that could be aligned or associated with this graze mark.

Two bullet holes and penetrating bullet paths were found in the interior of this vehicle. One entered near the center of the dash by grazing the on-off control for the radio/CD player and then proceeding on into the various components of the dash. The track of this bullet was plotted and the angular components diagrammed and photographed. Back-extrapolation of this bullet's path indicated that it is passed through an area just to the rear of the central area of the driver's side window (see photographs in this file). The bullet was ultimately recovered and impounded.

The second bullet hole in the interior of the vehicle was in the door of the glove box. This bullet perforated the glove box door after striking it at a shallow, near-horizontal incident angle. Upon opening the glove box door the bullet was observed embedded in the right interior side of the glove box. Back-extrapolation of this bullet's flight path (using the probe method as before) lead to an area just left of the central area of the driver's side window.

EVIDENCE RECEIVED AND EXAMINED

On July 11, 2005 at 4 pm the following items of evidence were received by the undersigned examiner at the central laboratory:

From Investigator Baker (U.S. Attorney's Office):

- the three .40 S&W caliber cartridge cases listed as having been recovered from the scene at 2909 S. 3rd Street
- one .45 Automatic cartridge case listed as having come from the driver's side floorboard
- a .40 S&W caliber Glock pistol, serial number ABC123 with left-hand holster and seven rounds of Speer *Gold Dot* ammunition
- a Model 1911A1 .45 Automatic pistol, serial number 654321 with 6 rounds of FMJ-RN ammunition
- a medium caliber JHP bullet in a medical examiner's vial and labeled as coming from John Johnson
- a large caliber, FMJ bullet in a hospital vial and labeled as coming from Officer Sam Smith.

Retrieved from the decedent's vehicle by this examiner on 7/11/2005:

- tempered glass samples removed from the upper molding of the driver's door
- tempered glass samples removed from the window carrier inside the driver's door
- pie-shaped sections of tempered glass removed from the lower interior of the driver's door

- a small copper bullet jacket fragment recovered approximately 1 in. from a bullet graze mark in the driver's side "B" pillar
- a fired bullet recovered from behind the radio/CD player
- a fired bullet recovered from inside the glove box.

OBSERVATIONS, TESTS AND RESULTS

Glock Pistol and Ammunition from Officer Sam Smith

This pistol was found to be fully operative and functioned as intended by the manufacturer. It was in a like new condition and showed no evidence of impact damage nor was any embedded glass found on this pistol.

Fired cartridge cases of the same brand and type ejected directly to the right with this pistol held in the usual shooting position (grip pointed down toward the ground, muzzle horizontal). The ammunition with this pistol is Speer *Gold Dot* loaded with 180 gr. JHP bullets.

Cartridge Cases—Pavement at the Scene

These three .40 S&W caliber cartridge cases bear the Speer headstamp. They were identified as having been fired in the submitted Glock pistol.

Cartridge Case—Driver's Floorboard

This cartridge is a WWII vintage .45 Automatic military cartridge comparable to the ammunition recovered in the decedent's .45 Automatic pistol.

Bullets from the Decedent's Vehicle

Both of these bullets are .40-caliber, 180 gr. Speer *Gold Dot* bullets possessing polygonal rifling engravings comparable to those found in Glock pistols.

Both of these bullets possess numerous particles of embedded glass and multiple "facets" on their ogives (nose portions).

Bullet Fragment from the Chrome Molding, Graze Mark Area of the Decedent's Vehicle

This item consists of a 5 gr. fragment of a copper bullet jacket. It is unsuitable for identification purposes but does possess a "land" impression from polygonal rifling.

Bullet from the Decedent, John Johnson

This bullet is a deformed but otherwise unexpanded .40-caliber, Speer *Gold Dot* bullet weighing 175 gr. and possessing polygonal rifling engravings comparable to those found in Glock pistols. It possesses obvious blue paint transfers on one side of its nose. There is also an area of missing bullet jacket material just aft of the blue paint transfers. *No* glass damage or embedded glass was present on this bullet.

Bullet from Officer Sam Smith

This bullet a .45-caliber, 230 gr. FMJ bullet fired from a firearm possessing conventional, 6-left rifling engravings. The nose of this bullet has a smooth, flattened area containing pulverized particles of glass. The flattened area is at right angles to the long axis of the bullet indicating a near-orthogonal impact with glass prior to striking Officer Smith.

Pieces of Tempered Glass from the Upper Molding of the Driver's Side Window

These pieces of diced tempered glass were examined under SW-UV light and found to fluoresce on the surfaces marked "E" for exterior.

Pieces of Tempered Glass Removed from the Window Carrier in the Driver's Door

These pieces of diced tempered glass were examined under SW-UV light and found to fluoresce on the surfaces marked "E" for exterior.

Two Pie-Shaped Sections of Tempered Glass from Inside the Driver's Door

Cone fracturing was found and photographed at the convergence points on these two sections of tempered glass.

These items were examined under SW-UV light and found to fluoresce on the surfaces possessing the cone fracturing.

SUMMARY AND CONCLUSIONS

Pie-Shaped Sections of Shattered Tempered Glass

These sections of tempered glass possess radial fractures and cone fracturing at the convergence point. These phenomena are the result of the first bullet striking and shattering this tempered glass window. SW-UV light examination of these sections of glass and known specimens collected from the decedent's vehicle revealed that this glass was made by the tin float method. This process causes one side of such glass to fluoresce when illuminated with SW-UV light and the opposite side to simply transmit or absorb the SW-UV light. From the observations on the evidence sections and the information derived from the known specimens taken from the 2002 blue Camarro, the bullet that shattered this window came from *inside* the vehicle.

Bullets Recovered from the Blue Camarro

Both of these .40-caliber *Gold Dot* bullets struck and perforated previously failed tempered glass as evidenced by the "faceting" on their ogives. Although they show extensive impact damage, they have retained their original weights.

John Johnson's Fatal Gunshot Wound

This .40-caliber *Gold Dot* bullet is associated with the grazing strike to the "B" pillar of the decedent's blue Camarro.

The reasons for this are the presence of blue paint transfers on this bullet and the missing 5 gr. of bullet jacket material. This missing area on the bullet corresponds to the 5 gr. bullet jacket fragment found trapped in the chrome molding immediately adjacent to the graze mark in the "B" pillar. The other two bullets fired by Officer Smith did not experience any weight loss.

The alignment of the graze mark with the area occupied by the driver places the shooter to the rear of the "B" pillar and firing along a forward and right-to-left path.

The total absence of any glass damage or embedded glass in this bullet indicates that the glass was no longer present in the window when this bullet entered the occupant area of the decedent's vehicle.

Bullet Recovered from Officer Smith's Gunshot Wound

This bullet is the first shot fired as evidenced by the smooth, even flattening of its nose in which was found pulverized glass. This finding is also supported by the results of the SW-UV light examinations of the sections of shattered side window glass. The nose-on flattening experienced by this bullet also establishes this shot as having been fired directly through and out of the driver's side window with Officer Smith located somewhere beyond (downrange) of the side window.

Trajectories of Officer Smith's Three Shots

The paths of the three shots fired by Officer Smith shows substantial changes or differences in their azimuth components. The bullet that struck the glove box came through the previously shattered glass of the driver's side window with a slight back-to-front angle. The shot that struck the radio/CD player entered the previously shattered driver's side window with an azimuth angle of approximately 45° along a back-to-front path. The fatal shot that first grazed the "B" pillar struck and entered the decedent's vehicle along an azimuth angle estimated to be 15°–25°.

These angles required the shooter to change his position between at lest two of these shots. This is contrary to the account given by the passenger, Jordan Jones, wherein he describes Smith as immediately outside the driver's window and firing into the vehicle.

This account does support Officer Smith's account insofar as his movements while returning fire.

Other Reconstructive Issues

The passenger, Jordan Jones, also described Smith as breaking the driver's side window with his pistol then thrusting his gun into the occupant area and firing all of his shots in very rapid succession.

This account is refuted for multiple reasons. The .45-caliber bullet from the decedent's pistol passed through and shattered the tempered glass in the driver's side window. Two of Officer Smith's bullets passed through failed tempered glass. His expended cartridge cases were found *outside* the vehicle, not inside the vehicle as one would expect if his pistol had been discharged in the manner described by Jones. The graze mark on the exterior of the "B" pillar also refutes the account provided by Jordan Jones. No impact damage or embedded glass was found on Officer Smith's pistol. No close range GSR or powder stippling were found around the decedent's entry wound when examined by the medical examiner.

ADDITIONAL TESTING

This report may be amended in the event additional testing is requested or carried out.

No such tests have been requested nor are any contemplated at this time.

DISPOSITION OF THE EVIDENCE

All of the submitted items were returned to their respective containers, resealed, marked, and transferred to the Evidence Impound Facility on July 15, 2005 at 4:55 pm.

signed,

_____

Attachments

1. CD containing a PowerPoint file of scene photographs, vehicle photographs, chemical test results on the vehicle, and reconstruction photographs taken by this examiner.
2. List of Publications relied on and/or authored by the examiner.
3. List of Trial and Deposition Testimony for the last 4 years.
4. Current Fee Schedule and CV.

The foregoing example may be more involved than necessary, particularly insofar as the attachments are concerned. This example was modeled along the lines of a civil case filed in Federal Court. Criminal cases in Federal and State Courts do not presently require such detail. Nonetheless, this report was for the purpose of exposing the reader to alternate ideas and perhaps a different style for writing reports dealing with reconstructive issues.

A suggested general outline for a reconstruction report is given below.

- Caption and Title Page
- Documents Received and Reviewed
- Case Overview
- Matters Not in Dispute
- Specific Reconstruction Issues
- Scene Description—Scene Processing
- Physical Evidence Received
- Observations, Tests, Results
- Summary and Conclusions
- Reservations/Additional Testing Suggested or Contemplated
- Disposition of the Evidence
- Attachments: List of Publications, List of Trial and Deposition Testimony, Fee Schedule.

Among the many things included in the "Appendix" to this book are some suggested direct examination questions for use at trial. It is expected that the interested reader will make copies of these and any other lists or illustrations in the Appendix for subsequent modification and use in the reader's jurisdiction or laboratory.

CONCLUDING COMMENTS

The classical definition of a science is *an orderly body of knowledge with principles that are clearly enunciated.* Additional requirements specify that *the subject be susceptible to testing and that it be reality-oriented.* This book has been an effort to organize the many aspects of shooting incident reconstruction into an orderly body of knowledge. The guiding principles have been unbiased analytical thinking and the application of *the Scientific Method.* The basis, the purpose and the various interpretations for each technique or procedure described within these pages have been described and references provided. Methods for documenting and preserving the evidence so that it can be reviewed by others and ultimately presented to a court or jury have been presented. All of the matters dealt with are reality-oriented. A ricocheted bullet removed from a body and found to contain particles of embedded asphalt is quite real.

The author has been very careful to use the phrases *shooting scene, shooting incident,* and *shooting incident reconstruction.* The phrases *crime scene* or *crime scene reconstruction* have been avoided for several reasons. Not all instances of firearms usage or even misusage are crimes. A police officer or armed homeowner defends himself against an armed assailant and the criminalist is requested to examine and reconstruct this scene. Neither shooter is charged with a crime. A hunting accident or the discharge of a dropped firearm may result in some form of civil litigation but no crime was ever charged. Whether a criminal or a civil matter, the ultimate goal of criminalistics is the same, and that is the reconstruction of the events. This is done from an examination and evaluation of the physical evidence with the purpose of determining what did and did not occur. This book has focused on the various characteristics of firearms and firearms-generated evidence and certain phenomena associated with them that have reconstructive properties. Eye witness accounts are notoriously untrustworthy. Participants in a shooting incident may have strong motives to misrepresent the facts. A careful, thoughtful, thorough, and well-documented reconstruction provides an objective and clear voice to this otherwise silent evidence.

Lucien C. Haag
May 1, 2005
Carefree, Arizona

REFERENCES AND FURTHER READING

Frye v. U.S. 293 Fed. 1013 (D.C.Cir. 1923).

Daubert v. Merrell Dow Pharmaceuticals, Inc. 509 U.S. 579, 113 S.Ct. 2786, 125 L.Ed.2d 469 (1993).

Kumho Tire v. Carmichael, 526 U.S. 137 (1999).

U.S. v St. Jean 45 M. J. 435 1996 CAAF Lexis 117 (1996).

Burrard, G., *The Identification of Firearms and Forensic Ballistics*, A.S. Barnes and Co., NY, 1962.

Davis, J., *Toolmarks, Firearms and the Striagraph*, Charles C. Thomas, Springfield, IL, 1958.

De Forest, P.R., R.E. Gaensslen and H.C. Lee, *Forensic Science—An Introduction to Criminalistics*, McGraw-Hill Publishing Co., NY, 1983.

Grzybowski, R.A. and J.E. Murdock, "Firearm and Toolmark Identification— Meeting the *Daubert* Challenge," *AFTE Journal* 30:1 (Winter 1998) pp. 3–14.

Hatcher, J.S., F.J., Jury and J. Weller, *Firearms Investigation, Identification and Evidence*, The Stackpole Co., Harrisburg, PA, 1957.

Kelly, G.G., *The Gun in the Case*, Whitcombe & Tombs, Ltd., New Zealand, 1963.

Kirk, P.L., "The Ontogeny of Criminalistics," *Journal of Criminal Law, Criminology and Police Science*, 54 (1963) pp. 235–238.

Kirk, P.L., *Crime Investigation*, 2, ed. J. Thornton, John Wiley & Sons, NY, 1974.

Mathews, J.H., *Firearms Identification*, Vol. I, II, III, Charles C. Thomas Publisher, Springfield, IL, 1962.

Moenssens, A., F.E. Inbau and J.E. Starrs, *Scientific Evidence in Criminal Cases*, 3rd edn, The Foundation Press, Mineola, NY, 1986.

O'Hara, C.E. and J.W. Osterburg, *An Introduction to Criminalistics*, 2nd edn, Indiana University Press, Bloomington, IN, 1972.

Saferstein, R., *Criminalistics—An Introduction to Forensic Science*, Prentice Hall, Englewood Cliffs, NJ, 1977.

Saferstein, R. ed., *Criminalistics—Forensic Science Handbook*, Prentice Hall, Englewood Cliffs, NJ, 1982.

Saferstein, R. ed., *Criminalistics—Forensic Science Handbook*, Vol. III, Regents/ Prentice Hall, Englewood Cliffs, NJ, 1993.

Svensson, A., O. Wendel and B.A.J. Fisher, *Techniques of Crime Scene Investigation*, 4th edn, Elsevier Scientific Publishing Co., NY, 1987.

Thorwald, J., *The Century of the Detective*, Harcourt, Brace and World, NY, 1964.

APPENDIX

SHOOTING INCIDENT EVALUATION AND RECONSTRUCTION DOCUMENTS SOUGHT OR REQUESTED

☐ Police/Sheriff's Report to include any Supplements, Witness Statements, Diagrams, Sketches, Scene Photographs, and Video Walk-Throughs

☐ Autopsy Report to include any anatomical diagram(s)

☐ Autopsy Photographs (particularly of the gunshot wounds before and after clean-up, with and without probes if used)

☐ Emergency Medical Records (E.R./Trauma Records)

☐ EMT Records

☐ A Copy of the Evidence Impound List

☐ Crime Lab Report(s) to include bench notes, worksheets, any diagrams, photos, and/or videotapes prepared by the examiner(s)

☐ Statement/Deposition of the shooter(s)

☐ Statement/Deposition of the victim(s)

☐ Statement/Deposition of any witnesses

☐ A Copy of the Complaint/Allegation/Pleadings (civil)

☐ Any pertinent Interrogatories or Depositions (civil)

☐ Any media photographs or videotapes taken at the time of the incident

☐ Other.

MATERIALS CHECKLIST FOR SHOOTING SCENE EXAMINATION

- ❐ Rubber Gloves/Protective Supplies
- ❐ Bullet Metal Reagents + BenchKote (Filter Paper) + Sprayers
- ❐ Lasers + Reflective Card (for laser photography)
- ❐ Tripods with Adapters for Cameras and Lasers
- ❐ High Intensity Flashlight(s) and Magnifier (Stereoscope)
- ❐ Digital Inclinometer
- ❐ Plumb Bob
- ❐ Angle Measuring Devices: Half- and Full Zero-Edge Protractors
- ❐ Pocket Calculator with Scientific Functions
- ❐ Compass
- ❐ Probe Kit or Dowel Rods
- ❐ Colored String Lines, Tacks, and Re-useable Adhesive
- ❐ Tape Measures/Scales (for photography)
- ❐ Distance Measuring Device
- ❐ Masking Tape and Double Stick Tape
- ❐ Marking Spray Paint
- ❐ Fingerprint Brush and Powder + Finger Print Cards
- ❐ Marking Pens
- ❐ Evidence Location Markers, Numbers/Letters, and Cones
- ❐ Videocamera + Still Camera with capture/storage devices, Gray Card
- ❐ Removable Strobe (Flash) Unit if Laser Photography is to be used
- ❐ Digital Calipers
- ❐ Digital Micrometer
- ❐ Large Calipers
- ❐ Tools (for disassembling car door panels, etc.)
- ❐ Dremmel Tool/Small Saw/Scalpel/Knife/Cutting Tools
- ❐ Assorted Evidence Collection Containers
- ❐ Worksheets/Note Pad/Photolog

Other Materials _____

_____

Case File/Phone # of Contact Person _____

Date/Time of Arrival _____

Persons Present:

Date/Time of Departure _____

VEHICLE DATA, MEASUREMENTS, AND BEHAVIOR WITH OCCUPANTS AND/OR MOVEMENT

Identification information: make _____ model _____

color _____

VIN _____ plate _____ engine _____

- ☐ Transmission type
- ☐ Front/rear wheel drive_____
- ☐ External dimensions
- ☐ Suspension characteristics (height changes with occupant loading)
- ☐ Tire sizes and description
- ☐ Height of vehicle body above ground level
- ☐ Height and dimensions of window and door openings
- ☐ Positions of windows/method of operation (manual/electric)
- ☐ Thickness of side/rear window glass (if struck)
- ☐ Documented samples of side/rear window glass collected (if struck)
- ☐ Angle of the windshield (if struck) relative to the horizontal plane
- ☐ Interior dimensions of occupant area
- ☐ Position of front seat(s), head rest, seat back(s), door locks
- ☐ Marking of vehicle position at the scene*
- ☐ Position of the front tires*
- ☐ Position of shifting lever*
- ☐ Position of steering wheel*
- ☐ Position of parking brake*
- ☐ Position of center arm rest/console hatch
- ☐ Glovebox locked/unlocked
- ☐ Ballistic "accessibility" of struck tires (if any)
- ☐ Manufacturer's diagrams of the vehicle
- ☐ Modifications to the vehicle (if any)
- ☐ Behavior of the vehicle in motion (if appropriate)
- ☐ The nature and/or effects of the scene terrain
- ☐ Other _____

* For vehicles struck at rest and not subsequently moved.

VEHICLE DIAGRAMS

Figure A.1

*Some Useful Forms for
the Examination of
Automobiles*

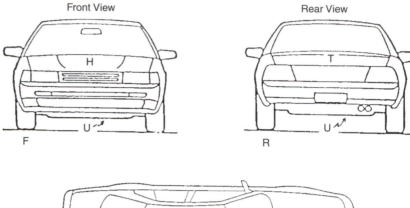

Front View

Rear View

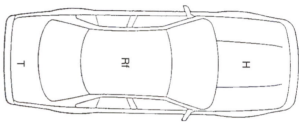

Figure A.2

*Some Useful Forms for
the Examination of
Automobiles*

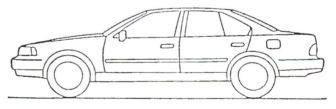

Left Side Profile View

D

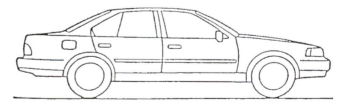

Right Side Profile View

P

LOGSHEET FOR PROJECTILE STRIKES TO VEHICLES

KEY: F = Front   H = Hood   R = Rear   T = Trunk Lid   D = Driver's Side
        Rf = Roof   P = Passenger's Side   U = Underside

Case # _____ Vehicle Description _____

Reference Point on Veh. _____

STRIKE	DISTANCE	HEIGHT	VERT. ∠	AZIMUTH ∠	COMMENTS

SHOOTING RECONSTRUCTION CHECKLIST

I. THE FIREARM

A. Shooter's/Witness' Explanation of the Pre-discharge Condition of the Firearm

1. Previous firings of this firearm (when, brand and type of ammunition, sample available?)
2. Storage and condition of the gun prior to the incident (previously cleaned, oiled, dirty, rusty, loaded, magazine separate from the gun, etc.).
3. Loading/preparation of the gun.
4. Source and type(s) of ammunition in the gun (brand, source, bullet weight, type).

Proposed laboratory evaluation: _____

_____

_____

B. Shooter's/Witness' Explanation of the Manner of Discharge

1. Manner in which the gun was held (e.g., one hand, two hands, arms extended, canted, etc.). [Measure approximate height above ground level and any upward or downward angles if possible.]
2. Means of discharge (e.g., double action, single action, dropped, "slam-fire", etc.).
3. Manner in which the gun was sighted. [Note the settings of any optical sight on the Firearm, mark, and secure these settings.]
4. Conduct of the shooter/participants relative to the firearm after the discharge of the firearm (e.g., gun recocked, reloaded, action cycled, gun dropped, tossed, magazine removed, etc.).

Proposed laboratory evaluation: _____

_____

_____

C. Location, Position, and Orientation of the Recovered Firearm

1. Record in the scene diagram and orientation photographs.
2. Take close-up photographs showing the configuration of the gun *before* the action is moved or opened. [*Note*: Ascertain if prints and/or DNA-containing cellular material are import and proceed accordingly.]

3. Documentation of:
   a. Position of the cylinder (revolvers)/position of the slide (pistols)/ position of the bolt/breech block (rifles or shotguns).
   b. Position and sequence of cartridges in the cylinder (revolvers) including a description of each cartridge.
   c. Position of the hammer (if present).
   d. Position of any mechanical safety mechanism.
   e. Magazine (if present) out or in. If in, was it fully seated?
   f. Number of live rounds remaining in the magazine/type and description.
   g. Live or expended cartridge in the chamber—yes/no. If yes, give description.
   h. Position of cock and/or load indicators (if present).
   i. Visible damage to the gun. Note location(s) and preserve any trace evidence associated with such damage.
   j. Location and description of any visible trace evidence on the gun (e.g., blood, hair, tissue, fibers, paint smears, impact damage, etc.).

Proposed laboratory evaluation: _____

_____

_____

D. Trace Evidence Considerations

1. On and/or in the firearm (e.g., powder particles in the bore and chamber(s), "flares" on the face of a revolver cylinder, bullet metal deposits in the bore, primer lacquer particles on the breechface; hairs, fibers, blood, tissue; impactive transfers of trace evidence such as wood, asphalt, concrete particles).
2. On the victim or other objects (e.g., gunshot residues, powder deposits, bullet "wipe"; ricochet/graze marks containing bullet metal; firearm contact/imprint marks, cylinder gap deposits).
3. Clothing of the gunshot victim.
4. Clothing of the shooter.
5. Acquisition of comparison specimens (like ammunition; materials from possible impact surfaces; comparison samples of blood, hair, fibers, etc.).

Proposed laboratory evaluation: _____

_____

_____

II. EXTERIOR/TERMINAL BALLISTIC QUESTIONS

A. Range Determinations

1. Information derived from the wound (stellate wound, atypical entry wounds, gunshot residue/powder pattern, penetration depth, etc.).
2. Directionality (bullet "wipe," cone fractures, lead "splash," deformation of struck surface or object, fracture lines in painted metal surfaces, pinch point, lead-in mark, etc.).
3. Gunshot residue/powder deposition and pattern.
4. Shot patterns (shotgun/shot cartridge shootings).
5. Wads, shotshell fillers, shot collars, shotcups, sabots (exterior ballistic characteristics).
6. Trajectory considerations (angle of departure, angle of fall, line of sight, midrange, and maximum ordinate height, sight picture, bullet flight time, "lagtime").
7. Performance characteristics of the responsible gun/ammunition combination (general operation of the gun, GSR production, muzzle velocity of the projectile, cartridge ejection pattern, etc.).
8. Appearance, visibility of the gun, its discharge (sound, muzzle flash, etc.).
9. Weather conditions at the time of the shooting (if applicable* and known).
10. Site description and MSL elevation (if applicable*).

* *Note*: The elevation of the site, terrain features, and meteorological conditions is of interest or importance in long range shootings.

Proposed laboratory evaluation: _____

_____

_____

B. Terminal Ballistic Phenomena

1. Penetration depth (as an expression of terminal velocity).
   [*Note*: Determined by measurement, radiographs ("X-rays"), interview of pathologist.]
2. Penetration path/angle (as an expression of the bullet's pre-impact path).
   [*Note*: Determine and/or document vertical and azimuth components of penetration path.]
3. Projectile deformation/trace evidence transfers (degree of projectile deformation related to impact velocity, the character of bullet deformation may be relatable to the incident angle in ricochets and the nature of the impacted surface; examples of transfers are fabric imprints, embedded bone, soil, adhering tissue, blood, fibers, etc.).

4. Bullet "wipe" (usually contains traces of the bullet's composition, primer-generated residue and possibly bullet lubricant).

5. Sequence of shots through intersecting radial fractures in plate glass, plastics, ceramics, skulls, etc. Sequencing of shots through mixed bullet types (revolvers, barrel residues, trace evidence methods).

Proposed laboratory evaluation:_____

_____

_____

III. EJECTED CARTRIDGE CASES/MISFIRED CARTRIDGES/UNFIRED CARTRIDGES

A. **Location(s) with measurements** (shooting scene diagram).

B. **Orientation Photographs**

C. **Close-up Photograph(s)**

D. **Cartridge Description** (headstamp, damage, adhering trace evidence)

E. **Nature and Description of the Surface upon which the Cartridge was Found**

F. **Nature of Any Nearby Surfaces that the Cartridge may have Struck** (walls, fences).

Proposed laboratory evaluation: _____

_____

_____

IV. SUPPLEMENTAL BACKGROUND INFORMATION

A. **Prior Use of the Gun** (when and where fired, number of shots fired, type of ammunition, persons present, fired cartridges picked up? cleaned afterwards?).

B. **Cartridges Previously Loaded or Fired in the Gun**.

C. **Source and Age of the Ammunition Actually Used in the Shooting Incident** (ammunition properties and characteristics—e.g., bullet weight, composition and design, powder type and charge, cartridge headstamp, etc.).

[*Note*: Obtain cartridge box(s) if possible (for lot number).]

Proposed laboratory evaluation: _____

_____

_____

V. MISCELLANEOUS TECHNIQUES AND COMPARISONS

A. Comparison of Ammunition Components (class characteristics, corresponding manufacturing marks, bunter marks, propellant type, propellant and/or primer chemistry, bullet weight, construction and composition, etc.).

B. Comparison of Bullet Cores and/or Bullet Fragments
   1. Physical matches of cores to separated jackets; fragments to fragments; weight considerations.
   2. Instrumental analysis of projectile fragments; comparison of results with fired and/or unfired projectiles of the same type.

C. Acoustical Evaluation and Comparison of Recorded Gunshots
   1. Description of equipment involved that recorded the gunshots and/or bullet passage/impact.
   2. Site conditions at the time the gunshots and/or bullet passage/impact sounds were recorded (meteorological conditions, wind speed and direction, sources of background noises, if any).
   3. Location and orientation of equipment involved.
   4. Distance(s) involved between the possible source of the shot(s) and the recording site.

Proposed laboratory evaluation: _____

_____

_____

SUGGESTED GUIDELINES FOR RECONSTRUCTION AND REENACTMENT OF SHOOTING INCIDENTS

1. Scope

These guidelines are intended to provide direction for the forensic scientist undertaking a reconstruction or reenactment of a shooting incident. These guidelines describe the:

1.1 Collection of data necessary to undertake the reconstruction

1.2 Documentation required in the report

1.3 Appropriate procedures and precautions for the presentation of a reconstruction or reenactment in court.

2. Definitions

2.1 *Reconstruction*: The determination of the sequence of two or more events in a particular incident utilizing information derived from the physical evidence, data from the analysis of physical evidence, recognized physical laws, and/or inferences drawn from experimentation related to the incident under investigation.

2.2 *Reenactment*: The demonstration of a reconstruction through the use of live actors or computer animation.

2.3 *Visual aid*: Any device used to demonstrate any aspect of a reconstruction or reenactment.

2.4 *Event*: A single occurrence, action, or happening.

2.5 *Incident*: A series of related events.

2.6 *Transfer evidence*: Evidence that is transferred from one object to another by virtue of the contact of the two objects.

2.7 *Trace evidence*: Physical evidence of a microscopic or submicroscopic size which, due to its small size, is deposited on or transferred to one or more objects without being manifestly apparent at the time of transfer or deposition. (*Note*: Trace evidence is differentiated from Transfer evidence in that contact is *not* required. Gunshot residues deposited on a victim or left in the bore of a firearm, for example, do not require contact.)

2.8 *Impression evidence*: Evidence produced by the static deformation or dynamic alteration of one object by another.

3. Significance and Use

The reconstruction or reenactment of an event or incident may attempt to establish one or more of the following:

3.1 The manner in which a firearm was discharged

3.2 The range from which a firearm was discharged

3.3 The position or orientation of a firearm at the moment of discharge

3.4 The location of the shooter at the time of discharge

3.5 The position of the victim at the moment of projectile impact and/or the discharge of a firearm

3.6 The number and/or sequence of shots in multiple discharge shooting incidents

3.7 The presence of a person or object at a shooting scene

3.8 Other exterior and/or terminal ballistic events that may have reconstructive value

3.9 Establish a time line for actions in an incident based on:

    3.9.1 The physical requirements for movement or actions of persons

    3.9.2 The considerations of operational requirements of mechanical systems or other objects or minimum rates of fire of the gun

    3.9.3 Acoustical information derived from audio recordings (911 tapes, videotapes, audiotapes, etc.).

4. Data Collection

The reliability of any reconstruction or reenactment ultimately relies on the validity and accuracy of the underlying data. The forensic scientist must determine that the data that is relied upon is sufficiently accurate for the purposes for which it is used.

4.1 *Scene data*

    4.1.1 The dimensions of the relevant area should be determined with reasonable accuracy.

        4.1.1.1 Data from interior rooms should include ceiling height, dimensions, layout (floor plan), and furnishings.

    4.1.2 Photographic documentation of the scene should be thorough, but should not be considered a substitute for adequate sketches.

        4.1.2.1 Overall photographic documentation should include views from several angles.

        4.1.2.2 Photographic documentation should include intermediate and close-up views of all items of physical evidence.

        4.1.2.3 Particular attention should be paid to ensure adequate photographic documentation of evidence which, by its nature, cannot be removed from the scene. Examples are bullet holes in some types of walls, plate glass windows, certain bullet impact sites, etc.

        4.1.2.4 Videotaping of the scene may provide a useful adjunct to but is not a substitute for adequate still photography. Such videotaping should include a factual narrative of the scene or subject matter being videotaped. The narrator should qualify observations with terms such as "—*appears to be* a bullet hole,—impact site,—ricochet mark, etc."

4.2 *Physical evidence data*

4.2.1 The locations of all items of physical evidence should be documented.

4.2.2 Critical scene information must be documented photographically and by such other methods as would allow another investigator who does not have access to the scene to use the photographs and data to evaluate the reconstruction. Some examples of common scene evidence which requires careful documentation are:

4.2.2.1 Projectile trajectories, bullet holes, and impact sites

4.2.2.2 Blood spatter patterns and blood trails

4.2.2.3 Expended cartridge cases, shotshell wads, gunshot residue deposits, etc.

5. Laboratory Testing

All data from laboratory examinations, analyses, or experiments should be recorded in a manner consistent with good scientific practice and in a manner that can be understood by another forensic scientist familiar with shooting incident reconstruction.

5.1 *Experimental verification in reconstruction*: Experimental verification of all stages of the reconstruction should be theoretically possible and, when possible, actually done.

5.1.1 Elements of the reconstruction that are not, at least in principle, experimentally verifiable should not be included in the report, testimony, or visual aids used at trial.

5.1.2 Any experiments should be designed to test specific hypotheses or accounts of an incident and should include appropriate controls.

5.1.2.1 The variables associated with an experiment should be explicitly identified and any assumptions made should be explicitly stated.

5.1.3 Experiments should be designed to reproduce those elements of the original circumstance that, in the opinion of the forensic scientist, are relevant to the purposes of the experiment.

5.1.4 Reliance on statements of witnesses or actors in the incident should be kept to a minimum and when utilized, so stated in any reconstructive effort.

5.1.4.1 If the forensic scientist relies on any witness or actor statement as a part of a reconstruction, it is highly desirable to design tests or experiments to evaluate such statements when possible. In the reconstruction the investigator should:

5.1.4.1.1 Evaluate whether the statement is reasonably correct and reliable

5.1.4.1.2 Specifically refer to the statement in any report or testimony.

5.1.5 If the reconstruction refutes the account given by a witness or actor, the specific basis of such refutation should be clearly stated in any subsequent report.

6. Presentation of a Reconstruction or a Reenactment

The purpose of the reconstruction or reenactment is to give the client, attorney, court, or jury a clearer understanding of what happened in the incident under consideration. It is also appropriate to state those matters and events that can be excluded as having occurred. Whether in a written report, by means of photographic or other documentation, or in testimony, it is incumbent on the forensic scientist to be certain that the recipient has a clear understanding of the nature of the investigator's opinion, including the limitations and uncertainties of that opinion.

6.1 *Written report*: A written report should convey to an untrained reader who is familiar with the basic facts under consideration the expert's view of what happened. This report should also contain enough data so that the technically sophisticated reader can evaluate the methods used by the expert to arrive at the conclusions expressed in the report, and to make a preliminary evaluation of the reliability of the opinions expressed in the report.

6.1.1 All documents reviewed should be explicitly identified in the report.

6.1.2 If any additional information has been used in the reconstruction, the source and nature of that information should be explicitly stated in the report.

6.1.3 All items of physical evidence received or examined should be listed in the report.

6.1.4 The source of any reference material used in experiments or tests should be explicitly described in the report. For example:

6.1.4.1 Tables of ballistic data

6.1.4.2 Tables or data relating to body dimensions and/or gunshot wounds in bodies.

6.1.5 Data from experiments or observations should be included in reports in such a way so as to provide the technically astute reader a basis for evaluating the opinions expressed in the report.

6.1.6 Computer programs used for the analysis of data or the production of information used for the reconstruction should be identified.

6.1.7 If photographic or videotape documentation is supplied with the original report, the photographs or videotape should be referred to in the report, and copies should be made available to interested parties.

6.2 *Testimony*

6.2.1 The purpose of testimony is to convey to the court or jury those pertinent elements of the reconstruction that can be stated with a reasonable degree of scientific certainty by the forensic scientist or witness.

6.2.2 The expert should not include in testimony or visual aids information that is not based on his technical evaluation of the evidence or circumstances of the incident.

6.2.3 The witness should provide the court and jury with an understandable assessment of the reliability of conclusions.

Such an assessment should include:

6.2.3.1 Statements of reasonable alternative possibilities

6.2.3.2 Estimates of the uncertainty (margin of error) in experimental results.

6.3 *Presentation of reenactments*

6.3.1 It must be recognized that it is not possible to determine all of the elements necessary for a reenactment of an incident with the same degree of reliability.

6.3.2 The courtroom presentation of a reenactment of a shooting incident will be at the discretion of the trial judge. It is incumbent on the forensic scientist who developed the reenactment to advise the court as to what elements of the reenactment can be verified by the witness and what elements are assumptions.

6.3.2.1 The expert must remember that many of the unverifiable aspects of a reenactment (e.g., facial expressions of participants, positions of hands and arms, positions of persons or objects prior to the earliest moment of the reconstruction, etc.) may be critical to arguments made by counsel and can have an undue impact on a jury.

6.3.2.2 While the determination that a visual aid or reenactment is "more probative than prejudicial" is up to the trial judge, the expert must be sure that the judge is aware of which aspects of the reenactment or visual aid can be defended scientifically and which cannot.

COAT WORKSHEET

Figure A.3

examiner _____   case # _____

Item # _____   date _____

Remarks: _____

_____

_____

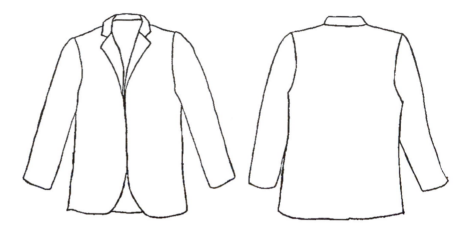

LONG SLEEVED SHIRT WORKSHEET

examiner_____   case # _____

Item # _____   date _____

Remarks:_____

_____

_____

Figure A.4

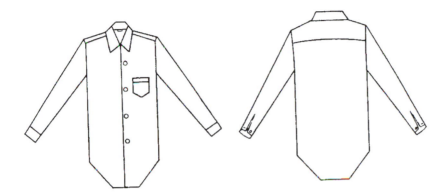

UNDER SHIRT/T-SHIRT WORKSHEET

Figure A.5

WORKSHEET

examiner_____   case # _____

item # _____   date _____

Remarks:_____

_____

_____

PANTS WORKSHEET

examiner_____   case #_____   *Figure A.6*

item #_____   date_____

Remarks:_____

_____

_____

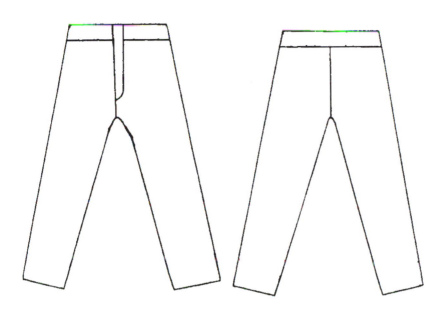

SOME CYCLIC RATES OF FIRE FOR
RECOIL-OPERATED SEMIAUTOMATIC PISTOLS
[MEASURED WITH PACT IV TIMER]

FORMULAS ASSOCIATED WITH RATES OF FIRE

Cyclic Rate, CR (in rds. per minute) $= (60 \div \text{shot-to-shot time in seconds}) - 1$

or

$$CR = [(X - 1) \div T_x] \times 60$$

where $X =$ the number of shots and $T_x =$ the time to fire X shots
Shot-to-shot interval, $I = 60 \div (CR + 1)$
Time to fire X shots, $T_x = I \times (X - 1)$

or

$$T_x = 60 \ (X - 1) \div (CR + 1)$$

Firearm	Average Cyclic Rate	Ave. Shot-to-shot Interval (sec.)	5 rds in— sec.
Glock 17 9 mm	245 ± 20 RPM (no target) = ca. 4 rds./sec.	0.24	0.98
	175 ± 50 RPM (directed fire) = ca. 3 rds./sec.	0.34	1.36
Glock 22 .40 S&W	208 ± 12 RPM (directed fire) = ca. 3.5 rds./sec.	0.29	1.15
S&W M39 9 mm	260 ± 32 RPM (no target) = ca. 4.3 rds./sec.	0.23	0.92
Starfire 9 mm	236 ± 16 RPM (no target) = ca. 4 rds./sec.	0.25	1.01
Ruger P85 9 mm	229 RPM (no target) based on 10 rds. in 2.349 sec.	0.26	1.04
SIG P226 9 mm	248 ± 15 RPM (no target) = ca. 4 rds./sec.	0.24	0.96
	78 ± 14 RPM (directed fire) = ca. 1.3 rds./sec.	0.76	3.00
Beretta 85F .380 Auto	218 RPM (no target) based on 7 rds. in 1.645 sec.	0.27	1.10
Beretta 92F 9 mm	237 RPM (no target) based on 6 rds. in 1.262 sec.	0.25	1.01
	173 RPM (directed fire) based on 6 rds. in 1.733 sec.	0.34	1.38
Beretta 96 .40S&W	248 RPM (no target) based on 5 rds. in 0.926 sec.	0.24	0.93
	160 ± 58 RPM (directed fire) = ca. 3 rds./sec.	0.37	1.49
Colt 1911A1 .45 Auto	253 RPM (no target) based on 7 rds. in 1.420 sec.	0.24	0.94

"No target" means to fire as fast as possible with no effort to aim the pistol.
"Directed fire" means that there was an effort to aim the pistol at a man-sized target for each shot.

REFERENCE

Haag, L.C., "Rates of Fire for Some Common Semi-Automatic and Full Automatic Firearms," *AFTE Journal* 32:3 (Summer 2000) pp. 252–258.

CONVERSION AND COMPUTATIONAL FACTORS

U.S. to Metric	Metric to U.S.

Length/Distance

1 in. = 25.4 mm (2.54 cm)	1 mm = 0.03937 in.
1 yd. = 0.9144 meters	1 cm = 0.3937 in.
1 ft. = 0.3048 meters	1 m = 1.0936 yd (3.2808 ft.)

Velocity/Speed

fps × 0.3048 = m/s	m/s × 3.2808 = fps
fps × 0.6818 = mph	m/s × 3.60 = kph
mph × 1.609 = kph	kph × 0.6214 = mph

Gravitational Acceleration (g)

32.17 f/s/s	980.7 cm/s/s

Weight/Mass

1 gr. = 0.0648 g	1 g = 15.432 gr.
1 oz. = 28.35 g	1 kg = 2.203 lbs.
1 lb. = 16 oz. = 454 g	
7000 gr. = 1 lb.	
lbs/g = slugs	

Energy

ft.-lbs × 0.1382 = kg-m	kg-m × 7.233 = ft.-lbs
ft.-lbs × 1.355 = joules	joules × 0.7375 = ft.-lbs
	kg-m × 9.805 = joules

K.E. (ft.-lbs.) = grains × velocity^2/450,380 where projectile velocity is in fps

Momentum

lb.-sec. × 0.454 = kg-sec.	kg-sec. × 2.203 = lb.-sec.

Momentum
(lb.-sec.) = grains × velocity/225,190 where projectile velocity is in fps

Pressure

1 bar = 14.5036 psi	780 mm = 29.92 in. Hg
1 atmosphere = 29.92 in. Hg = 780 mm Hg	1000 millibars = 29.53 in. Hg

Force

1 lb. = 4.45 N (Newtons)	1 N = 10^5 dynes = 0.2247 lbs
	1 dyne/cm^2 = 1.45 × 10^{-5} lbs/in.^{2}

Area

1 in.^{2} = 6.45 cm^2	1 cm^2 = 0.155 in.^{2}
Area of a circle = πr^2	

Volume

1 in.^{3} × 16.38 = cm^3	1 cm^3 × 0.0610 = in.^{3}
Gal. × 3.785 = liters	Liters × 0.264 = gallons
Vol. of a sphere = $4/3\pi r^3$	

Temperature

°F = 9/5C + 32	°C = 5/9(F − 32)

Speed of Sound (Sea Level & 15 °C)

1117 fps	340.4 m/s

Density of Water (20 °C)

8.4 lbs/gal	0.99823 g/cc

TABLE OF COMMON BULLET WEIGHTS
(L = LEAD J = JACKETED)

Cartridge	Weights (in gr.) and Basic Construction
.22 Short	27L 29L
.22 Long	27L 29L
.22 Long Rifle	27L 31L 32L 33L 36L 37L 38L 40L
.22 Magnum Rimfire	30J 33J 34J 40J 50J
.223 Rem. (5.56 mm)	40J 45J 50J 52J 53J 55J 62J 64J 69J
.25 Automatic	35J 45J 50J
.30 (7.65 mm) Luger	93J
.30 Carbine	110J
.32 Automatic	60J 65J 71J 73J 77J
.32 S&W	85L 88L
.32 S&W Long	83L 98L 100L
.32 Short Colt	80L
.32 Long Colt	82L
.32–20WCF	100L 100J
.32 H&R Mag	85J 95L 95J
7.62 × 39 mm	122J 123J 124J 125J
7.62 mm (.30) Tokarev	87J
7.62 mm Nagant	108L
7.63 mm (.30) Mauser	88J 93J
9 mm Luger	90J 95J 100J 105J 115J 120J 123J 124J 147J
9 mm Makarov	90J 95J
.38 S&W	145L 146L 200L
.380 Automatic	60J 82J 85J 88J 90J 95J 102J
.38 (Super) Automatic	115J 125J 129J 130J
.38 Short Colt	125L
.38 Long Colt	150L
.38 Special	95J 110J 125J 129J 130J 132J 140J 148L 147J 150L 158L 200L
.38–40 WCF	180J
.357 SIG	124J 125J 147J 150J
.357 Magnum	105J 110J 125J 140L 140J 142J 145J 150J 158L 158J 165J 170J 180J
.357 Maximum	158J 180J
.40 S&W	135J 140J 145J 155J 165J 170J 180J
.10 mm Auto	155J 170J 175J 180J 200J
.41 Short Colt	160L
.41 Long Colt	200L
.41 Rem.Mag	170J 175J 200J 210J 250J
.44 Russian	247L
.44–40 WCF	200J 225L
.44 Special	200L 180J 200L 200J 210L 240L 246L
.44 Magnum	180J 185J 200J 210J 240J 240L 250J 270J 300J
.45 Automatic	165J 185J 200J 225J 230J
.45 Auto Rim	230L
.45 Colt	200J 225J 225L 250J 250L 255L 300J
.454 Casull	240J 250J 260J 300J 350J
.455 Webley	262L 265L
.475 Linebaugh	400J
.480 Ruger	325J
.50 AE	300J 325J

Note: This list is not all-inclusive. Many uncommon and obsolete cartridges are not listed. Moreover, bullets of alternate and uncommon weights can be handloaded in most of these cartridges. New loadings for certain cartridges frequently appear on the market and without prior notice.

SOME GENERAL GUIDELINES FOR LASER PHOTOGRAPHY

BASIC EQUIPMENT

35 mm FILM camera with:
  "B" (bulb) shutter capability
  adjustable aperture (f-stop) and manual focus
  cable release with lock
  lens cap
ASA 100 (or slower) *film*
Sturdy Tripod
Removable Flash unit
Laser(s) and mounting apparatus
White cardstock

PREPARATORY GUIDELINES

- Establish and secure laser illustrating reconstructed projectile path.
- Set up camera at a right angle (profile view) to the desired portion of the laser path.
- If desirable, include a vertical or true horizontal reference line or object at the center point.
- For *outdoor* scenes: late evening, dusk or night. For *indoor* scenes: subdued lighting.
- For very dark conditions, try f/2.8 to f/5.6 shutter aperture. For dimly lit conditions, f/8 to f/16.
- Set camera on Manual Focus and focus at the laser line distance.
- Attach the cable release and loosen the cable lock.
- Note that the card handler should wear dark clothing if there is subdued lighting.
- If multiple laser lines are to be photographed on the same frame of film, use an opaque lens can and cover the open lens between setups.

PHOTOGRAPHIC GUIDELINES

- Position the card handler at one end of the laser line. Stand on the side of the line *opposite* the camera and face the camera. (Several practice walks are recommended.)
- Insert the white card in the laser beam with the card oriented at a 45° angle to the camera.
- Have the camera operator open the shutter with the cable release and lock it open.
- Walk the laser line at a slow walk (ca. 1 foot per second).
- Signal the camera operator to release (close) the shutter by unlocking the cable release as the card handler reaches the end of the laser line. If multiple laser lines are to be photographed, cover the open lens with the lens cap.
- Take multiple time exposures at differing f-stops.
- If the area is very dark and fill-in lighting is desired, use a hand-held flash unit and discharge it away from the open lens of the camera and toward the areas to be illuminated.

DIGITAL PHOTOGRAPHY OF LASER PATHS
(NIKON D100 OR EQUIVALENT)

- Position the card handler at one end of the laser line. Stand on the side of the line *opposite* the camera and face the camera. (Several practice walks are recommended.)
- Insert the white card in the laser beam with the card oriented at a 45° angle to the camera.
- Set focus to manual.
- For brighter conditions, set ASA to low settings (100 or lower); for darker conditions, set as high as 1000+.
- Adjust f stop to get depth of field (higher number [32] for greater depth of focus, smaller [5.6] to let more light in).
- Adjust as you go by viewing each shot.
- Have the camera operator open the shutter with the cable release and lock it open.
- Walk the laser line at a walk (ca. 1 fps). Faster for green lasers.
- Clear all people out of the field of view, and pop off any strobes for fill light. Never point the flash toward the camera when using fill lighting.
- If multiple laser lines are to be photographed, cover the open lens with the lens cap or dark cardboard held tightly against the lens in between runs.
- Dotted laser lines may be created by cycling the card in and out of the beam as you move along the laser line (see Figure A.8).
- Signal the camera operator to release (close) the shutter by unlocking the cable release as the card handler reaches the end of the laser line.

Figure A.7

Nighttime scene as it appeared. Visit books.elsevier.com/companions/9780120884735 for a full-color version of this figure.

```
Nikon D100 – 28 mm
f/10 – 7 sec.
ISO 800
Darkness
```

Figure A.8

Laser depiction of three shots by two shooters. Visit books.elsevier.com/companions/9780120884735 for a full-color version of this figure.

```
Nikon D100 – 28 mm
f/9 – 14.2 sec.
ISO 800
Twilight
```

The key and biggest advantage of working with a flexible digital platform such as the Nikon D100 is the ability to evaluate your image immediately. Bracketing settings (f stop, exposure time, number of fill flashes, ASA, etc.) is still the best and the safest way to obtain the best finished product.
Source: Digital procedures and photographs by Michael Haag and Joe Foster, Alququerque, NM Police Crime Laboratory, September 2005.

SOME BULLET CORE AND JACKET WEIGHTS

Cartridge	Mfg./Type	Bullet Wt.	Jacket Wt.	Core Wt.
.224 in.	Sierra FMJ-BT	55 gr.	16.5 gr.	38.5 gr.
.224 in.	Sierra JSP Blitz	55 gr.	11 gr.	44 gr.
.32 Auto	Rem. MJ-RN	71 gr.	16 gr.	55 gr.
7.62 × 39 mm	Russian JHP	123 gr.	33 gr. (Fe)	88 gr. + 1.5 gr. plastic plug
.308 in.	Sierra JSP	125 gr.	32 gr.	93 gr.
.310 in.	Hornady JSP	125 gr.	26 gr.	99 gr.
.380 Auto	REM. FMJ-RN	95 gr.	17 gr.	78 gr.
.380 Auto	WIN. FMJ-RN	95 gr.	17 gr.	78 gr.
.380 Auto	Fiocchi FMJ-RN	95 gr.	15 gr. (brass)	80 gr.
9 mm Luger	WIN.-U.S.A. FMJ-RN	115 gr.	21 gr.	94 gr.
9 mm Luger	FED Hi-Shok JHP	115 gr.	19 gr.	96 gr.
9 mm Luger	FED.-Am.-Eag. FMJ-RN	124 gr.	21 gr.	103 gr.
9 mm Luger	FED. JHP	124 gr.	21 gr.	103 gr.
9 mm Luger	FED.-Am.-Ea. FMJ-TC	147 gr.	23 gr.	124 gr.
9 mm Luger	WIN.(Law Enf.) JHP	147 gr.	24.5 gr.	122.5 gr.
.38 ACP	PMC FMJ-RN	130 gr.	26 gr.	104 gr.
.38/357M	WIN. JHP	110 gr.	15 gr.	95 gr.
.38 SPL.	WIN. SilverTip JHP	125 gr.	5.5 gr. (Al)	119.5 gr.
.38/357M	WIN. FMJ	130 gr.	23 gr.	107 gr.
.357M	WIN. SilverTip JHP	145 gr.	22 gr.	123 gr.
.38/357M	WIN. JSP	158 gr.	23 gr.	135 gr.
.38/357M	FED. JSP	125 gr.	14 gr.	111 gr.
.357M	FED. JHP	180 gr.	21 gr.	159 gr.
.357M	PMC Starfire JHP	150 gr.	15 gr.	135 gr.
.40 S&W	FED. HYDRA-SHOK	155 gr.	29 gr.	126 gr.
.40 S&W	FED. HYDRA-SHOK	165 gr.	31 gr.	134 gr.
.40 S&W	FED. HYDRA-SHOK	180 gr.	32 gr.	150 gr.
.40 S&W	FED. TACTICAL	165 gr.	47 gr.	118 gr.
.40 S&W	WIN –U.S.A. FMJ	180 gr.	29.5 gr.	150.5 gr.
.40 S&W	WIN-SXT/B.T. JHP	180 gr.	35.5 gr.	144.5 gr.
40/10 mm	HORNADY FMJ	200 gr.	24 gr.	176 gr.
41 Rem. Mag.	WIN. JHP	210 gr.	31.5 gr.	178.5 gr.
.44 Mag.	REM. – UMC. JSP	180 gr.	28.5 gr.	151.5 gr.
.44 Mag.	WIN. – U.S.A. JSP	240 gr.	36 gr.	204 gr.
.45 Auto	REM. Golden Saber JHP	185 gr.	45 gr.	140 gr.
.45 Auto	SPEER JHP	200 gr.	25 gr.	175 gr.
.45 Auto	REM. FMJ-RN	230 gr.	35 gr.	195 gr.
.45 Auto	WIN. FMJ-RN	230 gr.	36 gr.	194 gr.

Note: Lead cores removed by heating the pre-weighed bullet with a torch and melting and rapping the liquid lead out of the jacket.

GENERAL FOUNDATIONAL QUESTIONS FOR:_____

Would you please state your name and occupation?

Who is a Criminalist/Forensic scientist?

By who are you presently employed?

How long have you been employed as a criminalist/forensic scientist?

Have your past duties include the examination of firearms and firearms-related evidence?

Please describe the types of firearms examinations you conducted during your _____ years as a criminalist/forensic scientist.

Would you please take a moment and describe your educational background?

Has any of your training included the examination of firearms and firearms-related evidence? Please describe the nature of that training.

Have you also attended seminars on the subject of firearms evidence and examination? On more than one occasion?

Are you a member of any professional associations or societies?

Would you name them please?

Do any of these organizations deal with firearms evidence and examination?

Have you been called to testify as an expert witness in courts of law as a result of the various types of testing you have carried out on firearms and firearms-related evidence?

Are there certain aspects of investigations involving shooting incidents that are reconstructive in nature?

Would a *muzzle distance determination* based on a powder pattern on a victim's shirt be a common example of a shooting reconstruction?

Would a *range-of-fire determination* based on a pellet pattern from a shotgun be another example?

Are these fairly routine determinations for people in your field?

Is the reconstruction of crime scenes, accident scenes, and certain events that took place at such scenes one of the common, recognized objectives of criminalistics and forensic science?

Regarding shooting incidents, aside from the previous examples, can you give us some additional examples of the reconstructive aspects of such cases?

Sample answers

The manner in which a firearm was discharged (e.g., impact vs. normal discharge).

The approximate range from which a firearm was discharged (e.g., powder pattern).

The position of the shooter based on scene geometry and bullet path.

The position/orientation of a victim based on wound path and projectile-struck objects.

The position/orientation of a firearm at the moment of discharge.

The number and/or sequence of shots (e.g., two shots in tempered glass).

The direction of fire (bullet holes in glass, wood, sheet metal).

Do you have any special training in the reconstruction of shooting incidents? Please describe that training.

[Examples: 5-day, 40-hr. course by the author at the *Gunsite Training Facility*, Paulden, AZ]

Past workshops and/or presentations given at the annual training seminars of The Association of Firearm and Tool Mark Examiners.

Optional Q. If one or more special shooting reconstruction courses have been attended:

Did you successfully complete that training course?

Would you describe in general terms what is involved in attempting to reconstruct a shooting incident?

Is it appropriate to consider the accounts of the various actors, participants or witnesses in a given case?

Why?

How did you start your evaluation in this case?

SPECIFIC FOUNDATIONAL QUESTIONS

When were you first contacted in this case?

Who contacted you in this matter?

What were you asked to do?

Did you review any documents or other materials in this undertaking?

What did you study?

[List or Describe the documents]

Did you examine any physical evidence related to this case?

Could you, in very general terms, give us a brief description of the physical evidence you personally examined? (*Note*: Specific details regarding certain items to be gone into later.)

Optional Q.

Did you go to the scene?

When?

Who was present?

What was your purpose in going to the scene?

Would you give us a brief overview of your activities at the scene?

Did you carry out any ballistic or other type of testing at your laboratory, related to this case?

If "Yes," what sort of testing did you carry out at your laboratory?

When were these tests carried out?

What was their purpose in general terms?

[Specific questions to follow after consultation with the attorney(s).]

GLOSSARY

A

ACP Abbreviation for Automatic Colt Pistol. Normally used to designate a cartridge, as in the 45ACP.

AFTE Abbreviation for the Association of Firearm and Tool Mark Examiners.

AP Abbreviation for **Armor Piercing** ammunition.

API Abbreviation for **Armor Piercing-Incendiary** ammunition.

Accelerator® Cartridge A type of Remington centerfire ammunition utilizing a substitute bullet mounted in a nylon sabot.

Accidental Discharge The gun and/or ammunition is the source of the discharge, for example, a compromised safety system, inadequate safety system, high primer, broken firing pin protruding from the bolt, etc.

Acetic Acid A chemical reagent used in the modified Griess test for nitrite residue detection. Acetic acid forms nitrous acid in reacting with nitrites in gunpowder residues which in turn reacts with the other constituents in this reagent to form an azo-dye. Acetic acid may also be employed in certain situations with the sodium rhodizonate test to solubilize lead residues and make them reactive with this reagent.

Ammunition The material fired in and from any weapon or firearm such as cartridges, bullets, and shot; one or more loaded cartridges consisting of a primed case, propellant, and with or without one or more projectiles.

Angle of Deflection The angle formed between the path of the departing projectile subsequent to an impact and the pre-impact path of the projectile's flight.

Angle of Departure The angle formed between a horizontal line and the centerline of the bore at the moment the projectile leaves the muzzle of the firearm. *Note*: This angle is related to the initial angle at which the projectile departs relative to the surface of the earth.

Angle of Elevation   The vertical angle formed between the target and the axis of the barrel bore. *Note*: This angle is related to the initial angle the projectile departs relative to the Line of Sight (LOS) to the target.

Over level terrain between the gun and the target, this angle will also be the angle of departure relative to the horizontal plane.

Angle of Fall   In shooting scenes, this term is used to describe the arrival angle relative to the horizontal plane of a bullet descending from a long range flight. This angle is the same as the vertical angle component of any bullet path at a shooting scene.

Angle Finder   A device that is designed to measure or display vertical or azimuth angles or both.

Angle of Incidence   In ricochet events, angle of incidence is the intercept angle described by the pre-impact path of the projectile and the plane of the impact surface at the impact site when viewed in profile. The angle formed between the path of the projectile prior to impact and the plane of the impacted surface. *Note*: As used in this text, this definition differs from the NATO definition of incident angle. To convert from the forensic definition used here to the corresponding NATO angle, use the equation [90° − F.A.] = NATO angle.

Angle of Ricochet   Using the same coordinate system as for the angle of incidence, this angle is defined by the path taken by the ricocheted projectile as it departs the impacted surface with one additional qualification—the plane of the impact site is that surface *prior* to bullet impact even though in some situations the bullet is departing a much modified surface (e.g., water, sheet metal, soil).

The angle formed between the path of the departing projectile subsequent to impact and the plane of the impacted surface.

Note: The term "departure angle" or "angle of departure" has been used to describe ricochet angle but this is discouraged because these terms also apply to a gun barrel's launch angle relative to the horizontal plane.

Annulus   The ring-like space between the exterior surface of the primer and the primer pocket or battery cup on the base of a cartridge.

Antimony (Sb)   A metal frequently used to harden lead by alloying the two. Percentages of antimony in lead shot and lead bullets typically range from 0.5 to 3% Sb.

Antimony Sulfide (Sb$_2$S$_3$)   A component of most common priming mixtures that serves as a fuel.

Anvil   An internal metal component in a boxer primer assembly against which the priming mixture is crushed by the impact of the firing pin; a metal feature in the primer pocket of a Berdan-primed cartridge case, against which the priming compound is crushed by the impact of the firing pin; the breech end of the chamber in a rimfire firearm against which the rim is crushed by the firing pin's impact.

Anvil Mark A microscopic mark impressed on the forward face of the rim of a rimfire cartridge case as it is forced against the breech end of the barrel by the impact of the firing pin. These marks are characteristic of the breech under the firing pin and have been used to identify a cartridge case with a specific firearm.

Apogee Regarding exterior ballistics, the apogee is the highest point in a bullet's flight path. For trajectories in air, this point will always be slightly displaced downrange of the true midpoint of the trajectory.

Armor Piercing Bullet A bullet containing a hardened core composed of a substance other than lead or lead alloy; any bullet manufactured, represented, or designed to be metal or armor piercing.

Assault Rifle A rifle of intermediate caliber capable of both semiautomatic and fully automatic fire by means of a selector.

Assault Weapon A media and/or legislative term applied to most any military-style, semiautomatic rifle, or carbine. The legislative bodies of some states have defined certain firearms as *assault weapons* based on certain features such as rifles having pistol grips, extended box magazines, flash suppressors, etc., or they have simply prepared lists of specifically named firearms. There is no scientific or technical definition for the term *assault weapon*. Moreover, they are not fully automatic firearms.

Automatic Weapon Any firearm which discharges multiple shots with a single actuation of the trigger. In common usage, the term is often applied erroneously to describe what should be described as an autoloading, semiautomatic, or self-loading firearm.

Azimuth Angle An angle or bearing lying in the horizontal plane usually described on the basis of compass directions or with north, south, east, west descriptors. In shooting reconstruction, an arbitrary north–south or east–west reference line may be chosen as a reference for azimuth angles related to that line.

B

BB The designation of spherical shot having a diameter of .180 in. used in shotshell loads. The term BB is also used to designate steel or lead air rifle shot of .175 in. diameter. Although the two definitions cause some confusion, they have co-existed for many years.

BC Abbreviation for **Ballistic Coefficient**.

BHN Abbreviation for **Brinell Hardness Number**, a system of hardness measurements commonly used for measuring lead alloys (see also Brinell Hardness Number).

Back-Spatter The short-range ejection of small droplets of blood and possibly other biological debris back along the path of a penetrating projectile or shot charge. Such bio-matter is often deposited on and in the bore of the responsible firearm when such firearm is either in contact with the injury site or in very close proximity to the entry wound site at the moment of discharge.

Backthrust The force exerted on the breech block by the head of the cartridge case during propellant burning.

Ballistic Coefficient A mathematical expression of a bullet's ability to counter atmospheric resistance (aerodynamic drag), as compared to a specified "standard" reference projectile. It is also a form-fitting factor for a particular bullet's exterior ballistic behavior as compared to that of the standard bullet for which well-established performance data is known. Generally abbreviated as **BC**. **BC** values related to the G_1 standard bullet are utilized with nearly all contemporary exterior ballistic programs. BC is defined by the equation w/id^2, where w = mass in pounds, i = coefficient of form (i.e., a form factor) and d = bullet diameter in inches. The G_1 standard bullet weighs 1 lb., has a form factor of 1 and a diameter of 1 in. and therefore a BC of 1.00.

Ballistic Soap A glycerin-type soap especially designed to simulate muscle tissue for use in wound ballistics studies of projectile behavior; most commonly, penetration depth and projectile deformation or expansion.

Ballistic Tip® A trademark of the Nosler Bullet Company for jacketed rifle bullets possessing specially designed and mounted plastic tips of various colors. The particular color denotes the caliber of the bullet.

Ballistics The science and study of projectiles in motion. Usually divided into three parts: (1) Interior Ballistics which studies the projectiles movement inside the gun, (2) Exterior Ballistics which studies the projectiles movement between the muzzle and the target, and (3) Terminal Ballistics which studies the projectiles movement and behavior in the target.

Balloting (*bullet*) A bullet "ballots" when it moves through the bore of a firearm with a bumping, buffeting action. Balloting is a yawing motion of a bullet while traveling through the bore, resulting in incomplete, intermittent rifling impressions often extending onto the ogive of the bullet.

Ball Powder Any of a series of double-base powders originally developed by Olin in the 1930s, having a spherical or flattened spherical shape. Examples would include Winchester's 231, 748, or 760 powders. Such powders are now manufactured in many countries other than the United States.

Barium Nitrate A component of most priming mixtures which acts as an oxidizer of the particular fuel in such mixtures.

Barrel/Cylinder Gap The distance from the face of a revolver's cylinder to the face of the barrel. Normally, this gap is somewhere in the range of .003–.006 in., depending on the manufacturers' specifications. This location is the source of certain gunshot residues possessing important reconstruction properties.

Barrel Length   For shoulder arms and most handguns, barrel length is the distance between the muzzle of the barrel and the face of the breech block or bolt. For revolvers, it is the overall length of the barrel only, including the portion within the frame.

Barrel Time   See **IBT**

Battery   As applied to firearms, the position of readiness for firing. A firearm is referred to as being "in-battery" when the locking mechanism is fully closed and the action is ready to be fired, or the breechblock is fully forward against a chambered cartridge.

Bearing Surface   The area of a bullet which actually contacts the bore of a firearm during its passage through the barrel.

Belted Case   A case having a raised band, or belt, around the base just ahead of the extractor groove. Intended to provide positive headspacing on cartridges with long, sloping shoulders, the belt allows the cartridge to feed and function more reliably than a rimmed case. Contrary to the common misconception, the belt adds little or nothing to the strength of the case.

BenchKote®   A special absorbent paper with a plastic backing manufactured by Whatman.

Berdan Case/Primer   A primer/case system, designed by Col. Hiram Berdan, having two or more flash holes, and an anvil formed into the primer pocket of the cartridge case. Although widely used throughout the world, this system has never been popular in the United States, due largely to the difficulty in reloading Berdan cases.

Berdan Primer   An ignition component consisting of a cup, explosive mixture, and covering foil or paper disk. The anvil is an integral part of the cartridge case head in the bottom of the primer pocket. One or more flash holes are drilled or pierced through the bottom of the primer pocket into the propellant cavity of the case.

Bevel Base Bullet   A bullet possessing a beveled edge at its heel. This feature assists the seating of such bullets in the cartridge case at the time of manufacture or during a reloading process.

Billiard Ball Effect   The divergence of pellets from the axis of the wound channel, caused by collisions between pellets in a shot string as they move into and through tissue or organs. The resultant scatter of pellets can give the appearance of a distant shot when viewed in X-ray films.

Birdshot   A general term used to indicate any shot smaller than buckshot. Popular sizes range from 0.09 to 0.13 in. in diameter.

Birefringence   An optical property of materials that possess two or more refractive indices when they are viewed with a polarizing microscope.

Black Powder   A mechanical mixture of potassium nitrate, charcoal, and sulfur with the most common proportions being 75:15:10. For sporting arms use, various granulations are available. These are designated f_g, ff_g, fff_g and $ffff_g$, largest to smallest, respectively. Although obsolete for over a century, the propellant is still in popular usage with antique and replica firearms of the 18th and 19th centuries.

Black Talon® A trademark of the Olin-Winchester Corporation, denoting a group of high performance handgun projectiles possessing a black copper oxide coating. Upon expansion in tissue, the petals of the bullet's jacket have a talon-like shape that is characteristic and unique.

Blowback Operation In firearms, an automatic or semiautomatic firearm design that directly utilizes the breech pressure exerted on the head of the cartridge case to actuate the mechanism. In blowback operated semiautomatic pistols, the slide is not locked to the barrel or frame and begins its rearward movement as the bullet is being accelerated down the bore of such firearms.

Blown Primer A primer that is separated completely from the cartridge or shotshell after firing due to severe expansion of the primer pocket and head. This is usually the consequence of greatly elevated pressures during discharge.

Boattail Base Bullet A tapered section between a bullet's bearing surface and base, intended to reduce the effects of aerodynamic drag. This, in turn, gives the bullet a higher ballistic coefficient than a comparable flat-based bullet.

Bolt Action The working mechanism of a firearm in which the breech closure operates in line with the bore in a manually reciprocating manner to cock, load, and unload the firearm as well as extract the fired cartridge case from the chamber.

Bolt Locking Lug(s) The protrusion or protrusions from the surface of the bolt body which lock into mating recesses in the receiver, barrel, or barrel extension to resist rearward thrust of the chamber pressure.

Bore The inside portion of a barrel through which the projectile or shot charge passes.

Bore diameter The inside diameter of the barrel before the rifling is cut. In a barrel with an equal number of grooves, the bore diameter refers to the measurement from the top of one land to the top of the opposing land.

Bottleneck Cartridge A cartridge case having a main body diameter and a distinct angular shoulder stepping down to a smaller diameter at the neck portion of the case.

Bow Effect The flow pattern of abrasive materials in soil, sod, and/or sand around the nose, ogive, and/or bearing surface of a bullet generated during penetration into and ricochet from such materials. This characteristic pattern is uniquely associated with ricochets from soil, sand, or sod that have yielded to the bullet's impact and allowed the bullet to enter into the substrate to some depth before departing the substrate.

It is most noticeable on the ogive of the bullet, but may extend back along the bearing surface as well. This type of marking takes its name from the similarity of the flow pattern of water off the bow of a boat.

Boxer Case/Primer   A primer/case system, designed by Col. Edward Boxer, having one flash hole located in the center of the primer pocket and a separate anvil pressed into the primer cup. Due to its ease of reloading, the Boxer system is best suited to the handloaders' needs. Ironically, the system invented by an Englishman (Boxer) is most prevalent in the United States, while an American system (Berdan's) is commonly used in England and Europe.

Boxer Primer   An ignition component consisting of a cup, explosive mixture, anvil, and covering foil or paper disk, which, together, forms the completed primer ready for assembly into the primer pocket of a cartridge case. A central flash hole is pierced through the bottom of the primer pocket into the propellant cavity of the cartridge case. Used in modern commercial centerfire ammunition made in Canada and the United States, and it is now available from manufacturers in many other countries.

Breech   The part of a firearm at the rear of the bore into which the cartridge or propellant is inserted.

Breechblock   The locking and cartridge head supporting mechanism of a firearm that does not operate in line with the axis of the bore. Examples would include Sharps rifles, Martini-Henry rifles, and Spencer rifles.

Breech Bolt   The locking and cartridge head supporting mechanism of a firearm that operates in line with the axis of the bore, such as in common bolt action rifles.

Breech Face   That part of the breechblock or breech bolt that is against the head of the cartridge case or shotshell during firing.

Breech Face Marks   Negative impressions of the breech face of the firearm found on the head of the cartridge case, and/or in the primer around the firing pin impression, after discharge.

Brinell Hardness Number   A system of hardness measurement used for lead and lead alloys in which pure lead is BHN 4.

Buckshot   Lead pellets ranging in size from .20 to .36 in. diameter normally loaded in shotshells. These pellets are sometimes plated with copper or nickel.

Bullet   A non-spherical projectile fired from a firearm. A complete, loaded cartridge is not a "bullet," although a bullet is part of a loaded cartridge. The bullet emerges from the muzzle of a pistol or rifle.

Bullet Drift   The lateral deviation in a bullet's flight through the atmosphere due to rotational effects. A bullet fired from a right twist firearm will drift right, and left for a left twist firearm. This effect is only noticeable and significant in long-range fire (such as 1000 yd).

Bullet Drop   The vertical distance a bullet has fallen, under the influence of gravity, at any point in its flight path. The distance is measured from a point on its path to the straight line from axis of the bore to target.

Bullet Path   The vertical distance, normally expressed in inches, above or below a firearm's line of sight. The path followed by a bullet in its flight to a target.

Bullet Pull   The amount of pull, normally measured in pounds, needed to pull a bullet from the case mouth. Also referred to as "neck tension."

Bullet Upset   In Interior Ballistics: The change of bullet form due to chamber pressure. In Exterior Ballistics: The expansion of a bullet upon impact with target.

Bullet Wipe   The discolored area on the immediate periphery of a bullet hole, caused by the transference of residues from the bearing surface of the bullet. These dark gray to black residues typically contain carbon, lead, bullet metal, and possibly other constituents such as bullet lubricant and primer residues. Bullet wipe occurs at any range of fire so long as the bullet has not passed through some intermediate object.

Bullet Yaw   An instability caused by the eccentricity or imbalance of the bullet in flight. Yaw is usually the greatest in the initial portion of its flight in the atmosphere after which the bullet "goes to sleep" and becomes fully spin-stabilized. Yaw occurs after elongated bullets strike an object or when they enter media other than the atmosphere.

Bulged Barrel   A barrel with an abnormal enlargement in its bore.

Burning Rate   A term used to describe the relative quickness of deflagration of a given powder as compared to a known standard. Burning rate is extremely important in determining a powder's suitability for a given cartridge.

C

CUP   Abbreviation for "Copper Units of Pressure." This relates to firearm discharge pressure measured in a copper crusher testing system. There is no direct correlation between CUP and pressure expressed in pounds per square inch (PSI), and no conversion factor to extrapolate one from the other. CUP values provide a means of testing and evaluating peak pressure generated during the discharge of a suitably modified firearm.

Calcium Silicide   A component of some priming mixtures that serves as a fuel.

Caliber   The diameter of a projectile, commonly expressed in hundredths or thousandths of an inch in the United States, when discussing small arms, although it may also be expressed in metric units. Caliber may also refer to bore or groove diameter, again, in either inches or millimeters.

A term also used to designate the specific cartridge(s) for which a firearm is chambered.

Calibers May be used as a unit of measure. For example, a bullet can be described as three calibers in length, when its length is three times its diameter.

Cannelure One or more circumferentially cut or pressed grooves around the shank of a bullet. Cannelures provide an area into which the case mouth may be securely crimped and/or lubrication may be deposited. Cannelures may also have product identification value.

Canting The tipping or tilting of a gun to one side at the time it is fired.

Cap and Ball A muzzle-loading firearm (most commonly a revolver) using the percussion cap ignition system and firing round lead balls.

Carbine A rifle of relatively short length and light weight originally designed for mounted troops.

Cartridge A single, complete round of ammunition. See **Ammunition**. Modern cartridges normally consist of (1) a case, (2) a bullet, (3) a primer, and (4) a powder charge.

Cartridge Case The container for all the other components which comprise a cartridge.

Cartridge Cook-Off The firing of a cartridge without operation of the firing mechanism, due to extreme overheating in a firearm chamber. Usually associated with closed-bolt machine guns after prolonged bursts of fire. This event is believed to be due to one or more components in the primer (such as tetracene) undergoing thermal initiation at about 320°F (160°C).

Case Refers to **Cartridge Case**. The terms "shell," "brass," "casing," and "hull" have also been used to denote a cartridge case.

Case Cannelure One or more circumferential rings around a cartridge case typically used by manufacturers to denote a certain type of load or product line.

Case Head Separation or Rupture A generally circumferential separation in the side wall of a cartridge case. It may be complete or partial.

Cast bullet A bullet produced by pouring molten lead (or lead alloy) into a mould.

Celsius Temperature Scale The temperature scale setting the freezing point of water as 0° and the boiling point of water as 100° with equal divisions between and extended beyond these reference points. The equation used to convert Celsius temperature to Fahrenheit temperature is °F = 9/5C + 32.

Center of Gravity The point through which the resultant force of gravity on an object (projectile) passes. This is a fixed location for a particular object (projectile).

Center of Pressure The focal point of the sum of aerodynamic forces acting on a projectile in flight at any one moment in time. This is a fictitious entity that can be defined as that point where the observed normal force would have to act in order to produce the observed overturning moment. For nearly all spin-stabilized projectiles this point is forward of the center of gravity. Unlike the center of gravity, it is *not* a fixed location for a particular projectile as it moves through the atmosphere and loses velocity and/or experiences changes in stability.

Chamber Marks   Individual microscopic markings engraved and/or imprinted on a cartridge case by the chamber wall as a result of any or all of the following: (1) chambering, (2) expansion during firing, and (3) extraction. Chamber marks produced during the discharge process are generally most noticeable with firearms that utilize the blowback method of operation.

Chilled Shot   Lead shot containing more than 0.5% of an alloying metal, usually antimony, to increase its hardness. Also called **Hard Shot**.

Chisum Trail   An elongated transference of bullet metal at the departure end of low incident angle ricochet marks on smooth, flat, unyielding surfaces. This asymmetrical elongated transference will be on the left side of ricochet marks for bullets fired from left twist firearms, and on the right side for bullets from right twist firearms. It is caused by the right or left edge of a flattened bullet remaining in contact with the surface after the main body of the bullet has lifted off the surface. The author named this phenomenon after Criminalist Jerry Chisum, who first described it to the author.

Choke   An interior constriction at or near the muzzle end of a shotgun barrel for the purpose of controlling shot dispersion. Markings by U.S. manufacturers typically utilize the following symbols: Full Choke = FC Full (Greatest constriction.); Improved-Modified = Imp. Mod. (Less constriction than full.); Modified = Mod. (Less constriction than improved-modified.); Improved-Cylinder = IC, Imp. Cyl. (Less constriction than modified.); Skeet = Skeet, Sk (Less constriction than improved-cylinder.); Cylinder Bore = Cyl. (Least constriction or no constriction.). European markings are normally as follows: Full Choke = *; Improved-Modified = **; Modified = ***; Improved-Cylinder = ****; Cylinder = CL.

Chronograph (ballistic)   An instrument used in determining the velocity of a projectile. Most are based on the time taken by a projectile to traverse a known distance between two points monitored by some form of detection system.

Clip   A device which holds ammunition to be charged into a magazine. Clips may be inserted into the firearm and remain there during firing, as with the M1 Garand, or may be used only to aid in charging the magazine, as with the 1903 Springfield, M14, or M16 rifles. This latter type is referred to as a "stripper clip," while the former is called a "charger clip" or "en bloc" clip.

Cocking Indicator   Any device to indicate that a firearm hammer or striker is cocked.

Concentric Fractures Fractures or cracks in glass or other similar brittle or ceramic material which take a generally circular form around the bullet hole or impact site in such materials.

Cone Fracture The characteristic cone shape of the exit side of a projectile hole through a relatively brittle medium (e.g., glass, bone) caused by spalling around the exit.

Cordite An early extruded, smokeless, double-base propellant widely used in England, and in particular, in early .303 British cartridges. Cordite is distinguished by its length, which normally runs the full length of the powder chamber. Invented in 1889, cordite served as the bases for many of our currently used extruded propellants.

Corrosive Primer Any primer using potassium chlorate in its priming compound. When fired, a portion of this will become potassium chloride, similar to common table salt, and be deposited in the barrel, causing corrosion (rusting) upon standing. Cleaning using normal powder and copper solvents will not remove the corrosion causing residue left in the bore. These deposits can easily be removed by using warm water, followed by standard cleaning and oiling. Corrosive priming mixtures were used in most U.S. military ammunition prior to 1952.

Crack Rule Also known as the "T" Test referring to the stopping of the propagation of one or more radial fractures during a projectile's impact in plate glass or other similar material by a fracture from a previous shot.

Criminalistics An early definition was: "That science which applies the physical sciences in the investigation of crimes." Derived from the German word *Kriminalistik*, criminalistics has taken on a broader definition and would be more properly defined as, "That science directed to the recognition, identification, individualization and evaluation of physical evidence as it relates to some law-science matter. It also includes the reconstruction of events based on the analysis of physical evidence. It draws upon the physical and natural sciences to accomplish its mission."

Crimp A turning inward of a case mouth to increase its tension on a bullet. Crimping is necessary when loading for revolvers, tubular magazines, and some rifles with extremely heavy recoil.

Crimped Primer Refers to a primer which has been staked, stabbed, or otherwise crimped into the primer pocket. Commonly found on military cartridge cases. In reloaded cases of this type the remnants of this crimp will have been removed by swaging or reaming before the new primer is seated.

Critical Angle The incident or intercept angle at and above which the particular projectile at a given impact velocity no longer ricochets from the impacted surface.

Crown The point of the bore where the rifling terminates at the muzzle.

Cupronickel   An alloy of copper and nickel, also known as "German Silver." Cupronickel was once used extensively as a jacket material, despite its serious tendency to leave metal fouling in the barrel. In the United States it has been replaced almost entirely by gilding metal, a copper–zinc alloy.

Cut Cannelure   A smooth cannelure cut or formed in a jacketed bullet that lacks any knurling but otherwise serves the same purposes as other cannelures, namely the crimping of the cartridge case into the bullet and/or product identification.

Cylinder   The rotatable part of a revolver that contains the firing chambers.

Cylinder Flare   The circular gray to black deposit around the front margin of the chamber or chambers of a revolver composed of gunshot residues deposited during the discharge process. Also called "halo" or simply a "flare."

Cylinder Gap   In a revolver, the maximum space between the cylinder and the barrel. Also called the *cylinder–barrel* gap. The cylinder gap is a source of high-energy gunshot residues with unique reconstructive value. Cylinder gap values of 0.004–0.006 in. are normal.

D

Decibel   A unit of intensity of sound, equal to 20 times the common logarithm of the ratio of the pressure produced by the sound wave to a reference pressure. Abbreviation: dB and described by the formula

$$dB = 20 \log(P/P_0)$$

where $P_0 = 0.0002$ dynes/cm^2.

Deflagration   A rapid but controlled burning of a solid fuel or propellant producing large volumes of gas and heat.

A rapid combustion reaction that is propagated at a subsonic rate by heat transfer into the reacting material. This reaction is accompanied by a vigorous evolution of heat and flame. Deflagration is usually dependent upon having fuel and oxidizing agent in very close contact, either from having the fuel as a finely divided mixture with the oxidant (such as black powder), or by combining the two in the same chemical compound or mixture (such as nitrocellulose propellants). Deflagration exhibits a dependence upon the surrounding gas pressure in that increases in pressure increase the burning rate.

Deflection   (As differentiated from ricochet) a deviation in the projectile's normal path through the atmosphere as a consequence of an impact with some object. This term is further refined for two types of impactive events in a projectile's normal flight path.

1. *Deflection* as a consequence of a *ricochet* is used to describe any lateral component of the ricocheted projectile's departure path relative to the plane of the impacted surface as viewed from the shooter's position and with the plane of the surface normalized to a horizontal attitude. The angle formed between the path of the departing projectile subsequent to impact and the pre-impact plane of the projectile's path.

2. *Deflection* as a consequence of *perforating or striking an object* is used to describe deviations in *any* direction from the projectile's flight normal path as a consequence of perforating or striking an object rather than rebounding off of surfaces. For example, a bullet may be *deflected* by passage *through* a tree branch, a windshield, or a panel of sheet metal. These are not ricochet events. Since such deflection can occur in any direction (up, down, right, or left), the clock position of such deflection is used to describe this form of deflection. As viewed from the shooter's position (or position directly behind the projectile at impact), 12 o'clock will be taken as straight up relative to the horizontal plane at the location of the event, 3 o'clock to the right, 9 o'clock to the left, and so forth.

Density A physical property of all matter that is equivalent to the mass (sometimes weight) per unit volume. Example: pure lead has a density of 11.34 grams per cubic centimeter.

Departure Angle See **Angle of Departure**.

Design Defect The design of a product itself (all of them being made the same) that possesses a flaw or shortcoming. See also **Manufacturing Defect**.

Deterrent Coating A chemical coating applied to propellant particles, in order to bring their burning rates and characteristics into line with the manufacturers' specifications for that particular powder type.

Detonation An extremely rapid chemical rearrangement normally associated with high explosives resulting in the near-instantaneous production of large volumes of gas.

An extremely rapid exothermic decomposition reaction which proceeds at a rate greater than the speed of sound within the reacting material (unlike **Deflagration**). The normal mode of initiation is shock (such as a blasting cap or high level mechanical shock), or from initial combustion which, due to peculiarities of confinement or other circumstances, accelerates to such a degree that a shock wave is formed. Behind the shock wave is a reaction zone where material is converted to gaseous products at high temperature and pressure.

Dicing (of **Tempered Glass**) The characteristic failure in tempered glass that takes the general form of small squares to rectangular pieces.

Disconnector A device intended to disengage the sear from the trigger. (1) In a manually operated firearm, it is intended to prevent firing without pulling the trigger. (2) In a semiautomatic firearm, it is intended to prevent full automatic firing.

Disk-Flake Powder An extruded form of smokeless powder cut into thin circular disks that may have a central perforation (see **Perforated Disk-Flake Powder** and **Unperforated Disk-Flake Powder**). Such propellants are most commonly used in pistol and shotgun ammunition and may be of either single or double base formulation.

Dithiooxamide (DTO) A specific colorimetric reagent (also known as rubeanic acid) that reacts with copper ions to produce a dark greenish-gray product.

Doppler Radar A continuous-wave radar used chiefly to make precise speed measurements. It works on the basis of the Doppler effect, which is a change in observed wave frequency caused by motion. By measuring the difference in frequency, Doppler radar determines the speed of the object or projectile observed.

Double Action A handgun mechanism in which a single pull of the trigger accomplishes two events, the cocking of the hammer followed by the release of the hammer.

Double Base Powder A powder that uses both nitrocellulose and nitroglycerine as the propellant base, as opposed to a single-base powder which uses only nitrocellulose. Double-base powders generally have higher energy content, and as such possess higher flame temperatures and can be somewhat more erosive than comparable single-base powders. See also **Single Base** and **Triple Base Powder**.

Doubling The unintentional firing of a second shot. Doubling is usually associated with semiautomatic firearms in which the sear fails to capture or hold the hammer or striker.

Drag Coefficient An experimentally derived correction or fitting factor denoted as C_D in the drag equation, $F = \frac{1}{2}\rho V^2 A C_D$, necessary to make the drag force, F, fit the data for a bullet of cross-sectional area, A, traveling at velocity, V, in an atmosphere of density, ρ. C_D is not a constant and varies with velocity and atmospheric density.

Drag Force The force, F, in pounds in English units derived from the formula $F = \frac{1}{2}\rho V^2 A C_D$ where A is the bullet's cross-sectional area in square feet, V the velocity in feet per second, ρ the density of the atmosphere (in mass units), and C_D is the drag coefficient.

Dram Equivalent The traditional method of correlating relative velocities of shotshells loaded with smokeless propellant to shotshells loaded with black powder. The reference black powder load chosen was a 3 dram charge of black powder, with $1\frac{1}{8}$ oz. of shot and a velocity of 1200 fps. Therefore, a 3 dram equivalent load using smokeless powder would be with $1\frac{1}{8}$ oz. of shot having a velocity of 1200 fps, or $1\frac{1}{4}$ oz. of shot and a velocity of 1165 fps.

A $3\frac{1}{4}$ dram equivalent load might have $1\frac{1}{8}$ oz. of shot and a velocity of 1255 fps. Abbreviated Dram Equiv.

Draw Mark   A longitudinal scratch on a cartridge case caused by foreign material on either the draw punch or the die during fabrication.

Drop   See **Bullet Drop.**

Drop-Fire   The discharge of a loaded firearm as a result of an impact after being dropped. This may be the consequence of a design shortcoming, a compromised safety system, or the failure of the handler to engage the appropriate safety device.

Dry Firing   The releasing of the firing pin on an unloaded chamber of a firearm.

Dum-Dum   A term applied to some early expanding bullets for the .303 service cartridge loaded by the British arsenal at Dum-Dum, India, prior to 1899. Frequently used (incorrectly) by the media and others unfamiliar with firearms to indicate any expanding bullet.

Duplex Load
1. A cartridge case containing two projectiles or two sizes of shot with a single powder charge.
2. A cartridge case containing a single projectile with two types of powder.

E

Effective Range   The maximum distance at which a projectile can be expected to be useful in its intended purpose.

Ejection Port   An opening in the receiver to allow ejection of the fired cartridge case.

Ejection Port Marks   Indented or striated marks at one or more locations on a cartridge case as a result of striking one or more areas on the ejection port during egress. Such marks may bear reproducible patterns of striae.

Ejector   A portion of a firearm's mechanism that ejects or expels cartridges or cartridge cases from a firearm.

Ejector Marks   A small, impactively produced mark in the head of a cartridge case formed through violent contact with the ejector normally associated with semiautomatic and fully automatic firearms. These marks can be produced during the extraction-ejection process of manually operated firearms but this usually requires vigorous manipulation of the gun's mechanism.

Energy   The capacity for performing work. In ballistics, energy is normally expressed in kinetic units of "foot-pounds" in the American and English system of measurements and in kilogram-meters or joules of energy in the metric system. One foot-pound is equivalent to the energy required to lift one pound one foot against the force of gravity. To convert ft-lbs to kg-m, multiply ft-lbs by 0.1382 or by 1.355 to obtain the equivalent energy in joules.

Erosion   The wear, usually in the throat area of a barrel, caused by extreme heat and friction. Erosion occurs in all firearms but is aggravated by rapid fire, large case capacity, or the use of propellants with elevated flame temperatures.

Explosion   An extremely rapid chemical or mechanical action resulting in the very rapid production and expansion of gases.

Exterior Ballistics   The branch of ballistics that deals with the projectile's flight, from the time it leaves the muzzle of a firearm, until it makes impact with the target.

Extraction Groove   A groove cut or formed in the side wall of a cartridge case just forward of the face of the head and rim for the purpose of extraction.

Extractor   A mechanism for withdrawing the cartridge or cartridge case from the chamber of a firearm. This component usually has a hook-like shape that grasps the rim or extraction groove of the chambered cartridge case.

Extractor Marks   One or more small marks that may occur in several closely related locations on or near the rim of a cartridge. An *extractor override mark* on the rim of a cartridge case occurs during the chambering process. *Extractor gouge marks* can occur on the case wall immediately adjacent to the rim or in the extraction groove during the chambering process or during the discharge-extraction process. An *extractor bite mark* is normally associated with semiautomatic and fully automatic firearms and is the result of the violent contact between the extractor and the front face of the cartridge rim producing an indented toolmark at this location.

Extruded Powder   More properly called **Extruded Tubular Powder**.

Extruded Tubular Powder   A type of smokeless powder formed by forcing the dough-like nitrocellulose composition through a die of specific dimensions, and cutting it into particles of specified length. Extruded tubular powders are more or less cylindrical in shape, and may have one or more perforations running through its length. Common examples of extruded powders are IMR 4350, H4895, or Accurate 3100.

F

FMJ   Abbreviation for "Full Metal Jacket" bullet.

FMJ-BT   Abbreviation for "Full Metal Jacket-Boattail" bullet.

FMJ-RN   Abbreviation for "Full Metal Jacket-Round Nose" bullet.

FMJ-TC   Abbreviation for "Full Metal Jacket-Truncated Cone" bullet.

Facets (*on bullets*)   Multiple flat, squarish impressions on the nose and ogive of a bullet that has perforated previously shattered tempered glass. These facets are produced during the bullet's impact with the small, diced pieces of broken glass.

Far Zero   The second point at which the bullet path intersects the line of sight. This is commonly referred to as "zero" for a given firearm, at which the point of aim and the point of impact coincide.

Filler Wad   A cylindrical disk of fibrous material of various thicknesses used to adjust the volume of the contents of a shotshell.

Fireform   To alter the shape of a case by firing it, generally done to increase case capacity. Upon firing, pressure forces the existing case out to fit the larger chamber, creating the new dimensions desired. Fireforming is a common technique in making wildcat or improved cases.

Firing Pin   That portion of a firearm that strikes the primer of the cartridge, causing detonation of the primer composition and ignition of the propellant charge.

Firing Pin Aperture   The opening in the bolt or breechblock of a firearm through which the firing pin moves during the discharge process.

Firing Pin Drag Mark   The toolmark produced when a projecting firing pin remains in, or comes in contact with, the primer, cartridge case, or shotshell during the extraction–ejection cycle in certain firearm designs such as break open shotguns and semiautomatic pistols employing the Colt-Browning locking system.

Firing Pin Impression   The indentation in the primer of a centerfire cartridge case or in the rim of a rimfire cartridge case caused by the impact of the firing pin. Also called "**Firing Pin Indent**."

Firing Pin Scrape Mark   See **Firing Pin Drag Mark**.

Flare   The circular gray to black deposit around the front margin of the chamber or chambers of a revolver composed of gunshot residues deposited during the discharge process. Also called "halo."

Flash Hole   A hole, or holes, from the primer pocket to the powder chamber of a cartridge case; a hole in the end of a battery cup primer used in shotshells; a hole in the nipple of a percussion firearm.

Flash Suppressor   A muzzle attachment designed to reduce muzzle flash. Also called a "Flash Hider."

Forcing Cone   The section of a revolver or shotgun barrel just ahead of the chamber(s) that gradually reduces in diameter to bore or land diameter. The forcing cone serves to align the bullet or shot charge with the bore, while preventing deformation to the projectile(s).

Frangible   Susceptible to being broken up or shattered into small pieces or particles. This term is often associated with certain special purpose bullets that are designed to shatter into many small pieces upon impact with the hard metal backstop of a shooting range.

Frangible Bullet   A projectile designed to disintegrate upon impact on a hard surface in order to minimize ricochet or significant rebound of bullet fragments.

Freebore   Essentially, the throat area of a barrel. Normally, use of the term "freebore" indicates the rifle in question has an unusually long throat, as in the case in most of the Weatherby chamberings.

Free Fall Velocity   See **Terminal Velocity**.

Full Metal Jacket   A bullet having no exposed lead on the frontal portion. FMJs are non-expanding bullets used in both rifles and pistols. They are produced in several different configurations, i.e., round nose, spitzer, spitzer boat tail, etc., depending on their intended use. Also called "Full-Jacketed," "Full Patch", "Full Metal Case."

G

g   The symbol for the accelerative force of the earth's gravitation attraction. The average, sea level value for the earth's gravitational acceleration is 32.174 f/s/s or 9.807 m/s/s.

GSR   Abbreviation for **Gunshot Residue**.

Gage   An instrument or device for measuring or testing a parameter (such as a headspace gage or trigger pull gage).

Gas Check   A protective cup of copper, brass, or gilding metal placed on the base of a cast bullet. Gas checks are intended to reduce gas cutting and deformation of the bullet's base due to pressure or hot propellant gases.

Gas Operated   In firearms, a gun system that utilizes a portion of the gases produced by the powder's combustion to cycle the action. The U.S. military M1, M14, and M16 are all examples of gas-operated weapons.

Gauge(*shotguns*)   An archaic method of describing the diameter of the bore of shotguns based on the number of lead spheres just fitting the bore that equal 1 lb. For example, the bore of a 12-ga. shotgun is of such a diameter that 12 lead spheres, each weighing 1/12 of a pound, would just fit in the bore. A 20-ga. shotgun (possessing a smaller bore) would require 20 lead spheres of bore diameter to equal 1 lb. May also be spelled *gage* in some sources.

Gauge(*sheet metal*)   A system of thickness description. Contemporary automotive sheet metal is typically composed of 0.031 to 0.032 in. (0.79–0.82 mm) thick steel which is designated as 22-gauge sheet metal. The adjacent gauges of sheet metal have listed thicknesses of 0.0343 in. (21-gauge) and 0.0280 in. (23-gauge).

Gilding Metal   An alloy of 90–95% copper, and 10–5% zinc, now used extensively as a jacket material for bullets. Also termed Commercial Bronze. Neither name is recommended by Copper Development Association, Inc., but instead Alloy No. 220 and Alloy No. 210, respectively.

Gold Dot®   A trademark of the Speer ammunition company of Lewiston, Idaho, for a line of jacketed pistol bullets possessing a plated, hollowpoint jacket over a lead core. The manufacturing process leaves a characteristic "dot" of the jacketing material at the bottom of the hollow point cavity.

Golden Saber® A trademark of the Remington ammunition company for a line of brass-jacketed hollowpoint pistol bullets possessing canted skives on the bullet's ogive.

Grain A unit of weight (avoirdupois) equaling 1/7,000th of a pound. The most common unit of weight by which bullets and powder charges are measured. There are 7,000 gr. in a pound and 437.5 gr. in one ounce.

Gravitational Acceleration See **g.**

Griess Test A specific chemical test for the detection of nitrites. The Griess Test is typically used by the criminalists and firearms examiners in the laboratory to develop patterns of gunpowder residues (nitrites) around bullet holes.

Greenhill Formula A mathematical formula developed by Sir Alfred Greenhill to determine the twist necessary to stabilize an elongated bullet. The Greenhill formula states: the twist required (in calibers) equals 150 divided by the length of the bullet (in calibers).

Grooves The area between the lands in the bore of a rifled firearm. The grooves are cut or impressed into the surface of a bore.

Groove Diameter The major diameter in a barrel, which is the diameter of a circle circumscribed by the bottom of the grooves in a rifled barrel.

Gunshot Residue The total residues resulting from the discharge of a firearm. It includes both propellant and primer residues, carbonaceous material plus metallic residues from projectiles, fouling, and any lubricant associated with the bullets.

H

Hair Trigger A slang term or shooter's jargon for a trigger requiring very low force to actuate. There is no technical definition or quantified measure for a "hair trigger."

Half Cock The position of the hammer, when about half retracted and held by the sear, intended to prevent release of the hammer by a normal pull of the trigger. This can be the safety or loading position of many guns.

Half Protractor A protractor consisting of a 90° angle with gradations for each degree from 0–90, with a zero edge allowing bullet path measurements in corners and other tight spots inaccessible to full, 180° protractors. See **Zero-Edge Protractor**.

Halo See **Flare**.

Hammer Spur The knob or extension on an exposed hammer which acts as a cocking or decocking aid.

Hangfire A delay, sometimes quite noticeable to the shooter or listener, between the impact of the firing pin and the actual ignition of the cartridge. Such delays are typically less than one second in duration and, when they do occur, are usually on the order of 0.25 sec. or less. The causes are typically either a contaminated primer mixture or an improperly seated primer.

Hard Shot Also known as **Chilled Shot**. Lead shot that has been alloyed with antimony making it harder and less susceptible to deformation than pure lead shot of the same size.

Head As applied to cartridges, the base area of the case. This area encompasses the primer pocket, extractor groove, and the rim or belt, extending up to the beginning of the body of the case.

Head Separation A circumferential cracking around the body of the case, usually just above the web area. A complete head separation will normally leave the forward portion of the case in the chamber upon extraction. Generally caused by excessive chamber headspace.

Headspace (*cartridge*) The longitudinal dimensions of a cartridge that, when correct, properly position the cartridge in the chamber of a firearm that itself is of proper headspace.

Headspace (*chamber*) The distance from the face of the closed breech of a firearm to the surface in the chamber on which the cartridge case seats. Also, the amount of play between the case head and the breech face, in a fully closed action. Insufficient headspace will cause difficulty in chambering, while excessive headspace can result in cartridge head separations. Headspace problems may be the fault of the gun, the ammunition, or a combination of both. There is a necessary relationship between the headspace of the firearm's chamber and that of the cartridge in order for the cartridge to perform properly during discharge.

Headspace Gage An instrument for measuring the distance from the breech face of a firearm to that portion of the chamber against which the cartridge seats.

Headstamp A series of letters, numbers, or characters stamped into the head of a cartridge case to denote caliber, type, manufacturer, supplier, arsenal, and date of production, or other pertinent information.

Hydra-Shok® A trademark of the Federal Cartridge Company for a line of hollow point pistol bullets containing a unique central post of lead in the hollow point cavity.

High Primer A primer that has not been fully seated in the primer pocket, and extends slightly above the head of the case. High primers create a dangerous condition that can result in slam fires, particularly in semiautomatic firearms.

Hollow Base Bullet A type of bullet having a hollow cavity in its base designed to improve bore obturation during discharge. This design also moves the bullet's center of gravity forward compared to a bullet of the same caliber and dimensions with a solid base.

Hollow Point Bullet A type of bullet having an opening in the nose. Hollow points may be of either the hunting or the target styles. Contrary to popular opinion, hollow points are not always designed to expand on impact. Match grade hollow point target bullets, for example, rarely exhibit any expansion when fired into tissue or tissue simulant.

I

IBT Ignition Barrel Time The elapsed time from the contact of the firing-pin with a cartridge primer to the emergence of the projectile(s) from the muzzle of the firearm. Sometimes simply called "barrel time."

IMI Abbreviation for Israel Military Industries.

IMR Abbreviation for Improved Military Rifle. A series of single-base extruded tubular powders developed by DuPont®. Currently being manufactured by the IMR powder company.

Ignition Temperature The minimum temperature at which a combustible substance will ignite.

Ignition Time The time interval between the impact of the striker or firing pin on the primer, and a rise in pressure sufficient to start the bullet from its seated position in the cartridge case. The elapsed time from moment of firing-pin contact on the primer to the point on the x (time) axis equal to the point where the pressure–time curve indicates propellant burning has initiated.

Incendiary Bullet A bullet containing a chemical compound which ignites upon impact with the intended purpose of starting a fire.

Incident Angle The intercept angle, $(\angle I)$, described by the pre-impact path of the projectile and the plane of the impact surface at the impact site when viewed in profile.

Inclinometer A device for measuring or displaying the angle of a surface relative to the horizontal or vertical plane.

Inertia Firing Pin A type of firing-pin in which the forward movement is restrained until it receives the energy from a hammer blow. It is slightly recessed in the breech face before being struck by the hammer and is shorter in length than the housing in which it is contained. Upon hammer impact, it flies forward using only its own kinetic energy to strike and fire the primer.

Ingalls' Tables A set of ballistics tables computed by Col. James Ingalls, in which the drag characteristics of a "standard" projectile are used as a reference for comparison of other small arms bullets. The standard projectile for the Ingalls' Tables is the G_1 bullet. The ballistic coefficients of almost all U.S.

manufactured bullets can be referenced to Ingalls' tables, with only a slight degree of error.

Instrumental Velocity   The velocity of a projectile as registered on a chronograph. Instrumental velocity is the average velocity of the projectile as it traverses the distance between the "start" and "stop" screens of the unit. If an actual muzzle velocity is needed, the instrumental velocity must be corrected to the muzzle. With modern chronographs, given their short screen spacings and a "start" screen only a few feet in front of the muzzle, this is generally unnecessary, and the corrections rarely amount to more than a few feet per second.

Interior Ballistics   The branch of ballistics dealing with events occurring between the detonation of the primer and the projectile leaving the muzzle.

Internal Ballistics   See **Interior Ballistics**.

Involuntary Discharge   A situation where an activity carried out by one hand (such as grasping, struggling, pulling) results in an involuntary contraction of the fingers of the opposite hand and, when that hand is holding a firearm in the shooting configuration to include having one's finger on the trigger, a firing of the gun occurs. It has also been claimed to occur as the result of a startle reaction with the same requirements insofar as improper gun handling are concerned.

J

JHP   Abbreviation for Jacketed Hollow Point bullet.

Jacket   An outer sheath, covering the interior portion (core) of a bullet. Many different materials, including mild steel and cupro-nickel alloy, have been used in making jackets, but today, 95/5 gilding metal (Cu/Zn) is the standard for the industry in the United States.

Jacketed Bullet   A bullet having an outer jacket composed of a metal or metal alloy such as copper, gilding metal, brass, mild steel, cupro-nickel, or aluminum.

Jam   A malfunction of a firearm that prevents the action from operating; may be caused by faulty parts, ammunition, improper maintenance, or improper use of the firearm.

K

Kernel   An industry term for a single, individual particle of powder. Sometimes also referred to as a grain of powder, but must not be confused with the unit of weight. See **Grain**. Forensic examiners typically use the term "powder particle" rather than "kernel."

Keyhole An elongated bullet hole, indicating that the bullet was not traveling point-on or fully nose-forward at impact. Also, a keyhole may be a slightly "out-of-round" hole, or it may be a complete profile image of the bullet, where the projectile actually went through the target sideways. This is either the result of a stability problem or the consequence of a deflected or ricocheted bullet.

Keyholed Bullet A bullet that strikes or enters a medium in a yawed or destabilized orientation.

Kinetic Energy The energy of a body with respect to the motion of that body given by the formula $\frac{1}{2}mv^2$, where m is the mass of the projectile and v is its velocity.

Knurled Cannelure A cannelure with a series of small regular ridges or rectangles to help prevent slipping of the bullet while held in the cartridge case. The style and spacing of the knurling also relates to the source of the bullet's manufacture.

L

LCB Abbreviation for **Lead Core Bullet**.

LRN Abbreviation for **Lead Round Nose** bullet.

LUP Abbreviation for *Lead Units of Pressure*. This relates to the pressure measured in a lead crusher testing system. Most often used in low-pressure applications such as shotguns. There is no direct correlation between LUPs and pressure expressed in pounds per square inch (PSI), and no conversion factor to derive one from the other.

Lagtime The time difference between the sound of the arrival of the bullet (sound of impact or sound of passage) at a specific downrange location and the arrival of the sound of the shot at that location. This time interval is useful in calculating the range of fire when the approximate muzzle velocity and ballistic coefficient of the bullet are known.

Lamel Powder A type of smokeless propellant in which the individual particles are in the shape of thin square, diamond, and/or parallelogram-shaped flakes. In some samples the shape and dimensions of the particles are closely controlled, while others may show considerable variation. This type of powder is typically found in European and Scandinavian small arms ammunition and has also been available as an imported canister powder. Examples: Alcan 5, 7, and 8.

Laminated Glass A "sandwich" of glass layers which combines alternate layers of plastic material and single strength (plate) glass. The outside layer of glass may break when hit by an object, but the plastic layer is elastic and stretches. This holds the broken pieces of glass together and keeps them from flying in all directions. It is used in automobile windshields.

Land (s)   The raised portions of bore remaining after the cutting or forming of the grooves in a rifle barrel. Commonly referred to as "the rifling."

Laser Photography   As used in the text, the open shutter time exposure and recording of laser beams in darkness or subdued light.

Lead Core Bullet   A jacketed bullet having a lead or lead alloy core.

Lead Round Nose Bullet   A lead bullet with a radiused nose. Technically, the radius of the nose is $\frac{1}{2}$ the bullet's diameter.

Leade   The minute portion of a barrel's rifling which slopes from the unrifled throat to the full-depth rifling. Although frequently referred to as the throat, there is a definite difference between the two.

Leading   A build-up or accumulation of lead in the barrel of a firearm, caused by using cast or swaged lead bullets. This can be controlled to a considerable degree by using harder alloys, better lubricants, or lower velocities. Leading causes no permanent harm to a firearm, but is detrimental to accuracy and can be difficult to remove.

Lead-In Mark   A visible, thin, elongated deposition of bullet wipe transferred to a surface as a bullet first makes contact with that surface at a shallow incident angle. The lead-in mark is useful in establishing the direction of fire and travel of the projectile.

   The dark, elliptical transfer of material from a bullet as it makes its initial contact with a surface at low incident angle.

Lead Splash   The production and dispersal of vaporized and fine particles of lead as a result of impact. Lead splash is related to the bullet design and composition, the nature of the surface struck and the energy associated with the impact. The geometry of the deposition of lead splash can provide information on the direction of fire. The amount of lead splash is a function of impact velocity.

Lead Styphnate   An impact sensitive initial detonating agent that is used in common priming mixtures. Also known as trinitroresorcinate, it is one source of particulate and vaporous lead in gunshot residue and bullet wipe.

Line of Departure   A straight line projecting through the axis of the bore to infinity. The direction in which a projectile is moving when it leaves the muzzle of a firearm. While this is the initial direction of the bullet's velocity, it should be clearly understood that the bullet falls away from this line immediately upon leaving the muzzle. This is primarily due to gravity and other outside forces acting on the projectile.

Line of Sight   A straight line passing through the sights of a firearm to the target.

Locked Breech System   In more powerful firearms, the condition of the action in which the bolt or breechblock is solidly secured in a fixed relationship with the chamber so as to resist being driven back by chamber pressure.

Locking Lug(s)   The protruding lug(s) that engage the receiver to lock the action closed during firing. Locking lugs are normally situated on a firearm's bolt in a bolt action rifle, although there are exceptions.

Lock Time   The time interval between the sear's release of the striker or firing pin, and the subsequent impact on the primer.

Lot Number   A term generally used by American sporting ammunition manufacturers for denoting same day, conditions, and components used to load a batch of ammunition.

Lubaloy®   An Olin-Winchester process resulting in a thin copper plating on lead bullets.

Lube Grooves   See **Cannelure**.

M

M2 bullet   The U.S. military designation for the 150 gr. FMJ flat-based bullet loaded in .30–06 cartridges.

M80 bullet   The U.S. military designation for the 147 gr. FMJ-BT bullet loaded in 7.62NATO cartridges.

M193 bullet   The U.S. military designation for the 55 gr. FMJ-BT bullet loaded in 5.56 mm cartridges.

M855 bullet   The U.S. military designation for the 62 gr. FMJ-BT bullet loaded in 5.56 mm cartridges. This bullet also contains a hardened steel penetrator in its nose.

Machine Gun   An automatic weapon firing a full-size (rifle caliber or larger) cartridge, usually fired off a bipod, tripod, or other fixed mount. They may be clip-, magazine-, or belt-fed, depending on the design and intended use. They are most often employed as a crew-served weapon.

Mach Number   The number obtained by dividing the speed of the projectile by the speed of sound at the time and location of the shot.

Magazine   An ammunition reservoir from which cartridges are fed into a firearm's chamber. Magazines may be integral, as in the 1903 Springfield, or may be detachable, as in M14 and M16 series rifles. Although the terms are frequently used interchangeably, a clip and a magazine are not the same thing.

Magazine Follower   A spring actuated device to push cartridges in a magazine to the feeding position.

Magazine Safety   A feature in some semiautomatic firearms in which the removal of the box magazine renders the arm incapable of being fired by a normal pull of the trigger.

Magazine Lip Marks   Thin, often curvilinear marks on the body and/or rim of a cartridge case at about the 10 o'clock and 2 o'clock positions produced by the lips of a magazine as the cartridges are stripped from the magazine by the firearm's mechanism.

Magnum   A designation sometimes attached to a cartridge of greater capacity or power than others of similar caliber. This can be misleading, as magnum cartridges are not always the most powerful in their respective bore sizes. In rifles, the term usually refers to one of the belted cartridges, based on the original Holland & Holland magnums. Today, belts are used more for sales appeal than any true ballistic function. In shotshells, the term denotes a heavier charge of pellets.

Manufacturing Defect   An individual product that was either made or came out differently than the manufacturer's specifications. See also **Design Defect**.

Mass   A constant property of matter that reflects the amount of material present. It is related to weight through the formula $m = W/g$, where W is the weight and g is the gravitational accelerative force acting on mass m.

Meplat   The diameter of a flattened tip at the nose of a bullet. A term for the blunt tip of a bullet, specifically the tip's diameter.

Mercuric Primer   Any primer which uses mercury fulminate as a component in its priming compound. While no longer in use, surplus U.S. military, old commercial ammunition, and some foreign ammunition may still be encountered that is loaded with these primers.

Mikrosil   A two-part silicon rubber casting material designed specifically for forensic use in casting toolmarks, the bores of gun barrels, and a variety of other surfaces and objects.

MIL   The angle created by one unit at 1000 units of distance.

Mild Steel   A carbon-steel containing a maximum of 0.25% carbon.

Minute of Angle   A unit of angular measurement equaling 1/60th of a degree. One minute of angle works out very close to one inch per hundred yards, making it a convenient measurement for shooters to use in describing accuracy, sight elevation, or windage deflection. Also referred to as "MOA" or "minutes." One minute of angle = 1.0472 in. @ 100 yd.

Misfire   The complete failure of a cartridge to fire after being struck by the firing pin or striker. The cause may be due to a number of sources, to include a contaminated primer and/or contaminated powder.

Momentum   Expressed in American units of "pound-seconds," momentum is a quantity of motion. Momentum is obtained by multiplying a bullet's mass with its velocity (i.e., mv). In some instances, momentum may be a better indicator of a bullet's injury-producing potential and its penetrative ability than kinetic energy.

Momentum Transfer   The transference of momentum of one object as it collides or impacts another object. When a projectile of mass m and velocity V strikes and comes to rest in an unrestrained object of mass M, the conserved momentum is given by the expression $mV = (m + M)v$. This is useful and applicable to claims of bullets knocking gunshot victims over or spinning them around.

Muzzle The end portion of a firearm's barrel; the point from which the bullet exits.

Muzzle-Brake (Compensator) A device at the muzzle end usually integral with the barrel that uses the emerging gas behind a projectile to reduce recoil.

Muzzle Pressure The gas pressure remaining as the bullet exits the muzzle. High muzzle pressures are associated with greater muzzle blast and higher peak decibel values.

Muzzle Velocity The initial velocity of a projectile as it exits the muzzle.

N

2-NN 2-nitroso-1-naphthol; a colorimetric reagent for copper residues in suspected bullet holes and bullet impact sites.

Neck The parallel-sided portion of a case that grips the bullet. In a bottle-necked case, it is the area immediately ahead of the shoulder.

Nitrocellulose Also known as *cellulose nitrate* and *cellulose hexanitrate*, the principal ingredient of single-base, double-base and triple-base propellants.

Nitroglycerin *Glycerol trinitrate* A high explosive and ingredient of double-base and triple-base propellants.

Non-Corrosive Primer A primer which contains no potassium chlorate or similar compounds in its primer mixture. See **Corrosive Primer** and **Mercuric Primer**.

Non-Mercuric Primer A primer which contains no fulminate of mercury or other mercuric compound in its priming mixture. A mercuric primer may or may not be corrosive, depending on whether or not it contains potassium chlorate. See **Mercuric Primer** and **Corrosive Primer**.

Nyclad® Bullet A nylon-coated lead bullet; a Smith and Wesson original trade name which has been purchased by Federal Cartridge Corporation.

O

OAL Abbreviation for Over All Length. The total length of a loaded cartridge.

Obturation The sealing of a bore and chamber by pressure. During the firing process, pressure swells the case within the chamber, preventing gas from leaking back into the action. The same pressure, applied to the base of the projectile, causes it to swell or upset, filling and sealing the bore.

Ogive Literally, a French word meaning "pointed arch." In bullet design, the ogive is the radiused portion between the bearing surface and the meplat or tip of the bullet. This radius is often measured in "calibers," where "caliber" is a unit of measure based on the diameter of the particular bullet.

Out of Battery Discharge   A discharge that takes place when the firearm's locking mechanism is not fully closed. Unlike a slam-fire, an out of battery firing is normally the result of the shooter intentionally pulling the trigger. Upon firing, the unsupported case may rupture and vent gasses back into the action. This is a very hazardous situation for the shooter, and can damage or destroy the rifle.

Over Powder Wad   The wad between the propellant and other components in a shotshell.

P

PPK   From the German *Pistole, Polizei, Kriminal* (criminal police <plainclothes> pistol).

PSI   Abbreviation for *Pounds per Square Inch*.

Parabellum   From the Latin "for war." This term is typically associated with the 9 mm Parabellum cartridge.

Partition bullet   A bullet designed for controlled expansion having a jacket which is divided into two chambers which enclose the forward and rear cores of the bullet. It is designed so that the first chamber expands and the rear chamber holds together for improved penetration.

Patched Ball

1. For modern cased ammunition, the term "patched ball" refers to an FMJ bullet.
2. For muzzle-loading firearms, the term refers to round or conical lead projectiles that utilize cloth or other material which acts as a gas seal or a guide for the projectile.
3. Early fixed ammunition using paper as a gas seal for the projectile (a paper-patched bullet).

Pattern   The distribution of shot fired from a shotgun. Among firearms manufacturers, pattern is generally measured as a percentage of pellets striking in a 30 in. circle at 40 yd. Some skeet guns are measured with a 30 in. circle at 25 yd. A *blown* pattern is one that displays an erratic distribution of pellets.

Perforated Disk-Flake Powder   An extruded form of smokeless powder cut into thin, circular disks and possessing a small, central perforation to modify its burning characteristics.

Phantom Safety   A situation where the handler incorrectly senses or believes that a manually operated safety system has been engaged. This claim or theory usually arises with traditional single action revolvers and postulates that the trigger sear was perched rather than seated in the safety or quarter-cock notch of the hammer.

Physical Match   The examination of two or more objects either through physical, optical, or photographic means, which permits one to conclude whether the objects were either one entity or were once held or bonded together in a unique arrangement.

Pierced Primers   A primer which, upon firing, has been pierced by the firing pin. This allows gas to flow back into the action, and can be injurious to the shooter. A potentially dangerous situation, normally indicating excessively high pressures, but not unique to this cause.

Pinch Point   In painted metal surfaces, a small area of surviving paint that was pinched between the initial contact point of a low incident angle bullet and the painted metal surface. The pinch point establishes the entry side of an impact or ricochet mark and thereby the bullet's direction of travel.

Plate Glass   A flat glass having an extremely clear, smooth surface. Also known as single strength glass. Common flat glass lacking any special treatments or construction.

Plinking   Informal shooting, not following any organized rules of competition or at any designated distance. Plinking is shooting "just for fun."

Plumb Line   A simple device to indicate vertical direction, made by suspending a small mass of lead or other heavy material, free hanging and still, on a string line.

Point Blank Range   The range to which a shooter can obtain a hit in the vital zone of a particular target or game animal, without holding over or under the target.

Point of Aim   The place or point on a target which intersects the straight line generated by the alignment of the front and rear sights of a firearm.
The exact point on which the shooter aligns the firearm's sights.

Point of Impact   The point at which a projectile hits a target or other down-range object.

Polygonal Rifling   A rifling system wherein the lands and grooves have rounded profiles with no distinct driving or trailing edges as contrasted to the sharp edges between lands and grooves found in traditional rifling systems. The most common firearm utilizing polygonal rifling is the various models and calibers of Glock pistols.

Port Pressure   Applies only to gas-operated firearms. The amount of pressure remaining in the bore as the bullet passes the gas port. If port pressures are too high, damage can result from the violent cycling of the action. It is important to understand that this can occur, even when chamber pressures are within acceptable limits. Port pressure can be controlled by proper powder selection.

Powder Patterning The orderly process of preparing powder patterns at selected standoff distances on some form of witness panel material with a specific gun and ammunition combination.

Powder Stippling Small hemorrhagic marks on the skin produced by the impact of gunpowder particles, or, in inanimate objects, small pits or defects in the object caused by the impact of unburned and partially burned powder particles.

Powder Tattooing The embedding of partially consumed and unconsumed powder particles in the skin with accompanying hemorrhagic marks associated with living skin.

Primer A cartridge ignition component consisting of brass or gilding metal cup, priming mixture, anvil, and foil or paper disk, which ignites the propellant in the cartridge when struck with sufficient force.

Primer Cratering A circumferential rearward flow of primer metal surrounding the indentation of a firing-pin in a fired primer cup.

Primer Setback The condition when a primer moves partially out of its proper location in the primer pocket of a metallic cartridge or shotshell during firing.

Print-Through A condition where the impressions of the lands in a jacketed bullet print through the jacket and can be seen on the separated lead core.

Proof Cartridge A special high-pressure load used to test the strength of a newly manufactured or rebuilt firearm. Also referred to as a "blue pill" load, pressures in these rounds may run as much as 40% higher than standard for a given cartridge.

Pyrodex® A Hodgdon trade name for a black powder substitute with similar burning characteristics. Pyrodex comes in several granulations for use in percussion revolvers, small and large caliber rifles, and shotguns.

Q

QE Abbreviation for quadrant elevation. A military term for the elevation of the muzzle of the firearm above the horizontal plane, usually expressed in mils.

R

RF Abbreviation for Rimfire (see **Rimfire**).

Radial Fractures The fractures or cracks that radiate out from an impact site in non-crystalline materials such as glass, ceramics, bone, and certain plastics.

Recoil The rearward movement of a firearm as a result of a discharge.

Recoil Operation Short recoil: A firearm mechanism (action) in which the breechblock remains locked to the barrel only while the pressure is high. This involves a barrel travel of only about $\frac{1}{2}$ in. The device locking the breechblock to the barrel is then released and the two components separate. The barrel may remain stationary and await the return of the breechblock, but in most modern designs, the barrel has its own spring and goes forward into battery. Long recoil: A system in which the bolt and barrel recoil a greater distance than the length of the unfired cartridge. The breechblock is then held to the rear while the barrel is driven forward by its own spring. When the barrel is fully forward it trips the catch, releasing the breechblock, which then feeds the next cartridge into the chamber.

Reference Ammunition Ammunition used in test ranges to evaluate test barrels, ranges, and other velocity and pressure measuring equipment. May also be used as a control sample by which other characteristics are compared, such as accuracy, pattern, etc.

Rib Marks A series of raised and depressed features on the edges of radial and concentric fractured in bullet-struck glass and other comparable materials that, in the case of radial fractures, start at right angles to the backside (exit) surface of the fracture and turn toward the source of breaking force. Rib marks on concentric fractures start at right angles to the front side (entry) surface.

Ricochet The glancing rebound of a projectile after impact with a surface.

Rifling The series of spiral grooves, cut or pressed into the bore of a firearm, designed to impart spin to a projectile.

Rifling Twist The direction (right or left) and rate of turn of the rifling helix usually expressed as one turn in x inches or y mm.

Rimfire Any cartridge having its priming mixture contained within its rim. For all practical purposes, rimfire cartridges are non-reloadable.

Ringed Barrel A barrel from a firearm that has been fired while containing an obstruction. The resultant excessive radial pressure causes a circumferential bulge in the barrel.

Round Military terminology for a single, loaded cartridge.

S

SAAMI Abbreviation for Sporting Arms and Ammunition Manufacturers Institute.

SEE Abbreviation for Secondary Explosive Effect. SEE has been described as a condition that can occur when slow-burning tubular powders such as IMR 4831 are used at greatly reduced charge weights in large capacity bottle-necked cartridges. Rather than burning in a normal fashion, the powder detonates, as though it were a severe overload. Also known as a *high pressure excursion*.

SS109 A European designation for the 62 gr. M855 FMJ-BT bullet loaded in 5.56 mm cartridges. This bullet also contains a hardened steel penetrator in its nose.

SWAT   Special Weapons and Tactics/Training or Special Weapons Assault Team.

S&W   Smith & Wesson.

SWC   Abbreviation for **Semi-Wadcutter Bullet** (see **Semi-Wadcutter Bullet**).

SXT®   Abbreviation for Supreme Expansion Technology. A Winchester trademark for certain pistol bullets.

Sabot   Literally, a French word meaning "shoe." In weapons systems, sabots are a device used to center a subcaliber projectile in a bore for firing. The sabot normally disengages from the projectile shortly after it exits the muzzle, falling to the ground a short distance in front of the gun.

Safety Mechanism   A device on a firearm intended to help provide protection against accidental discharge under normal usage when properly engaged.

Such a mechanism is considered "On" when the position of the safety device is set in a manner to provide protection against accidental discharge under normal usage. Such a mechanism is considered "Off" when it is set to allow the firearm to be discharged by a normal pull of the trigger.

A *manual* safety is one that must be manually engaged and subsequently disengaged to permit normal firing.

An *automatic* safety is one that goes to the "On" position when the action of the gun is opened. A *passive* safety is in place (or "On") until the trigger is pulled. An example would be the transfer bar system in some revolvers.

Schlieren Photography   From the German word *schlieren* (streaks). A special type of photography using a point (spark) source of illumination that exposes a bare sheet of film mounted on a flat surface. Objects and disturbances in the atmosphere between the spark discharge and the film are recorded as shadows. Schlieren photography is used to capture the supersonic shock wave and/or air turbulence created by bullets in flight.

Sear   An internal component in a firearm that is designed to retain the hammer or striker in the cocked position. When released, it permits the hammer to fall or the striker to fly forward firing the cartridge in the chamber.

Sectional Density   A bullet's weight, in pounds, divided by its diameter in inches squared. High sectional density is essential to producing a good ballistic coefficient and deep penetration.

Selective Fire   The capability of some automatic weapons to fire in either the automatic or semiautomatic mode at the firer's discretion. Theses weapons normally have a switch or selector lever to facilitate the operator's choice.

SemiAutomatic Firearm   *A* firearm which fires, extracts, ejects, and reloads once for each pull and release of the trigger. Also called Self-loading or Auto-loading firearms.

Semi-Jacketed Bullet   A bullet with a partial jacket, exposing a lead nose.

Semi-Wadcutter Bullet   A projectile with a distinct, short truncated cone at the forward end.

Shockwave   The disturbance of air surrounding and behind a supersonic bullet caused by a compression of the air column directly in front of the bullet.

Shot Collar   A plastic or paper insert surrounding the shot charge in a shotshell to reduce distortion of the shot when passing through the barrel. The most common shot collars are the plastic inserts found in some Winchester shotshells.

Shotcup   A wad, or shot protector. Various designs of shotcups are made of plastic and designed to reduce pellet deformation during barrel travel.

Sight Picture   The visual image observed by the shooter when the firearm sights are properly aligned on the point-of-aim.

Sight Radius   The distance between the rear sight and the front sight on a firearm.

SilverTip®   A trademark of the Winchester ammunition company most frequently associated with a line of jacketed hollow point pistol bullets possessing either nickel-plated gilding metal jackets or aluminum jackets.

Sine Function   In a right triangle, the ratio of the side opposite a given angle and the hypotenuse.

Single Action Firearm   A firearm with an action requiring the manual cocking of the hammer or striker before sufficient pressure on the trigger releases the firing mechanism. In such firearms, pulling the trigger accomplishes the singular act of releasing the hammer or striker from its fully cocked position.

Single Base Powder   Any smokeless propellant which uses nitrocellulose as its only explosive base (see **Smokeless Powder, Double-Base powder**, and **Triple-Base powder**).

Single Strength Glass   See **Plate Glass**.

Skid Marks   Rifling marks formed on the bearing surface of bullets as they enter the rifling of the barrel before being fully aligned and gripped by the rifling. Skid marks have the appearance of a widening of the land impressions at their beginnings. Skid marks are typically associated with revolvers because of the bullet's movement from the unrifled cylinder into the forcing cone of the barrel.

Skive   A small slit or cut in the ogival portion of a jacketed bullet for the purpose of improving expansion.

Slamfire   An accidental discharge that occurs during the feeding cycle, with no manipulation of the trigger on the part of the shooter. Most frequently associated with semiautomatic firearms in combination with poorly assembled ammunition. The most common cause in handloaded ammunition is a high primer, improperly set headspace (insufficient resizing), or a combination of both. Other causes include incorrect primer, a broken and protruding firing pin, or a badly worn sear. These are extremely serious conditions that can damage or destroy the firearm and injure the shooter.

Slide   A member attached to and reciprocating with the breechblock in a semi or fully automatic firearm.

Slide Drag Mark   A *Slide Drag Mark* occurs in a semiautomatic pistol when there is a live cartridge at the top of the magazine and the slide is retracted. The lug on the underside of the slide *drags* across the 12 o'clock position on the cartridge case leaving a striated mark. This mark can*not* occur when a cartridge is manually inserted in the firing chamber and the slide is released or when a loaded magazine is inserted in the firearm with the slide locked back.

Slide Scuff Mark   A *Slide Scuff Mark* occurs in a semiautomatic pistol when there is a live cartridge at the top of the magazine, and the retracted slide, upon release or forward movement, impacts the 12 o'clock edge of the cartridge head producing a small, indented mark at this location. This mark also occurs during the normal chambering of a live cartridge during the firing process. This mark can*not* occur when a cartridge is manually inserted in the firing chamber and the slide is released.

Slide Stop   A device to retain slide in an open or rearward position.

Smokeless Powder   A propellant powder composed primarily of nitrocellulose (single-base), or nitrocellulose and nitroglycerine (double-base). There are triple-base powders as well, but they are not used in reloading powders in the United States. Smokeless powder comes in several forms, such as tubular, ball, and flake.

Snap-Fire   The discharge of a loaded firearm through the inadvertent release of the hammer or striker from a partially retracted position. Such discharge may be the consequence of a compromised safety system, the lack of a safety system for such an event, or the failure on the part of the handler to engage the appropriate safety system.

Sodium Rhodizonate   A chemical reagent used to detect lead by converting it to lead rhodizonate.

Spall   A crater formed from chipped or fragmented material as a result of projectile impact in brittle or frangible materials such as concrete, cinder blocks, sheetrock, etc. It more often describes such a crater on the *exit* side of a bullet impact site but has also been used to describe an impact crater in otherwise hard materials.

Spark Photography   See Schlieren Photography.

Speed of Sound   The rate of speed at which sound travels in air; at standard sea level conditions, approximately 1115 feet per second but varies with air temperature, altitude, barometric pressure, and relative humidity.

Spitzer   Literally, German for "pointed." In firearms terminology, spitzer refers to a pointed bullet. A Spitzer bullet has a sharp pointed, long ogive, usually of seven calibers (i.e., length to diameter ratio of 7 to 1) or more.

Sprue   The opening in a bullet casting mold that permits entry of the molten metal. Also the name given the waste piece cast in this opening.

Sprue Cutter Marks   The toolmark left on a cast bullet which results from the cutting off of the sprue.

Squaring (*of a vehicle*)   A rectangular system of reference for diagramming and measuring projectile strikes to automobiles.

Squib Load or Discharge   A cartridge or shell which produces projectile velocity and sound substantially lower than normal. This may result in projectile and/or wads remaining in the bore.

Steel Core Bullet   A jacketed bullet containing a core, usually composed of mild steel, and frequently centered or secured inside the jacket with lead.

Steel Jacketed Bullet   A bullet with mild steel used as jacket material, and often coated or plated with a thin layer of copper as a corrosion inhibitor.

Steel Penetrator   A hardened steel component within a jacketed bullet designed to improve penetration in "hard" targets.

Step Base Bullet   A Jacketed bullet having a small, flat recess or "step" in its base.

Stippling   See **Powder Stippling**.

Stove-Piping   A failure to eject where the fired case is caught in the ejection port by the forward motion of the slide or bolt. The cartridge case protruding out of the ejection port is said to resemble an old-fashioned stove pipe.

SubMachine Gun   An automatic or selective fire weapon chambered for a pistol cartridge. These firearms are normally compact, and intended to be used at close combat ranges.

Swage   To form metal under pressure. Normally done in a press, using a punch or die.

Swaged Bullet   A bullet that has been formed by pressing and forming the bullet material in a die.

Sympathetic Discharge   See **Involuntary Discharge**.

Sympathetic Firing   The simultaneous firing of two or more cartridges chambered in the cylinder of a revolver, one of which is in battery.

T

TC   Abbreviation for Truncated Cone Bullet (see **Truncated Cone Bullet**).

TMJ   Abbreviation for Total Metal-Jacketed Bullet (see **Total Metal-Jacketed Bullet**).

Tangent   In a right triangle, the ratio of the side opposite a given angle to the adjacent leg of the triangle.

Tartrate Buffer   An aqueous solution of sodium bitartrate and tartaric acid formulated to provide a pH-2.8 environment for the sodium rhodizonate test reagent.

Tattooing   See **Powder Tattooing**.

Tempered Glass   A single piece of glass that has been given a special heat and cooling treatment that results in much greater strength. It is used for glass doors, shower doors, arcadia doors, as well as side and rear automobile windows. Upon failure it shatters and dices into small fragments.

Terminal Ballistics   The branch of ballistics study that deals with the projectile's impact on a target.

Terminal Velocity   The final velocity reached by a free-falling object in a specific medium, usually air.

Throat   The unrifled portion of the bore immediately ahead of the chamber, and before the leade (see **Freebore**).

Time of Flight   The time taken by a projectile to traverse two points, or a specific distance. Time of flight is a critical factor in a number of exterior ballistic calculations.

Total Metal-Jacketed Bullet   A bullet made by copper plating a lead slug to create a jacket that completely encases the core. This jacket is much thicker than cosmetic copper plating.

Tracer Bullet   A bullet containing a chemical compound that is ignited during the discharge process and that produces visible or infrared illumination during its flight. Selected colors such as red, orange, yellow, or green are achieved through the incorporation of certain metallic compounds in the composition of visible tracer bullets.

Trajectory   The arched path that a bullet follows in flight (see **Bullet Path**).

Trajectory Rod   A straight probe or rod often with centering cones and constructed of inert, brightly colored materials specifically designed for tracking and illustrating the nominal path of a projectile through one or more materials.

Trigger Pull   The amount of force which must be applied to the trigger of a firearm to cause sear release. It is measured with hanging weights or an appropriate scale touching the trigger at a point where the trigger finger would normally rest, and with the force applied approximately parallel to the bore axis.

Triple Base Powder   A propellant composed of nitrocellulose, nitroglycerine, and nitroguanidine. Generally used in large caliber military ammunition.

Truncated Cone Bullet   A bullet with a tapered or conical nose profile that has been terminated (by design on the part of the manufacturer) before reaching a point.

Twelve-Twenty Burst   A dangerous situation where a 20-ga. shotshell is inserted into the chamber of a 12-ga. shotgun followed by a 12-ga. shotshell. The 20-ga. shell represents a serious obstruction at the forcing cone when the 12-ga. shell is discharged and will produce a "ringing" of the barrel in this area at the minimum or a bursting of the barrel in this area.

Twist   The direction (right or left) and rate of turn of the rifling helix.

Twist Direction   The direction of rotation of the rifling, as determined by studying the inclination of the rifling at the top of the bore as one looks through a barrel.

Twist Rate   The rate at which a firearm's rifling turns within the bore. This is normally expressed as the distance required for the rifling (and projectile) to make one complete revolution. Depending on the origin of the firearm, this may be written in inches or in millimeters. Examples: 1 turn in 12 in. or 1 turn in 305 mm.

Tubular Powder   See **Extruded Tubular Powder**.

U

UMC   Abbreviation for Union Metallic Cartridge Co.

Unintentional Discharge   The handler of the gun is the source of the discharge. Examples: improper let-down of a hammer, involuntary/sympathetic discharge during a struggle, slip and fall, etc.

Unperforated Disk-Flake Powder   An extruded form of smokeless powder cut into thin, circular disks of selected diameters and thicknesses. Examples are Bullseye and Unique.

V

Vertical Angle   In shooting scene reconstruction, it is the vertical component of a projectile's reconstructed flight path. This angle is given a minus sign if the path followed by the projectile is downward and a positive sign if upward. A flight path that parallels a level surface has a vertical angle of 0.0°.

Vector® Ammunition   A unique line of illuminating projectiles manufactured by the Hornady company that utilized an igniter composition followed by a fine zirconium wire centered in the lead core of these open-based pistol bullets. These special cartridges are no longer available.

W

WCF   Abbreviation for Winchester Center Fire. Designates a center fire cartridge designed or produced by Winchester. Examples would include the .30 WCF (.30–30), .38–40 WCF, and .44–40 WCF.

WRA   Abbreviation for Winchester Repeating Arms.

W-W   Abbreviation for Winchester-Western.

Wadcutter   A bullet having a full-caliber flat nose, intended to cut a clean hole in the target for easier scoring.

Web The solid portion of a cartridge case between the primer pocket and the powder chamber. The primer pocket and powder chamber are joined by the flash hole or vent in the web.

Weight A property of matter that depends both on the mass of an object or material and the effects of gravity on that mass represented by the formula $W = ma$, where W is the weight, m is the mass, and a is the accelerative force of gravity (32.174 f/s/s or 9.807 m/s/s average earth values at sea level). *Note*: In some formulas the accelerative force of gravity is denoted by the letter g.

Windage Lateral correction of a firearm's sights to compensate for the projectile's deflection by wind or drift. This term also refers to the space between an undersized spherical lead ball and the bore of a muzzle-loading rifle.

Witness Panel Any one of a variety of materials such as thin cardstock or poster board positioned and mounted in such a way so as to "witness" or record the position and orientation of a perforating bullet or bullet fragments. Cardstock witness panels are also used to record pellet patterns from shogun discharges at selected ranges. The patterns of gunshot residue deposits are also recorded on witness panels of selected materials for this purpose.

Work Hardened A change in the grain structure of a metal as a result of repeatedly stressing it. In cartridge cases, work hardening most frequently occurs in and around the neck area, from the stresses of repeated firings and resizings. This causes brittleness, and leads to cracking and splitting of the case.

Wound Ballistics A special case in Terminal Ballistics dealing with the behavior of projectiles in tissue and tissue simulants. It includes bullet performance (threshold velocity to achieve penetration, bullet expansion, fragmentation, deformation, path deviation, and yaw point), penetration characteristics, and velocity loss as a consequence of the perforation of tissue and tissue simulants.

X

X-ray Magnification The magnification effect for radio-opaque images on X-ray films that result from the conical beam of X-rays from the X-ray source. The effect is that the physical dimensions of the radio-opaque object on the film are always larger than the actual object within the body or subject being examined.

Y

Yaw The rotation of a bullet at an angle (usually very slight) to its line of flight. Some yaw is almost always present when a bullet is fired, but this usually dampens out within 200 yd if the bullet is properly stabilized and well balanced. The angle between the longitudinal axis of a projectile and the line of the projectile's trajectory. Yaw is usually considered to exist before a bullet achieves full gyroscopic stability.

Yaw Card A size and thickness of cardstock selected to faithfully record the outline and orientation of a perforating projectile. A yaw card serves the same or similar function as a witness panel.

Z

Zero The adjustment of a firearm's sights in order to obtain impact at a desired point in relation to a specific point of aim at a given range.

Zero-Edge Protractor A protractor with a zero line or edge that lacks any tabs or ears, thus allowing this edge to be paced directly next to the struck surface and the medium used to represent the projectile's flight path.

INDEX